THE MOST BEAUTIFUL MOLECULE

THE MOST BEAUTIFUL MOLECULE

THE DISCOVERY OF THE BUCKYBALL

Hugh Aldersey-Williams

JOHN WILEY & SONS, INC.
New York • Chichester • Brisbane
Toronto • Singapore

This text is printed on acid-free paper.

Published by John Wiley & Sons, Inc.

Library of Congress Cataloging-in-Publication Data:
Aldersey-Williams, Hugh.
The most beautiful molecule : the discovery of the buckyball / Hugh Aldersey-Williams.
p. cm.
Includes bibliographical references and index.
ISBN 0-471-10938-X (cloth : acid-free)
1. Buckminsterfullerene. I. Title.
QD181.C1A43 1995
546′.681—dc20 95-12422

Printed in the United States of America.

10 9 8 7 6 5 4 3 2 1

For the chemistry masters of Highgate School in 1974 and still: Humphry Barnikel, Peter Knowles, and Andrew Szydlo, and most of all to Michael Morelle, who taught us the chemistry of Borodin, Tom Lehrer, and Flanders and Swann as well as of Lavoisier, Davy, and Mendeleev.

CONTENTS

ACKNOWLEDGEMENTS

First, I owe a debt of gratitude to Peter and Phillipa Knowles who presented me with the idea of writing this book.

I should like to thank all those scientists who gave most generously of their time in allowing me to interview them. They are credited individually in the relevant chapters, but I must here single out Harry Kroto and Rick Smalley who were exceptionally generous in this regard despite their equally exceptionally busy schedules. David Walton, too, bore with patience and good humour some of my more tedious requests. Many of these people were press-ganged into reading drafts of chapters of the book; I thank them, too, for their comments and suggestions. In addition, the following read portions of the manuscript: Yuan Liu, Gillian Beer, Ed Applewhite, Ian Smith, Simon Friedman, Raymond Schinazi, Peter Buseck, and Alan Mackay. John Lynch, Bonnie Goldstein, Philip Ball, and David Bradley provided additional research material, while Linda Rings, *née* Smalley, and Margaret Kroto recalled the events of 1967. I am grateful to them all.

The resources and staff of the Cambridge University Library and of several other libraries of that university proved invaluable to my researches, as did those of the Science Reference and Information Service of the British Library.

Friendly neighbourhood scientists, Jim Smith and Fiona Watt, offered constant encouragement, and John and Judith Aldersey-Williams gave practical advice when it was most needed.

Finally, I thank those who now know far more about one particular molecule than they will ever find useful: my agent, Caroline Davidson, who switched on rather than off at the mention of a book on chemistry; my editor, Piers Burnett, who kept the ball rolling; and, most of all, Moira whose commitment to my project on the third form of carbon deserves a gift of one of the other forms.

INTRODUCTION

The longest word in this book is also one of its most abundant. I make no apology for this, except to state the hope that I have compensated in other respects by keeping scientific jargon in check. The word is buckminster-fullerene.

As its odd name begins to imply, buckminsterfullerene, the chemical discovery that is the subject of this book, is a rather unusual substance. It has a shape that is unique in all of chemistry. Most molecules are compounds of two or more chemical elements. But a molecule of buck-minsterfullerene is made up of sixty atoms of just one element – carbon. Each atom occupies an exactly equivalent site and each is joined in an exactly equivalent way to its neighbours. The chemical bonds between these atoms follow the pattern of the seams on a soccer ball that hold the leather pentagon and hexagon facets together to form the familiar inflated sphere. This spheroidal cage has the highest symmetry of any known molecule. It is, quite simply, the most beautiful molecule.

We are on intimate terms with diamond and graphite, forms of carbon which have been known for millennia. Buckminsterfullerene, or the buckyball as it was quickly nicknamed, is a third stable form; it was discovered only in 1985 and made in quantities sufficient to be able to explore its possible uses only in 1990. It was named by its discoverers after the maverick architect Richard Buckminster Fuller who popularized geodesic domes, and has since been found to be one of a family of spher-oidal molecules known collectively as the fullerenes.

By virtue of its shape and symmetry, buckminsterfullerene possesses not only beauty but also novelty and likely utility. For all these qualities, buckminsterfullerene and its kin featured as the subject of nine of the ten most cited chemistry papers in 1991, as reported by an industry journal that monitors such things. In 1992, it was the subject of all ten. One paper a week appeared between 1985 and 1990. Once it was discovered how to make it in quantity, the pace accelerated to close to one a day.

Chemistry is supposed to be one of the more developed sciences. Within chemistry, carbon is the element about which we are supposed to know most. It is the element around which life is built. The remarkable thing about the discovery of buckminsterfullerene, then, is not how clever we are to have found it, but how unobservant and unimaginative we have been not to have found it sooner. The discovery is a potent reminder that science is not, as it is sometimes portrayed, on the verge of reaching a state of absolute knowledge but that after three centuries of modern chemistry we have only just begun the exploration. It highlights not how much we know, but how very little.

The story of the discovery of this unique molecule illustrates much about the way the science professions work – and occasionally fail to work – together. In its early stages, scientists – chemists, physicists, astronomers, mathematicians – collaborated fruitfully, although often without funding, driven by the excitement of a growing puzzle. Later, as pressure grew to exploit the discovery and put it to good use, the tale serves as a parable illustrating the increasing demands being placed on pure scientists to do more "relevant" work or to play a part in the development and commercial application of their new fundamental knowledge.

The discovery of buckminsterfullerene already ranks with some of the greatest moments in chemistry. As and when practical uses are found for it, its fame can only increase. The sheer beauty of the molecule has done much to revitalize chemistry teaching in schools. Equally important, it has boosted the morale of the core physical science professions with a conspicuous and inspiring success after years spent in the shadow of molecular biologists and cosmologists.

Public interest in science centres on two big questions: what is the origin of life, and what is the origin of the universe? Publishers and other popularizers of science have exploited this fascination by producing a host of books on genes and the cosmos; the physical sciences, in contrast, have been largely ignored. The startling discovery of buckminsterfullerene provides an opportunity to redress this imbalance. This tale of discovery is a celebration of the intimate world of physical science and its practitioners. It is on the face of it a self-contained tale: a molecule is born. And yet that birth is attended by the muses of many disciplines, not only the sciences but also the arts. As we shall see, this story, too, touches upon cosmology and biological science – and much else besides.

Another stereotype of "popular science" portrays research as a relentless race of one single-minded boffin against another to win the Nobel Prize. To fit this story into that mould would be to travesty a complex narrative that concerns many scientists who are humans first and – perhaps – geniuses second. (A rider to this stereotype is that the boffin is always male; this stereotype, sadly, I cannot dispel in this account.) Nevertheless, there are episodes where personal rivalry and even enmity have added fuel to the fire, and there is indeed every chance that the discoverers of buckminsterfullerene will one day be awarded the Nobel Prize.

The stereotype usually demands that serendipity play a part, an element of luck that leads to something being found that was not being sought. No one set out explicitly to discover a spheroidal sixty-atom molecule of carbon, and so to that extent we may claim that our story too has serendipity. Certainly, no one set out to overturn so completely our understanding of the physics and chemistry of carbon. Contrary, perhaps, to popular belief, it is the resolve of the modest and proper scientist *not* to seek to upset the applecart at every bump along the road. But to lay too much emphasis on serendipity is to do little justice to the skill and ingenuity of those involved, to their scientific method and to their occasionally unscientific method, and to their plain, dogged hard work. All of these qualities are of more merit than "serendipity". Serendipity, in short, is something the good scientist is prepared for, and takes advantage of.

There was no *Eureka!* moment in this story, though history may later demand that one be conjured up. Rather there were many small realizations and a creeping feeling, at first of uncertainty because things were somehow not as they should be, and then of certainty that some remarkable new phenomenon lay behind it all.

In this discovery, as in others, the scientists benefited from the occasional lucky break; but they were also hindered from time to time by an inability to break free of received wisdom and the orthodoxies of their specialisms. The story of buckminsterfullerene has its element of serendipity but also its share of these blocks. It is perhaps they which better illustrate the working of the scientific mind. The stubbornness that causes some to cling to a conventional wisdom even as evidence is presented to contradict it is not so very different from the stubbornness that advocates a radical new idea in the face of disbelief and opposition from one's peers. Both attitudes have no official place in the manual of scientific method, yet both were present in abundance in the aftermath of the 1985 discovery. The discoverers of buckminsterfullerene believed they had what they had before they could really prove it; their detractors continued to disbelieve in the face of evidence that was eventually to become overwhelming.

In parallel with this debate, evidence grew, too, for the other "fullerenes", spheroidal carbon molecules not just with sixty carbon atoms, but with seventy, seventy-six, eighty-four and other numbers. All are less perfect in their symmetry than buckminsterfullerene itself, but the structure of each suggests its own curious properties. In 1991, the canvas was broadened still further with the discovery of long "bucky-tubes" and then huge fullerenes, made up not of a few dozen carbon atoms, but of hundreds and thousands of atoms, yet still with the same beautiful latticework shells.

When the structure of buckminsterfullerene was finally proven to everyone's satisfaction, the prospects that such an unusual molecular architecture would have unusual properties quickly led corporations such as AT&T, Du Pont, Exxon, and IBM to invest millions of dollars in order to investigate potential applications. The most apparent characteristic is the hollowness of the cage shape, large enough to trap an atom of any element of the periodic table, and offering the prospect of fascinating modifications to the properties of the element so trapped. The exterior shell of the fullerenes is also interesting as a potential site for chemical reactions. Finally, this unique family of molecules suggests a whole microworld made up of structures of interconnecting domes and tubes with clear potential in the emerging field of nanotechnology, the skill of building machines invisible to the naked eye that will be able to perform tasks too fiddly for human hands. The ultimate goal becomes the possibility of building custom carbon architectures; joining modules of spheres, tubes and other shapes in a controlled fashion, working to a designer's blueprint.

Early conjectures as to what this new class of carbon molecules might be good for were optimistic indeed, but they were soon surpassed by reality as buckminsterfullerene derivatives quite unexpectedly showed themselves able to perform as electrical superconductors and even to exhibit biological activity against the human immunodeficiency virus that causes AIDS. Yet, as the development effort intensifies, it is important to remember that the original discovery was made in the pursuit not of commercial gain or social improvement but in the name of knowledge. At root this is a story of classic "bootleg science". No company or foundation specifically funded any search for an agent that would possess these properties. The work was done on the back of other, funded, projects and when time would allow. Yet its commercial implications are probably immense.

The truth is that there is very little correlation between the extent to which a project is funded and the importance of the resulting discoveries; or between the extent of funding and how quickly a discovery is put to the service of humankind. Because it is foolish to expect such a discovery to be put to general use very soon, I have limited the discussion of potential uses of buckminsterfullerene to the final chapters and have concentrated on the story of the discovery and its aftermath.

In the book as a whole, Chapters Two and Eight serve as tent poles across which the narrative is stretched. These chapters respectively describe the discovery in 1985 and the breakthrough in 1990 when it became possible to make useful quantities of buckminsterfullerene for the first time. These chapters are like the scherzos in a Mahler symphony, boisterous and fast-paced, the scenes of frenetic activity. Around them are more rambling, exploratory movements, the adagios of the piece. The period between 1985 and 1990 was one of frustration as scientists strove to learn more about the new molecule but could still not make enough of it to gain much headway. Chapter Three chronicles efforts to prove that buckminsterfullerene was indeed the novel molecule that it was claimed to be, efforts that were hampered by a continuing inability to produce sufficient quantities for easy experimentation. The subsequent hiatus provides us with an excuse to digress during the course of Chapters Four to Seven into topics more loosely connected with buckminsterfullerene before we rejoin the fray in Chapter Eight. Chapter Nine describes a final proof of the structure of the molecule, a proof quite as elegant and visual in its appeal as the molecule itself; Chapters Ten and Eleven conclude with a discussion of the likely uses for buckminsterfullerene and the implications of its discovery.

Chapter One sets the scene for what is possibly the greatest chemical discovery of this century. It is episodic in nature, interleaving some basic chemistry with snapshots of the obsessions of the scientists involved. Although their interests initially appear separate, they gradually converge, each person contributing an expertise that will make possible the discoveries to come. By chance, the two scientists who were principally responsible for bringing buckminsterfullerene into the world share a particular episode from their past that may have sparked their scientific creativity. It is this episode that is the subject of the following Prologue.

PROLOGUE

THE DOME OF DISCOVERY

Nineteen sixty-seven was the year of the Summer of Love. It was also Canada's centennial. To mark the anniversary, the Canadian, Quebec, and Montreal authorities threw a party for fifty million guests – the Montreal EXPO. Seventy-five countries or more spread their wares over nearly one thousand acres across two largely manmade islands in the St Lawrence river. Even as international expositions go, the scale of the event was almost beyond imagination.

It was Napoleonic France that launched the idea that a nation should periodically show off the fruits of its industry, but the first truly international exposition was held in 1851 in London. Other cities jumped on the bandwagon. Before Montreal, perhaps the most powerful image of an EXPO was of the Soviet and German pavilions facing each other starkly across the main axis of the exposition site at the Palais de Chaillot in Paris. In 1937, the fascistic architecture was an effective presage of war.

The mood was very different in 1967. Technological rationalism took the place of neoclassical bombast. Many buildings at the EXPO chose to express their nations' pride in a geometric rhetoric. The Canadian pavilion was a vast inverted pyramid. The Venezuelans constructed a group of brightly coloured cubes. Monaco favoured a forest of cylinders, while the Austrian pavilion was made up of nested tetrahedra. Over the EXPO site as a whole, the effect was as if a child had chosen to blow the entire contents of a chemical crystal-growing set in one orgy of geometric excess.

In the year that Christiaan Barnard conducted the first human heart transplant operation, belief in scientific progress was near absolute. Two years later the decade would culminate in man's landing on the moon, an event that was amply trailed in many of the exhibits in Montreal. Indeed, the strides being made in science and technology at this time were so great that, a quarter of a century later, the promoters of the Seville EXPO resorted to the same come-ons: "At EXPO'92 you'll get a foretaste of the year 2000. Space probes, satellites, high resolution TV." At Montreal, it had been the SECAM system of colour television in the French pavilion while the Soviet Union demonstrated the conditions "under which a cosmonaut will walk on the moon".

Montreal caught the spirit of the time as few other cities would have done or could have done. It drew on the wealth and confidence of the North American continent and added to it that peculiarly French combination of vision and *dirigisme* that can imagine *grands projets* and then bring them into glorious existence. The fusion brought into being an International EXPO on a scale that has not been equalled since.

In the heat of this summer, two young scientists separately made the long journey by car up from New Jersey where they had jobs with two of America's major high-tech corporations.

Rick Smalley explored the crystalline cityscape of the EXPO site in the company of his sister. With short, dark hair and blue jeans, he had the look of a typical clean-cut all-American young man. Only his unusually sharp blue eyes roving and focusing on the sights around him hinted at the single-mindedness that he brought to his research. Both physicist and chemist, he was possessed by the excitement of developments in his chosen sciences. His sister, Linda, was a teacher and his senior by just over a year. Living many hundreds of miles apart, they had not met since perhaps last Christmas or Thanksgiving. The EXPO was a good pretext for them to catch up on each other's lives.

Smalley was working for Shell Chemical Company in the quality control department of its polypropylene plant in Woodbury. A workaday job perhaps, but it was an enormous plant and the technology by which the polypropylene was made had its own excitement. It was Smalley's job to troubleshoot the temperamental production process. The plant ran twenty-four hours a day and Smalley was one of the few staff who came in at night, using the valuable quiet time to develop new analytical techniques all geared to monitoring the plastic production more effectively. Smalley was enjoying working there, but nevertheless, his routine fell rather short of his high school wish to "play at the frontiers".

Linda had flown up to Philadelphia and Rick had collected her from the airport. They were unaccustomed to upping sticks and being tourists together and this was their first visit to Canada. It was also their first time in Philadelphia. As they checked off the essential sights in the cradle of America's liberty, they began the slow process of reacquainting themselves before pressing on.

The Montreal reunion was a generally happy one. But Rick was tired from his work and the long drive, the modest hotel room they had booked was cramped and lacked air-conditioning, and under the broiling Canadian sun their conversation occasionally turned to squabbling. As they did so, a spirit of competition bubbled to the surface that had lain pretty much dormant since the two were rivals in chemistry class at high school.

On another day during the same month, Harry and Margaret Kroto, a young English couple, were pushing their infant boy along in an elaborate white pram of the kind that is now obsolete. Harry was also a scientist, a physical chemist to be exact, but with an artist's eye that was as happy in contemplation of the merits of one typeface over another as in analysing the fine structure of a molecular spectrum. Casually dressed in shirt, slacks and shades, he had been living and working in North America for nearly three years and betrayed little of that characteristic awkwardness of the English abroad.

He and his wife were relaxed and happy as they watched their son clambering over the modern plastic furniture in the cafés, although it was not without regret that they were counting down the days before they were due to leave North America for their return trip to Britain. Between touring the various national pavilions, they found time to stop and throw pebbles into one of the lakes that dotted the EXPO islands.

In the United States, Harry was working for a spell in Murray Hill at the Bell Laboratories, famous for their pioneering role in the invention of the transistor and the laser. He was using laser light to study the properties of molecules such as benzene and carbon disulphide, shining the light so that it would reveal something of how the molecules interacted with one another. It was not the sort of work that yields an easy answer for those who demand that scientific research should have an immediate commercial benefit. He was simply doing what scientists have always done, adding to the stock of knowledge about the natural world. Margaret, meanwhile, was housebound, looking after the addition to the family.

The trip to Montreal was the principal vacation of the year, and their last in North America before Harry was to return to England, drawn by an odd mix of patriotism, a wish to be closer to where the Sixties were really Swinging, and an invitation to begin an academic research career at the brand new University of Sussex.

The two couples set about touring the EXPO rather differently. The Krotos' strategy involved a map and a schedule. Their interest was as much in the cultural and creative aspects of each nation as in their demonstrations of scientific prowess. They stood in line to enter the Czechoslovakian pavilion, where Margaret was captivated by a display of contemporary glassware in cool greens and blues. Harry was more interested in the work of the film animators for which the country is famous.

The Smalleys' approach was more haphazard but ultimately more exhaustive. Rick and Linda's principal goal was to eat in as many of the pavilion restaurants as possible. They struck out at random, but by the end of their time, they had seen all the major pavilions and sampled the cuisine in many of them. Diverted momentarily from their culinary crusade, Rick was impressed by the sweeping cantilevered pavilion of the Soviet Union, and the rare glimpse it offered into a world normally hidden from view during those Cold War days. It had been the launch of the Sputnik satellite during his high school days that had stirred Rick's interest in science. Now, Rick wanted to inspect the latest in Soviet space technology.

Across a channel of water from the Soviet pavilion, linked to it by a pedestrian bridge known as the Cosmos Walk, lay the pavilion of the United States, the giant geodesic dome designed by Richard Buckminster Fuller that was the boldest architectural statement of all. The dome stood 76 metres tall. Its base was smaller than its diameter so that it appeared almost to be a complete sphere bobbing along on the surface of the Ile Sainte-Helène. Viewed from the platforms of other pavilions, the sphere seemed smooth. There was no way to gauge its size without referring to nearby trees or flagstaffs. Its alien shape and its scalelessness made it quite unlike any other building.

The outer frame of the dome was made up of steel struts linked into a repeating pattern of triangles. Six triangles came together at each point where the struts were joined, except for a number of special positions across the sphere where only five triangles met. If one were to stop focusing on the triangles, one would see the surface of the dome as a honeycomb of hexagons with, on closer examination, a few pentagonal irregularities.

Amid the ordered chaos of different nations' pavilions striving for effect, the dome had an undeniable serenity. So it was with a sense of awe that visitors pierced the tracery of tensioned steel and entered this cathedral consecrated to the theme "Creative America".

The spectacle was perhaps still more remarkable from the inside. The space was in fact quite different from that in a cathedral. The long axis of a cathedral is its nave. Here, the views stretched in all three dimensions. Light poured in from all sides of this structure without sides. There was little sense of being enclosed. Heavy platforms hovered impossibly in the vast space, setting off the feeling of weightlessness of the dome itself.

Both scientists were struck by the grandeur of the conception, although both, too, viewed it as a largely impractical one-off piece of technological bravura and not, as Fuller might have hoped, as the prototypical shape of shelter for the future. Rick Smalley was stunned by the contradiction that so transparent a structure could have such a dominating presence.

Like many Americans, he was aware of the name Buckminster Fuller, he knew what a geodesic dome was, but nothing more. Harry Kroto was more familiar with the maverick architect's work. He had seen published some of Fuller's more utopian schemes, but the Montreal dome represented the most tangible example of his principles and deserved closer inspection. Kroto negotiated his child's pram up and down the escalators and ramps that link the exhibit platforms. He drew away from the exhibit levels, and the pop-cultural clutter of Elvis Presley's guitar and paintings by Andy Warhol, to get a closer look at the structure of the dome. As he did so, hexagons dissolved into triangles and triangles into the individual components of the structure. Neither Kroto nor Smalley ever noticed the handful of pentagons insinuated among the hexagons of the gigantic orb.

In later years, both scientists' recollections of their visit to the American pavilion would become clouded. As part of his duties as an American father, Smalley made repeated visits to Disneyworld. After a time, perhaps, it was the larger dome of EPCOT that he recalled more readily. Since his student days Kroto had kept an extensive archive of magazines of the visual arts, and it was the photographs and diagrams of Fuller's EXPO pavilion in publications such as *Life*, *Graphis*, and *Paris Match* that he remembered as much as the real thing.

Lodged there by whatever means, the image of the geodesic dome that these two young scientists saw at EXPO '67 was a persistent one. It was an image that would come to assume a great importance when they first met eighteen years later.

1

MAKING MOLECULES

A molecule is a messy thing. It has a gangling skeleton whose bones are chemical bonds and whose joints are its component atoms. The bonds are often of different lengths and strengths. They poke out at odd angles. This gaunt portrait is messy enough but still only a crude sketch. Bonds are not lines and atoms are not points, though they are often drawn like this for convenience. Both fill space with clouds of electrons. These clouds flesh out the molecular skeleton to Michelin Man proportions.

Of course, a molecule no more looks like a Michelin Man than it does a stick figure. Both are models. These and other models have their uses, but none can said be to an accurate representation of the truth. Because of its size, well below the wavelengths of the visible light that we use to see, there is no way of saying what exactly a single molecule looks like. Pictures of atoms and molecules obtained by techniques beyond the reach of human physiology, such as X-ray diffraction and electron microscopy, are just that: pictures, subject to the limitations and ambiguities of any picture. They no more represent what atoms and molecules "look like" than someone who has been blind from birth can grasp what something looks like by simply feeling it.

Nevertheless, models and pictures are the key to our understanding of events that we cannot witness directly. Niels Bohr's model of the atom sticks in the mind – a sort of mini solar system in which negatively charged electrons are held in orbit around a positively charged "sun" made up of protons and neutrons, not, like planets, by gravitational force, but by electrostatic attraction. The picture has an enduring appeal and is still useful in some circumstances. It has been largely superseded, however, by a newer model in which electrons no longer describe neat orbits but are to be found within a three-dimensional volume known as an atomic orbital. Some of these orbitals are spherical, centred on the atomic nucleus; others have more complex shapes. An electron may travel anywhere within its assigned orbital volume like a fly trapped in a jam jar. The model has its limits. The walls of this jam jar are not in fact solid; they are elastic. There is a chance that the electron will be found outside the walls. The physical drawing of such an orbital is inadequate: it appears to show a hard border beyond which its electrons may not go, but in fact this border merely represents a fuzzy boundary within which those electrons will be found for a certain proportion of the time.

One picture of the bonding between atoms stems directly from the atomic orbital model. Each atom has a sequence of orbitals: most of these orbitals are of low energy and have their full quota of electrons; the ones that are not full, which may be conveniently thought of as the outermost orbitals, are higher in energy and are the ones that participate in chemical bonding. Atoms typically combine to form molecules when their incompletely filled atomic orbitals can overlap (another metaphor) to form a new, filled orbital. These overlaps are good if a number of conditions are met: the number of electrons in the unfilled atomic orbitals must be such that they are able to pool their resources to form a filled molecular orbital; the orbitals must be directed towards each other as much as possible so that their overlap is great and a strong bond is formed; where the orbitals cannot overlap directly then their respective sizes and mutual orientation must be such as to maximize what overlap there is.

If we could grip such a molecule – let it be one of ethyl alcohol – in a pair of tweezers and hold it up as we might a maggot to visual scrutiny, we would see its shape, a spine of three heavier atoms, two of carbon and one of oxygen, each with light hydrogen atoms attached to it in number according to the rules of chemical combination. It is in turn the number of electrons in the outermost, incomplete orbitals of combining atoms that determines these rules. Carbon has four electrons available to participate in the formation of bonds. So the first carbon atom in our skeleton has three hydrogen atoms attached to it since its fourth electron contributes to the bond with the second carbon atom. This second carbon atom has just two hydrogen atoms, bonded as it is to the first carbon and to the oxygen at the end of the spine. Oxygen has just two electrons available for bonding so only one hydrogen is attached here.

Bringing up a magnifying glass to our specimen of ethyl alcohol, we would notice that the carbon-carbon bond length is different from the carbon-oxygen bond length and that the carbon-hydrogen bonds likewise differ from the oxygen-hydrogen bond. Upon closer examination, we would notice that the carbon-hydrogen bonds, although very similar, are not identical on the two carbon atoms. We would notice that, if we regard the bonds as arms projecting outwards from the carbon atoms, then each of these arms points almost, though not exactly, along the lines that would join the atom to the corners of a tetrahedron (a regular three-dimensional shape with four equilateral triangular sides) whose centre is the atom itself. This indicates that each bond finds its equilibrium when it is positioned as far as possible from other bonds radiating from the same atom.

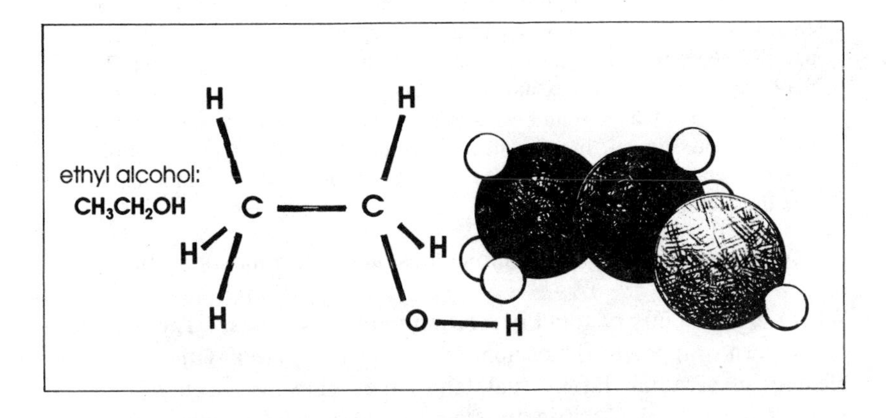

Molecular structure may be represented in a number of ways to indicate the position of atoms and overall shape

It would be difficult to hold a ruler and protractor up to our molecule in order to make these measurements. At all but the lowest temperatures, it squirms and wriggles, each of its bonds constantly bending and stretching at its own distinctive rate. If our grip slips, the molecule also begins to rotate. All molecules have just two or three axes of rotation but the number of ways they can vibrate increases as the number of constituent atoms grows. Ethyl alcohol, a molecule made up of just nine atoms, can rotate about three axes but it already has twenty-one distinctive modes of vibration. Each molecule has its own portfolio of characteristic rotations and vibrations and their detection and measurement are often invaluable in chemical identification.

All this and we have not yet even brought our molecule into contact with another that might react with it. If we do this now, carefully taking the right proportions of each reacting ingredient, we can obtain a virtually pure product. This is what organic chemists call synthesis. We have some control over the reaction. We determine the ingredients and we can judge the amount of each that we wish to react. But over the rate at which these ingredients react and over their mutual orientation and angle of attack, we are, as the chemist and author, Peter Atkins, pointed out, still barely in control:

> Chemists currently strive to achieve the fine control exercised by nature. To do so, they still stir, pour, heat, and distil, just as they have done ever since their intellectual ancestors vainly sought a reaction that would produce gold from lead. These crude processes seem to be ways of bending

> matter to our will and forcing it to undergo specific change. Modern chemists, though, use these techniques to direct reactions more precisely and rationally than alchemists, cooks, and Faraday's contemporaries. They may seek to build complex molecules, and to do so proceed by stealth and subtlety.... They have found ways to emulate (and sometimes improve on) nature by mixing, stirring, and heating in such a way that they do not break asunder what they have already joined – even though they cannot manipulate the atoms directly.[1]

Chemistry is the science of molecules, and it is a messy science.

Where does chemistry stand in relation to other sciences? The ambiguity of the verb is apposite. The sciences do not stand like pavilions and temples in an arcadian intellectual landscape. They are seen to stand in another sense, ranked vertically from the highest to the lowest, from the hardest to the softest. Many distinguished scientists acknowledge a hierarchy of sciences.[2]

It is a hierarchy organized according to mathematical content. First comes mathematics itself. Then comes physics, followed by chemistry, with the mathematical or theoretical branches of these subjects placed ahead of the experimental branches. These are the hard sciences, so called not because their problems are necessarily hard to solve (though they often are) but because they better withstand rigorous inspection of the way they conduct their investigations. Next in the sequence come biology, geology, metallurgy, medicine. Bringing up the rear, and ruled beyond the scientific pale by some, are anthropology, psychology, economics, and sociology. The gamut runs from the devisers of formulae to the users of formulae to those who observe, catalogue, and record nature and human nature. "Taken in its entirety, the hard sciences lord it over the soft sciences, the whole dreary hierarchy inhibiting the development of a coherent understanding of the multifaceted, complex aspects of human experience", according to one writer.[3]

The hierarchy seems at times to exist almost independently of the contributions that the sciences within it may have made.[4] It stigmatizes qualitative observation in favour of the quantitative. Fitting numbers to observed phenomena can reveal order within them, but calculation is never a substitute for observation.

Yet there is justification for the hierarchy. Mathematical analysis of natural phenomena is demanding as well as rewarding. Physicists *are* on average brighter than biologists, and theorists *are* better qualified than experimentalists.[5] Moreover the fact that as sciences develop they tend to become more describable by numbers and equations rather than less so lends support to the notion that all sciences aspire to the condition of mathematics. The hard or mathematical sciences can be regarded as members of an exclusive club who from time to time admit a new member. Chemistry was admitted by degrees: in the seventeenth century when Robert Boyle reconciled it with natural philosophy (as physics was then called); during the nineteenth century when John Dalton and Amedeo Avogadro established that atoms and molecules combine in fixed proportions; and through the twentieth century with the application to the subject of quantum theory. The biological sciences are undergoing similar "elevation" now.

The tide is unstoppable but it does not, as some suppose, threaten to sweep away the inherent character of the scientific disciplines as we presently understand them. Some of the best scientists – those few who are given column inches to express their thoughts as well as to report their findings – have felt compelled to state as much. The comments of two Nobel laureates will suffice. Peter Medawar warned: "We are mistaking the direction of the flow of thought when we speak of 'analysing' or 'reducing' a biological phenomenon to physics and chemistry."[6] And Roald Hoffmann thundered: "Chemistry is not reducible to physics."[7]

It is easy to see why physics can feel itself superior to chemistry and the other experimental sciences. It holds dominion over many more orders of magnitude (powers of ten) – of time, distance and energy. Indeed, on closer examination, the physics community can be seen to have its own caste system. Physicists specialize by dividing nature into categories according to the scale of the phenomena they study, from astrophysics through solid-state physics to quantum, nuclear, and particle physics. The first three of these categories concern us here, and of these astrophysics is something of a special case. It has a romantic public image but is also seen by some physicists as the black sheep of the family whose hypotheses are not always testable and whose wilder theories not only cannot be proven, as Karl Popper famously suggested is impossible for any scientific theory, but also, embarrassingly, cannot readily be disproven either.

Chemistry, too, has its demarcations. *Organic chemists* work with carbon, the element of life. Organic compounds are all those compounds of carbon other than its gaseous oxides and the carbonate minerals such as limestone. The term arose when it was thought that such compounds could only be derived from natural organisms or their remains, but was rendered obsolete in 1828 when Friedrich Wöhler made urea, previously known only by isolation from animal matter, from inorganic salts. Wöhler's synthesis notwithstanding, the distinction between organic and inorganic persists in the division of chemists' labour. Yet nature recognizes no such frontier; materials move freely from the inorganic realm to the organic one when a tree absorbs carbon dioxide from the atmosphere during photosynthesis to form sugars, and back again when that tree is struck by lightning and its organic content undergoes full combustion to carbon dioxide.

Organic chemists typically take small carbon molecules, such as our ethyl alcohol, or acetone or benzene or thousands of others, and combine them to build other molecules. Theirs is a creative, constructive business. Organic chemists are the chefs and architects of the chemistry professions. What they do cannot be reduced (the mathematicians and physicists might say elevated) to a matter of equations.

Inorganic chemists seem on the face of it many times superior to their organic colleagues. They have the other ninety-one naturally occurring chemical elements within their purlieu. But it is too much. There is no central theme, no shared goal, unless it be the audacious one of understanding as much of all these as we do of carbon. For here lies the difficulty. Inorganic chemistry is a miscellany. Inorganic chemists may be specialists in silicon or sulphur, devotees of the transition metals, those elements whose electronic structures lead them to form beautifully coloured salts, or lovers of the extroverted halogens, fluorine, chlorine, bromine and iodine. But these people will quite likely consider that they have less in common with one another than with some organic chemists.

Physical chemists toil in the service economy of chemistry, measuring, analysing, and quantifying the nature of molecules and the heat and speed of their interaction. They are in effect surveyors, accountants, doctors, spies, photographers and film-makers. They are the closest to the physicists, but theirs is a physics confined to modest orders of magnitude, to the span of the size and influence of molecules, to the duration of their existence and transformations, and to the energies with which these transformations occur.

One of the most important tools available to this chemical physics is spectroscopy, which is the study of the interaction of light and other forms of electromagnetic energy with matter. Spectroscopy is in fact not so much a tool as a tool kit. Different forms of spectroscopy at different wavelengths of light may be used to elucidate different aspects of atomic and molecular structure when these wavelengths coincide with the energies of electronic transitions in atoms or molecules or with the frequency of molecular rotation and vibration. So powerful and varied are the various forms of spectroscopy that they warrant an entire chapter later in the book.

At the top of the tree are the *theoretical chemists*. Their apparatus need amount to no more than a comfortable armchair – and, these days, we allow, a powerful computer. They have no call for lab coats, no cause to stain their fingers with foul reagents. For they can predict the existence or non-existence of a chemical species and its fundamental properties using mathematics alone.

The theoreticians use the methods of quantum mechanics. Quantum theory sets out the quantitative relationships that govern atomic and molecular structure by means of, for the most part, simple rules based on the fact that at this scale energy is quantized, exchangeable only in certain fixed amounts, or quanta. These rules lead in principle to yes or no answers that determine the prospects for the interaction of energy such as light waves or microwaves with matter, or of one atomic or molecular species with another. For example, the transition of an electron in a high-energy orbital of an atom or molecule to a low-energy orbital may be accompanied by the emission of electromagnetic radiation; conversely, the promotion of an electron from a low-level to a high-level orbital will usually require the absorption of such energy. Similar processes occur when a molecule makes the transition between one quantum frequency of vibration or rate of rotation and another; or when an atom is ionized – that is, stripped of one or more of its electrons, transforming it into a positively charged ion.

For such transitions to occur they must meet the necessary energy requirements. For example, an electron in a low-energy atomic orbital cannot be promoted to a higher energy level if the energy required to make the step up is greater than the energy of the photons of the light wavelength being used to irradiate the atom, no matter how long or how intense the burst of radiation. They must also meet certain other conditions to do with the symmetry of the species involved. In the jargon of the field, events are then either allowed or forbidden. If one wishes to calculate, for example, whether two simple molecules will react, it is sometimes possible to do so with a knowledge of their shape and the energies of their molecular orbitals. Theories based on these principles can then predict the preferred orientation during this coupling and the product that should result.

But our ability to make accurate predictions with the techniques of quantum mechanics quickly diminishes when we try to deal with heavier atoms, which contain more protons and neutrons in their nuclei and more electrons in their orbitals, or when we bring atoms together as molecules. Two bodies may be predicted with confidence to behave in a certain way. Thus, we have a complete and accurate quantum picture of the hydrogen atom with just one proton and one electron. But a carbon atom, for example, has a nucleus of six protons and six neutrons with six electrons spread between four different atomic orbitals. Consider one of these six electrons moving around the carbon nucleus. In isolation, we would have a picture much like that for the hydrogen atom. But the electron we have selected is influenced slightly by the presence of the other electrons and the other electrons by it. What was a problem of the mutual influence of two bodies is now a problem of seven bodies. Our selected electron is said to have been perturbed by the presence of the others.

Already we are losing precision in our attempt to provide a quantum portrait of carbon, which is after all one of the smaller elements. The picture becomes still more blurred with even the simplest molecules. Now we must consider not only the interaction of a number of electrons with each other and with their parent nucleus but also the mutual influence between nuclei and between their sets of electrons. Even the simplest molecules pose problems. Other chemists joke that the theoreticians' ideal molecule is the ion that contains two hydrogen nuclei, single protons, which share a single electron in the bond between them – which, to the chemists, is no molecule worth the name. Even this problem of three bodies is frequently simplified by the pretence that it is only the electron that moves under the influence of the far heavier protons and that there is no reciprocal action of the electron upon the protons. As for ethyl alcohol with its nine rather simple atoms, this certainly cannot be delineated in quantum terms with total accuracy.

Computers have enabled theoreticians to progress in recent years, but their methods still rely upon the making of judicious approximations before a powerful number-crunching iteration can be set going to produce an estimate of a given molecular quantity. Different theories demand different approximations. There are many such variants to the basic quantum theory which can lead to wildly different "solutions" from similar starting conditions. Theory and experiment are often out of step, and if an observation is in want of a theory to explain it, then one of these modifications is almost sure to oblige. To the sceptical chemist, it sometimes seems that a theory can always be found that will fit a given result. Alternatively, theory may predict a certain result, but the technique is lacking in order to do the experiment that will corroborate (or refute) it. Some theoreticians snootily maintain that they have no interest in the wet and smelly world of real chemistry. But they secretly know that chemistry is about experiments, and that they depend upon experimental verification for their own professional standing.

It has been argued that the rise of "mathematical observation" is a principal and long-standing cause of the growing alienation of science from other aspects of culture.[8] The split goes back to Galileo's inquisition. Newton used mathematics in part to preserve knowledge from those who would misconstrue it. He used Latin for the same reason, despite the fact that Shakespeare and Dante had long before written in the vernacular.

As science has been seen to grow apart from other modes of cultural expression, so individual sciences have also become more mutually isolated in consequence of the addition of layer upon layer of detail to our knowledge, which has led, in turn, to an assumption that there is a need to correct this drift. In fact, it has become virtually axiomatic that interdisciplinary science is superior to unidisciplinary science.[9] Yet science which is not interdisciplinary must have something going for it. This, after all, is largely the science that has evolved through humankind's systematic investigation of nature over several centuries, driven by curiosity and, increasingly, by the demands of commerce. The focus derives from the need to divide complex subjects, and, more prosaically, from the employment of techniques special to each subject. These constraints in turn dictate that when interdisciplinary science is purposely practised it is not by Renaissance men acting out of a spirit of idealism. Interdisciplinary science usually arises as a pragmatic collaboration when a specialist in one area feels that an expert in another will have something to contribute to the solution of a particular problem, or when these two are brought together by external forces. For this reason, the building of interdisciplinary teams is more a feature of industrial research, where problems are more determinate and where managers direct the activities of researchers, than of academia. Such research is a liaison of convenience. Knowledge of a discipline other than one's own need only extend to the point where it is felt that the other discipline might contribute to the solution of a problem in hand. It need hardly be encyclopaedic. Nevertheless, scientists often lack even basic awareness of each other's disciplines, and where they do have knowledge or curiosity they often play it down for fear of appearing to be jacks of all trades.[10]

Although interdisciplinary science is not the idealistic pursuit that casual use of the phrase sometimes suggests, some scientists feel that it is in the uncultivated margins of their fields and in the hedgerows between them that the best discoveries are to be made. For them, it is worth the effort to acquire special expertise that does not fall within the conventional definition of one or another discipline.

Richard Smalley is such a scientist. He came to Rice University in Houston in 1976 as a physical chemist and now holds full professorships in both chemistry and physics as well as the chairmanship of the Rice Quantum Institute. Smalley today appears the archetype of the scientist as magnate. He drives a black Audi with a manual shift and a telephone. His office, on the top floor of the Space Sciences building on the Italianate Rice campus, is plush and spacious compared to the rabbit hutches occupied by most academic scientists. Its walls are neatly lined with chemical and physical journals. In discussion with other scientists he holds sway as if at a corporate meeting, swivelling in his chair, bringing in first one person, then another. His unblinking grey-blue eyes focus sharply on each interlocutor. The cut of the white beard that traces a precise line along his jawbone gives him an almost military air, an impression that is enhanced as he strides through the windowless laboratory that adjoins his office and clambers among the metal plate of his apparatus like the captain of a submarine. His dual professorships are clumsy conventional recognition of the fact that Smalley's interests lie on the border between traditional physics and chemistry.

Chemistry is the science of molecules, but molecules are not the only entities to populate the mezzanine between atoms and their smaller constituents (the domain of particle physics) and bulk material (that of solid-state physics). Chemical bonds between atoms take many forms. The strong bonds that characterize well-found organic molecules such as ethyl alcohol or benzene are one type. These bonds involve the sharing of electrons from the atomic orbitals in new molecular orbitals and are known as covalent bonds. As we have noted, an orbital is the volume of space available to an electron, in effect its three-dimensional orbit. In atoms, electrons occupy atomic orbitals. In molecules, these overlap to form molecular orbitals. Atomic orbitals have different shapes which are assigned letters *s*, *p*, *d*. The *s* orbitals are spherical with the atomic nucleus at their centre. The *p* orbitals come in sets of three pairs of lobes that protrude from the nucleus along the three orthogonal axes; one pair sticks out top and bottom, another back and front, the third left and right. (The shape of the *d* orbitals is more complex and does not concern us here.) Molecular orbitals which arise from the overlap of *s* and *s* atomic orbitals and of *p* and *p* orbitals are correspondingly known as sigma (σ) and pi (π) orbitals. Where atoms come together, they may form bonds of this type or of certain other types if their respective electronic structures so permit. If the electronic structures are not in favourable conjunction, then the species may remain discrete. Sometimes, however, they may form less stable entities, entities that do not exist in nature, entities whose bonds can be made to form only briefly and under special conditions, entities beyond the ken of chemistry and, by default, more a part of physics.

In particular, atoms of a single element rarely form stable molecules. There are diatomic gases – hydrogen, nitrogen, oxygen, and the halogens – and oxygen occurs in triangular threesomes as ozone; four-fold phosphorus is a tetrahedron and sulphur forms rings of eight. But in most cases the aggregation of like atoms is a somewhat arbitrary affair. Such groupings do not earn the right to the title molecule and must be content with the status of "clusters".

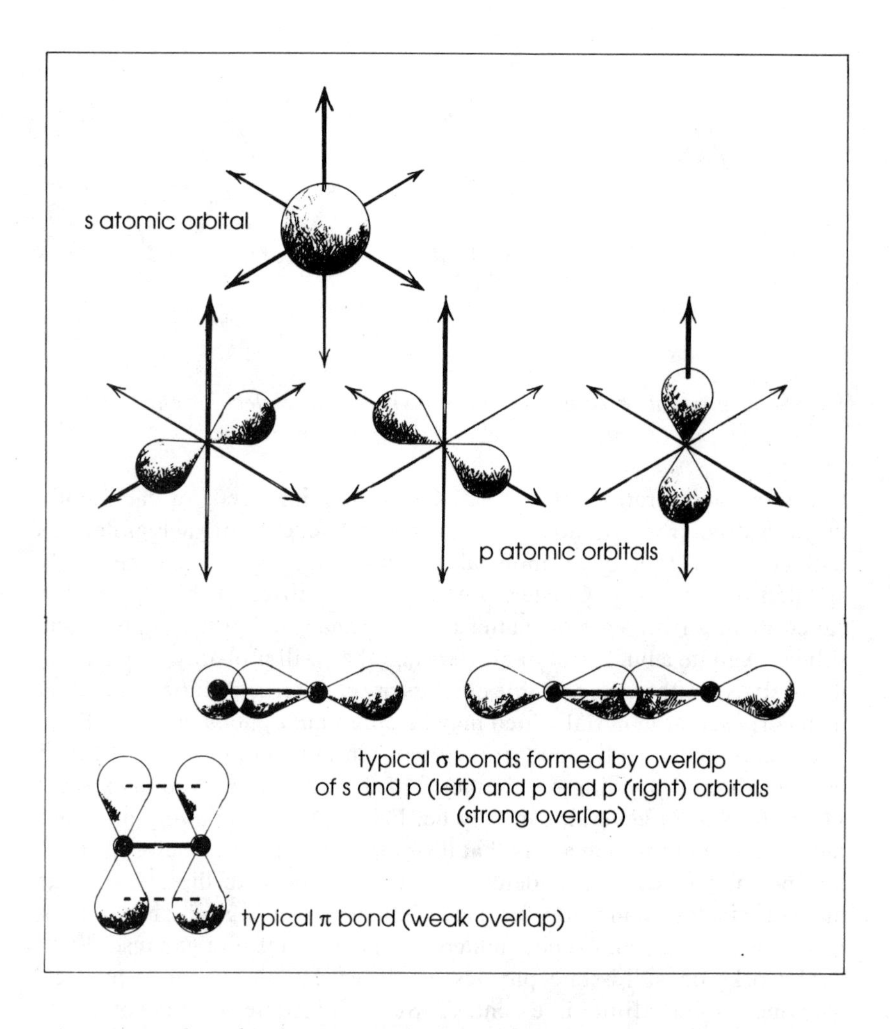

Atomic s *and* p *orbitals (top) overlap strongly to form sigma bonds (centre) and more weakly to form pi bonds (bottom)*

Any assembly of atoms might be called a cluster. What matters is how long and how strongly they cohere and in how fixed a shape. Molecules are merely a stable special case among the possible ways of combining atoms. It is in this poorly mapped but vast new territory, outside traditional chemistry and physics and at the limit of both theory and experiment, that Rick Smalley has chosen to pitch his camp.

Stable molecules are combinations of atoms held together by chemical

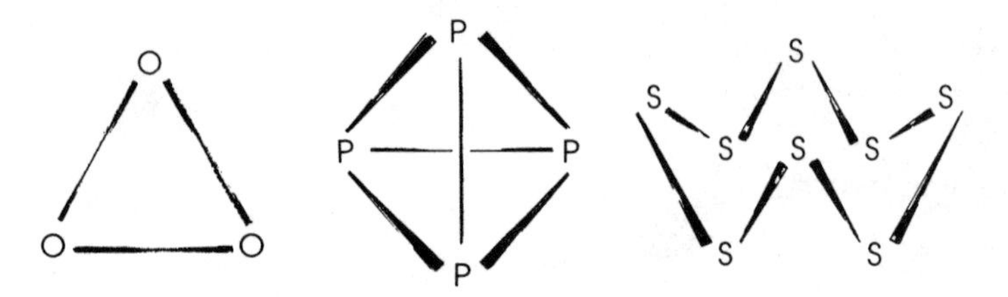

Some elements exist not as single atoms, but as molecular clusters. Ozone, phosphorus, and sulphur are shown

forces in configurations which satisfy the bonding needs of each atom; they are structures with no ends left loose. A bucketful of such molecules will generally remain as molecules unless that bucket is excessively agitated in some way. Clusters are sufficiently different that some have called them a fifth state of matter (after plasmas and gases, liquids, and solids). Agitate a bucketful of clusters and they will probably reconstitute themselves as other, more stable clusters or set to form a continuous rather than corpuscular material which may be either amorphous or crystalline.

Clusters form, split, and form again in this promiscuous fashion because the full bonding requirements of their constituent atoms are not satisfied. Nor, in general, can they be. Each atom has bonding positions radiating from it in such a way that it is impossible to form a closed structure no matter how many identical atoms are added to the cluster. The only satisfactory condition for such elements – among which are numbered most metals and semiconductors – is as an infinite expanse. Standard works on solid-state physics are littered with prefatory phrases, "assume a solid infinite in extent...." Such an infinite solid is composed of atoms happily bonded to each other – whether by covalent or by other types of bonds – in a stable array. But no solid is infinite, and for anything less than the infinite, there are always edges or surfaces from which half-bonds reach out hungrily in search of new atoms. What happens at these edges and surfaces is often the key to change in the bulk.

Scientists only really began to speculate about clusters in the 1960s. Theoreticians made the early running but, as so often, their predictions were wildly at odds. It only became possible to put theory to the test in the 1970s when Smalley and others succeeded in building an apparatus that could generate clusters upon demand and preserve them long enough for their physical properties to be probed.[11] In this apparatus, a sample is heated to produce a vapour which is blown at supersonic speeds into a vacuum. This cools the vapour and effectively freezes the action of cluster formation. The chilled clusters are then steered into an instrument that measures their mass.

The case of chromium provides a clear illustration of the confusion of theories at this time. Like all the transition metals, the chromium atom has five atomic orbitals denoted *d* in its outer, third, electron shell which is partly filled with electrons. There are six available electrons in the chromium atom. Each $3d$ orbital may take a pair of electrons. Five electrons half-fill the five orbitals, but because pairing demands a certain amount of energy the sixth prefers to occupy the next higher ($4s$) orbital rather than to pair with one of these five. The theoreticians could not decide whether the bond between two chromium atoms should be a strong, short sextuple bond using all six electrons in each atom, or a long, weak single bond using just the $4s$ electrons, or whether in fact any bond could be formed at all. Using his apparatus, Smalley was able to confirm that the chromium dimer has a sextuple bond.[12]

Knowledge of the element chromium in its pure state now covered the single atom, the diatomic cluster, and the "infinite" bulk. However, from there it was still impossible to infer the properties of other clusters simply by interpolation. *En route* from two atoms to many atoms, properties may not vary in smooth progression but may seesaw between extremes. Such irregular behaviour suggests intriguing goings-on. All the chemical elements are poorly understood at this scale, where atoms are present by the half-dozen, the dozen, or the score. Our ignorance is especially conspicuous in the case of carbon which is in other respects the most studied element by far.[13]

The study of the physics and chemistry of carbon has a long and distinguished history. In the mid-nineteenth century, Michael Faraday took carbon as the subject for some of his series of Christmas lectures for young people at the Royal Institution in London. His lectures on "the chemical history of a candle" shed light on many aspects of physical science:

> There is not a law under which any part of this universe is governed which does not come into play and is touched upon in these phenomena. There is no better, there is no more open door by which you can enter into the study of natural philosophy, than by considering the physical phenomena of a candle.[14]

Much of the drama of carbon chemistry – its role in combustion and in plant and animal life – unfolded during Faraday's lifetime, helped along in no small measure by his own researches. This understanding was achieved without certain knowledge of events at the atomic level which has been added in our century. Peter Atkins brought Faraday's discourse up to date by considering the chemistry that might befall a single carbon atom in the flame of the famous candle:

> Its future might be merely to link to other carbon atoms and form a particle of soot, which would imply at least a brief hiatus in its progress through the world, but its future might also be to form carbon dioxide and to join the gases of the air. There its mobility gives it wings to travel into different forms – perhaps to be absorbed into the oceans, perhaps to lie literally landlocked as limestone until the liberty brought by erosion sends it on its way again, and perhaps to be harvested from the skies by photosynthesis to make that awesome transition from inorganic to organic. One particular atom might in due course become part of a human brain and contribute to the awareness of its own past and future potential....
>
> The atom at the focus of our interest – carbon – is unique in the sense that it has stumbled into life: the reactions it can undergo, the liaisons it can enter into, are mundane, and they are shared, to a greater or lesser degree, by all the other elements. Carbon's kingliness as an element stems from its mediocrity: it does most things, and it does nothing to extremes, yet by virtue of that moderation it dominates nature. The pursuit of the carbon atom will show us many of the reactions that other elements can undergo, but nowhere in so well poised, so restrained a way.[15]

Atkins went on to describe in quantum-mechanical terms the formation of his carbon atom's first bonds, with hydrogen from the tallow, with oxygen from the air, and the complex assembly of various bonds to form large, stable molecules. But what if we limit our discussion to carbon only? What lies behind carbon's "merely" linking to other carbon atoms to form that smut? This is our direction.

Before considering the combination of carbon atoms, we should pause briefly to consider the various types of carbon atom that exist. Most elements occur in a number of types called isotopes with the same number of protons (and electrons) but varying numbers of neutrons. Carbon is unusual in that each of its major isotopes plays an active role in modern science. Carbon-12, which comprises nearly 99 per cent of all carbon, has been chosen as the standard by which all other atomic weights are measured.

Carbon-13, the only other stable isotope which makes up almost all the remaining one per cent, has six protons in its nucleus like carbon-12 but one extra neutron making a total of seven neutrons and hence a total of thirteen nuclear particles in all and an atomic weight of thirteen. Atomic nuclei with an odd number of protons and neutrons spin about their own axis. This is true of hydrogen with a single proton as its nucleus and also of carbon-13. By placing molecules of unknown structure containing these atoms in a magnetic field, it is possible to learn what other atoms are bonded to these atoms and hence to piece together the chemical structure. This is the basis of a form of spectroscopy that is one of the most powerful analytical tools in modern chemistry.

Carbon's best known isotope, however, is radioactive carbon-14. It is this form of carbon that is used to date archaeological artefacts. All these isotopes, the first two naturally occurring and the last made by the impact of cosmic rays upon atoms of nitrogen in the atmosphere, are present in any carbonaceous material. The presence of carbon-14 in organic matter is dictated in part by the extent to which the matter absorbs the isotope from the atmosphere (for example in the form of carbon dioxide during photosynthesis), and in part by the extent to which the isotope has decayed back to nitrogen, which it does with a half-life of 5730 years. Since any living matter stops absorbing carbon when it dies, it is thus possible to work out the approximate date of death of a specimen.

The nuclear and electronic nature of the carbon atom is well established. But everyday pure bulk carbon remains a surprisingly dark horse. Even such basic physical properties as its melting and boiling points remain matters of some ambiguity.[16] It was learned that single atoms quickly form bonded pairs during combustion. Then three-atom species were found in carbon vapour and in comets. Speculation is still rife concerning the role of both these species in combustion and other carbon transformations. Larger clusters are even less well understood.[17]

Detailed comprehension of carbon only recommences, as Rick Smalley recounted in a review paper, with the two forms familiar to all schoolchildren:

> In chemistry class in high school, most of us can remember hearing for the first time of one of the most vivid examples of chemical change. The pencil in our hands, we were told, contained elemental carbon in the form of graphite. As we dragged this pencil across a white page of paper, the black line left behind was, apparently, just a trail of carbon. In fact all the writing in most every book we'd ever read was, we were told, made from carbon black – a nearly pure form of graphite carbon....
>
> But we were then told there is a second form of pure carbon, one where the carbon atoms are connected in a tetrahedral lattice. If the common, dirty-looking graphite is simply pressed hard enough at a very high temperature, the chemical bonds can be rearranged so that the material is converted into this strikingly different form of carbon – it is converted into diamonds. This hard, smooth, brilliantly refractive gem stone, also familiar in our youthful experience from many touches of our mother's wedding or engagement ring, seemed so totally alien from sooty graphite that it was

> hard to believe the chemistry professor's claim that it was composed of just the same old carbon atoms. We were told that this equivalence could be proven simply by burning the diamond in oxygen and demonstrating that the only product was just CO_2. And not a few of us wondered if we had the guts to actually burn one to find out![18]

The elucidation of the structure of the two forms of carbon marks another remarkable episode in the history of science. In one of the first applications of the new technique of X-ray diffraction, the father and son William Henry Bragg and William Lawrence Bragg discovered the structure of diamond in 1913. The structure of graphite was resolved a decade later by J.D. Bernal among others.

Carbon has four electrons in its outer, second, electron shell, the electrons that allow it to form four bonds. Two of these electrons come from the spherical *s* orbital; two from the lobed *p* orbitals. However, when bonding, four new "hybrid" orbitals, known as sp^3 hybrids, are formed from the old *s* orbital and all three *p* orbitals, and each takes up one of these electrons. These hybrid orbitals have lobes pointing to the corners of an imagined tetrahedron. Upon repetition in a lattice, this tetrahedral bonding network builds up the rigid structure that explains the hardness of diamond. In the case of graphite, the hybridization is incomplete, producing three sp^2 hybrid orbitals with lobes directed towards the corners of an equilateral triangle, which leads to the characteristic honeycomb pattern of a graphite layer. One electron remains in each last unhybridized *p* orbital whose lobes protrude as before, at right angles from the plane of the graphite layer. This orbital is able to participate in the formation of a weaker type of bond, called a van der Waals bond, with its opposite numbers on adjoining parallel layers of graphite some distance away. Like sticks of candy floss stood too close to one another, these *p* orbitals merge to form a continuous electron cloud that extends over the entire surface of the graphite layer. The bonding power of this cloud of electrons is insufficient to prevent the honeycomb layers of the graphite from slipping past one another when an external force is applied. It is this structure that gives graphite its characteristic greasy feel. The electrons of these *p* orbitals are free to move within this cloud; they are no longer local to one atom but "delocalized". This mobility enables graphite to conduct electricity whereas diamond cannot.

The geometries that stem from the nature of carbon bonding in diamond and graphite lie at the root of almost all organic chemistry. Many organic compounds – ethyl alcohol among them – kink and branch at roughly tetrahedral angles like fragments of diamond. Other organic compounds have the planarity of a layer of graphite. Still others combine

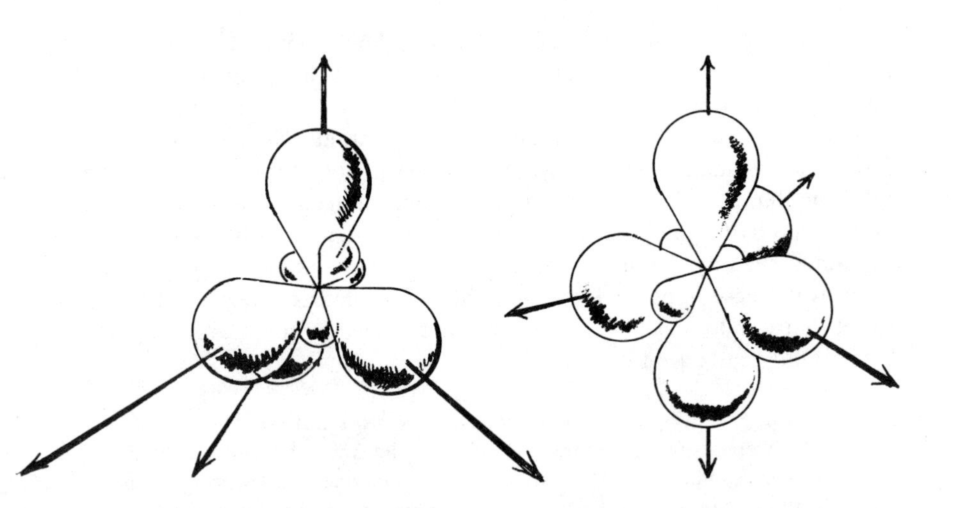

Carbon s *and* p *orbitals form hybrids. The* sp^3 *hybrid (left) gives a diamond-like structure;* sp^2 *(right) is planar*

aspects of the two geometries within the structure of a single molecule, but there are few other fundamental structural possibilities for organic compounds. The diamond lattice and the graphite sheet represent the two paradigms for carbon architecture.

Diamond and graphite, the rare and the commonplace, the clear and the dark, the tough and the yielding, provide ready literary images but are nevertheless of little practical interest to most chemists. Although organic molecules contain chains of carbon atoms variously like those running through both diamond and graphite, these pure *allotropes* of the one element provide a poor starting material for practical chemistry.

The revelation of the morphologies of pure carbon lagged well behind many of the basic discoveries of organic chemistry. Michael Faraday's great chemical discovery, made in 1825, was the isolation of benzene by fractional distillation of the residue left from the heating of whale-oil to produce domestic gas, then just coming into widespread use for lighting. Dalton's theory that atoms are solid and indivisible and that they combine in fixed proportions without transformation of the atoms themselves was current at this time, but there was no model that described the bonding between atoms in a satisfactory way, and so Faraday's discovery was of limited value without an idea of the molecular structure of the new compound from which it would be possible systematically to investigate its chemical reactivity.

A generation later, and still without the benefit of any electronic theory of bonding, August Kekulé deduced that carbon atoms must be able to form four bonds, and that they could bond readily with one another and not only with atoms of other elements. In 1859, he further postulated that double bonds could form between two carbon atoms. This explained the formulae of some compounds where carbon did not appear to bond to the full complement of four other atoms. Some molecules, however, have carbon and hydrogen present in proportions that apparently could not be explained even by double bonding. Benzene, with six atoms of each element, was a case in point. It was not until 1865 that Kekulé published the result that was to make him famous – the benzene structural formula. He recalled his realization thus:

> I was sitting, writing at my text-book; but the work did not progress; my thoughts were elsewhere. I turned my chair to the fire and dozed. Again the atoms were gambolling before my eyes. This time the smaller groups kept modestly in the background. My mental eye, rendered more acute by repeated visions of the kind, could now distinguish larger structures, of manifold conformation: long rows, sometimes more closely fitted together; all twining and twisting in snake-like motion. But look! What was that? One of the snakes had seized hold of its own tail, and the form whirled mockingly before my eyes. As if by a flash of lightning I awoke; and this time also I spent the rest of the night in working out the consequences of the hypothesis.[19]

Benzene, then, is not a carbon chain like many organic compounds but a ring. The ring structure with single and double bonds alternating around it explains the apparent shortfall of hydrogen atoms. The structure may be drawn in two ways: imagine placing one carbon as the downward point of the hexagon to be drawn out on a page; the first bond drawn clockwise from this six o'clock position may be single or it may be double, provided in each case that the subsequent bonds continue to alternate in the required fashion to produce the complete molecule. Quantum mechanics determines that we cannot know which of these two structures we have in any given instance. Both are equally likely and they are said to be resonant hybrids.

The story of the snakes chasing their tails is perhaps chemistry's most enduring myth. Whether Kekulé really dreamed his dream will probably never be known. Accounts of it date from celebrations of the twenty-fifth anniversary of his discovery, long after it had been put to highly profitable use and he had become famous. True or not, it will endure because it satisfies the human wish for serendipity.[20] More significant than the serendipity, however, is the fact that Kekulé had studied architecture at university. The discovery of the structure of benzene is a rational

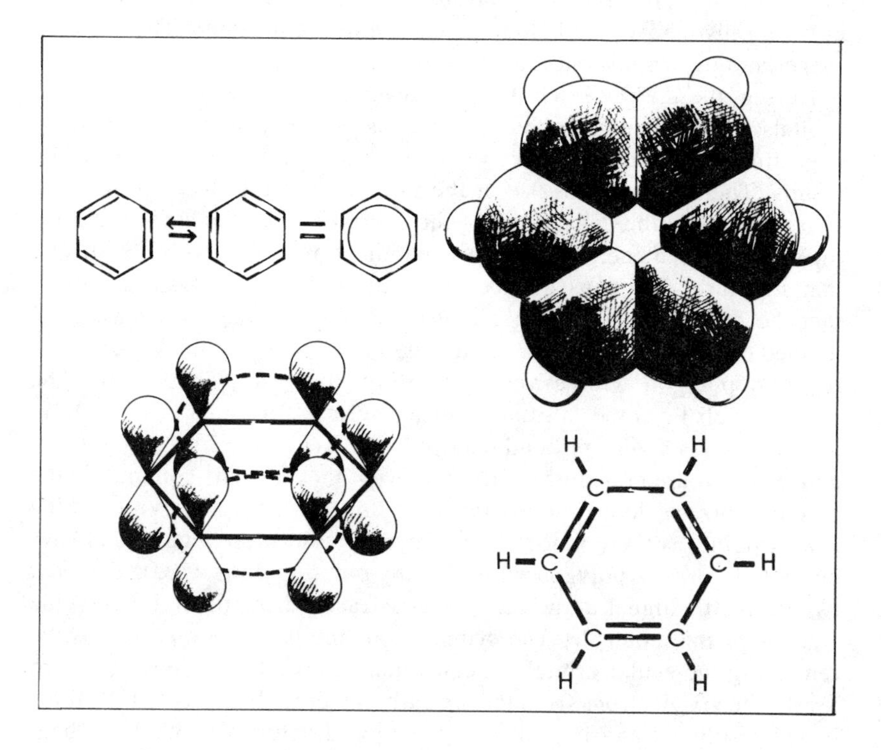

Representations of benzene. The ring within the hexagon signifies that the pi bond electrons may circulate

extension of Kekulé's earlier working out of the four-fold valency of carbon and of his dogged efforts to resolve the form of various hydrocarbons with equal numbers of hydrogen and carbon atoms – acetylene and styrene as well as benzene – that apparently flouted this tetravalency. Kekulé had a facility for visual thinking and devised many graphic structural systems for these molecules. Even when he published, Kekulé could not make a final decision between the ring and a double triangle structure which were his two surviving candidates whittled down from many contenders.[21]

In benzene as in graphite, it is sp^2 hybrid atomic orbitals that lead to the formation of the hexagonal carbon ring. These three-lobed hybrid orbitals overlap to form the six carbon-carbon sigma bonds in the ring and also contribute to the bonds with the six hydrogen atoms. The six p orbitals at right angles to the plane of the ring can be thought of as pairing off to form the three double bonds or pi bonds. A more accurate picture, however, portrays the electrons delocalized in two annular clouds hovering just above and below the carbon hexagon frame. These are no longer atomic orbitals but molecular orbitals.

During the 1930s Erich Hückel developed a theory of molecular orbitals for organic compounds such as benzene that was uniquely appealing both in the elegance of its working and in the sureness of its results. The theory makes use of the symmetry of the structure in question to simplify the calculation of the energies of its bonding electrons, the sum of which energies provides an indication of the likely stability of that structure. One consequence of the graphic nature of this theory is that the energy levels of these molecular orbitals may on occasion be divined directly from the structural diagram of the molecule. We draw the required polygon with a vertex pointing downward once more. The energy levels lie at the "altitude" of each vertex. The equator of the polygon represents a zero pi-bond energy level. Below the line lie the energetically favoured bonding orbitals; above it they are anti-bonding. In the case of benzene, there is one molecular orbital whose energy lies at the lowest point (as there is for all such cyclic molecules since it is always possible to draw a polygon with a vertex pointing downward). The next two lie a little higher at the same level as each other, still well below the equator of the molecule. The symmetry of the hexagon then places the remaining molecular orbitals at equivalent energy levels above the zero level. The six electrons from the six carbon *p* orbitals neatly fill the three bonding orbitals and it is this low-energy structure that explains benzene's stability. A similar planar ring with four or eight carbon atoms will not be as stable as benzene because two of its molecular orbitals which must take in electrons lie right at the zero level. Knowing in this way the energy levels of the molecular orbitals and the number of electrons available to fill them it is possible to gain a good idea of the stability and reactivity of any molecule of this kind. In general, Hückel showed that planar cyclic molecules with six, ten, fourteen ... carbon atoms are often stable whereas rings with four, eight, twelve ... carbon atoms are not.[22]

The cyclic, or ring-shaped, molecules predicted to be stable by Hückel's rule are called aromatic. The term predates modern understanding of molecular electronic structure and was originally applied to diverse molecules extracted from spices and resins. Aromaticity in a molecule is frequently an indication of particularly versatile chemical behaviour that can lead to the creation of chemical compounds with interesting properties and uses, and the chemistry of aromatic compounds has become a flourishing part of organic chemistry.

Hückel's theory also permits remarkably precise calculation of certain physical properties. Because we know the energy levels of the various

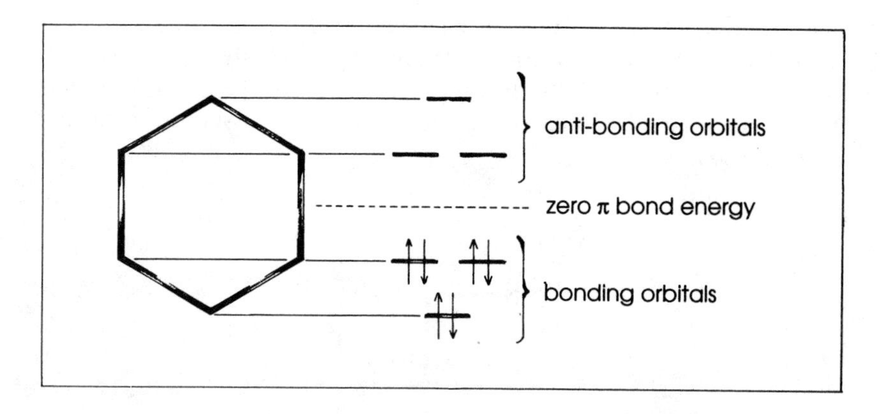

The position of the molecular orbitals of benzene on the energy ladder are given by its symmetrical structure

molecular orbitals, we can, for example, deduce the frequency of electromagnetic radiation required to excite one electron from its position in the highest occupied molecular orbital to the next available level, the lowest unoccupied molecular orbital. We can also calculate likely ionization energies, the energies required to remove entirely from the molecule one electron after another from successive highest remaining occupied molecular orbitals. The fact that we can do these things means that spectroscopy and other experimental probes are rendered that much more useful as diagnostics of the structure of these organic molecules.

Much of organic chemistry is devoted to the synthesis of molecules destined for practical use as drugs, dyes, perfumes, and plastics. Benzene and its derivatives are widely employed in this work. But there is a school that finds fulfilment by setting itself altogether more abstract targets.[23] The adherents of this school say that if a molecule can be drawn in the manner pioneered by Kekulé, as a carbon framework of atoms and bonds on a piece of paper, then it ought to be possible to make the thing.

The organic molecules that are conjured on paper in this way often have more visual than chemical interest. They are often molecules of high order or symmetry on paper but of such contortion in practice that they lend themselves neither to easy synthesis nor to systematic nomenclature.[24] It is ironic that these molecules often end up with highly descriptive names: cubane has eight carbon atoms at the corners of a cube; their mutual bonds form its edges. Fenestrane resembles a child's archetypal drawing of a window. Other oddities which have been synthesized include propellane with three cyclohexane rings sharing an edge from which radiate three hexagonal "blades". Paddlane is a similar molecule with benzene rings attached to a hub like a waterwheel. Molecules yet to leave the drawing-board include the fanciful carbon frames called Israelane and Helvetane, respectively a Star of David and a Swiss cross.

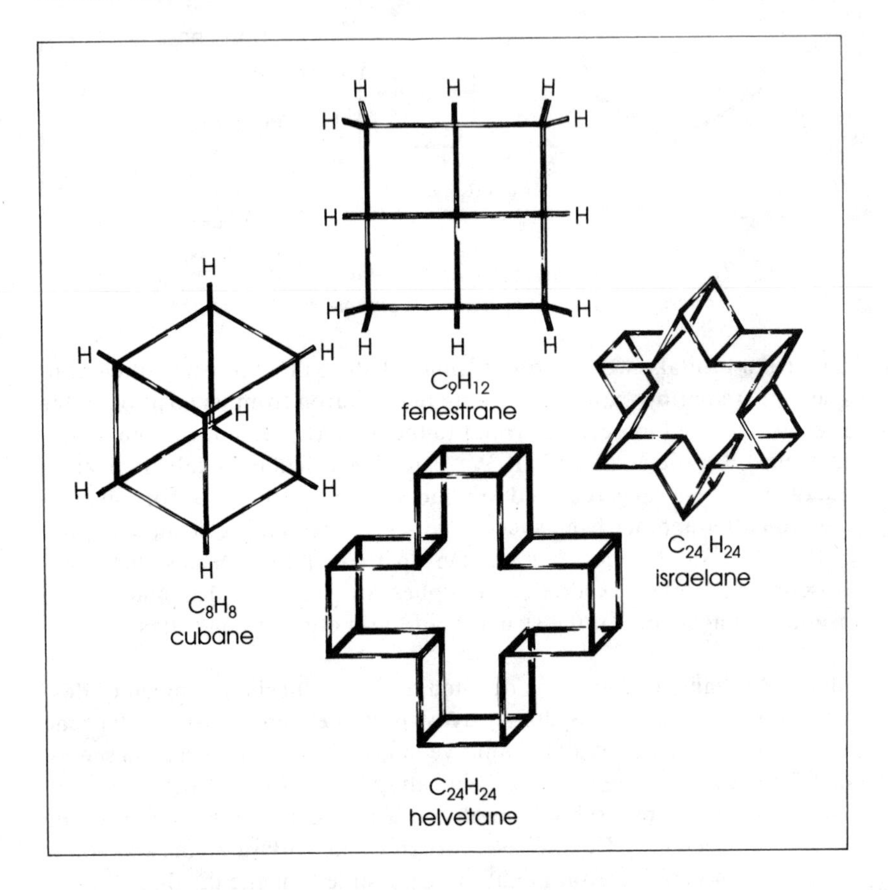

A miscellany of "designer" hydrocarbons. Cubane and fenestrane have been made; more complex ones have not

A few of these curiosities have more than just a passing aesthetic appeal. Adamantane, synthesized in 1933, is in essence a ten-carbon-atom fragment of diamond; the surface bonds that would continue the diamond lattice are tied off with hydrogen atoms. Diamantane and trimantane, larger derivatives of adamantane, extend the diamond lattice a little, adding carbon atoms, replacing carbon-hydrogen bonds with carbon-

carbon bonds, and pushing the hydrogen atoms further out from the centre of the growing lattice. Further extrapolation of this series would continue to tip the balance of the composition in favour of carbon, eventually creating a bulk material of pure diamond whose only hydrogen atoms would be those bonded at the advancing front, an insignificant proportion of the total confined to a surface contaminant layer.

An analogous family of molecules can be regarded as precursors to graphite. These begin with the benzene hexagon. Two such hexagons fused along a common edge produce naphthalene, another highly useful organic molecule and the ingredient of mothballs. Next comes anthracene, with three hexagonal rings in line. If we join the three rings to form a sequence that is kinked rather than straight, we form phenanthrene with the same number of carbon and hydrogen atoms but in a different arrangement, in other words an *isomer* of anthracene. In all cases the carbon frame is planar and rimmed with hydrogen atoms.

Then we continue: naphthacene has four rings in line, pentacene five, hexacene six. These molecules appear respectively orange, blue and green, which is to say they absorb blue, yellow and red light respectively. In doing so, they provide dramatic evidence for Hückel's theory. With each new ring in a molecule, there are more molecular orbital energy levels and hence a closer spacing between them. Thus, it requires progressively less energy (obtained here from the absorption of light of progressively longer wavelength and lower frequency) to promote an electron from the highest occupied to the lowest unoccupied orbital. Hence the sequence of colours.

As we assemble ever larger molecules, we can tile the hexagon units in many more patterns. Pyrene, for example, employs four hexagons not in line as in naphthacene but in the most compact arrangement of a lozenge. One of the most elegant of these molecules, known collectively as polycyclic aromatic hydrocarbons, is the poetic coronene with six hexagons forming a symmetric circlet around a seventh. Extension of this series by the addition of ever more hexagon units progressively pushes the hydrogen atoms out to the rim to yield a single plane of increasingly pure graphite.

By the early 1980s, the zenith of this art of academic synthesis was the achievement of a molecule called dodecahedrane by Leo Paquette and his colleagues at Ohio State University.[25] Dodecahedrane is a three-dimensional carbon frame with twelve pentagonal faces, a carbon atom at each of the twenty vertices linked to three other carbon atoms and each to a hydrogen atom – a Platonic playmate for cubane.[26]

These organic chemists might seem rather like those enthusiasts who build scale models of cathedrals out of matchsticks. They themselves

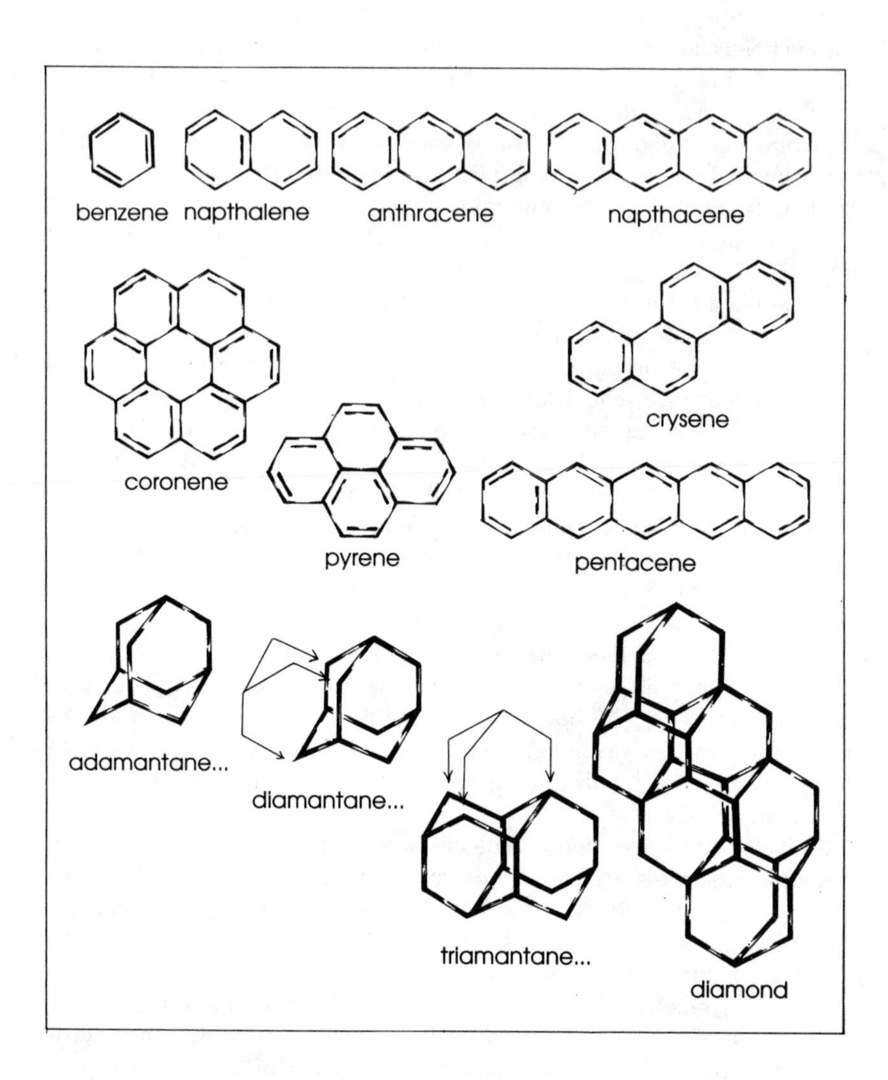

Adding hexagonal rings to planar hydrocarbons leads to graphite. Similar additions in three dimensions yield diamond

prefer analogies with chess, and offer a spurious justification for their obsession. Organic synthesis, say the synthesizers, is a cascade. Each step is a fork in the road, each intermediate leads somewhere new. The target molecule is just one of many that will be created if the design and exploration of a complex synthetic route proves successful. But no amount of protestation disguises the fact that the appeal really lies in the

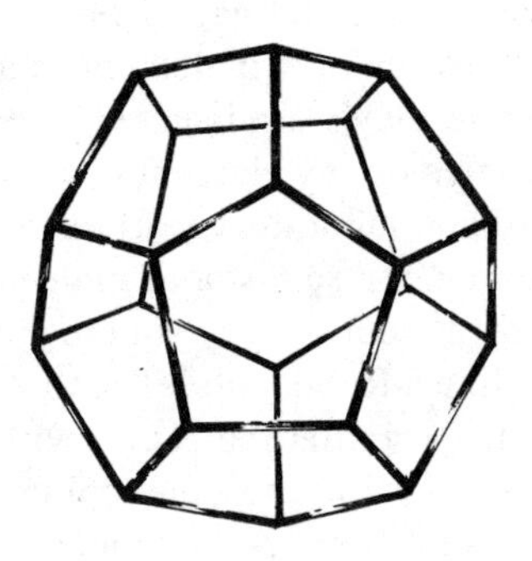

Dodecahedrane, $C_{20}H_{20}$. It is customary to show only the carbon framework; the hydrogen atoms are omitted

intellectual elegance of finding a path that leads from a few readily available starting compounds to some chemical Helen.

There are other puzzles that chemists can set themselves. Let us return to the mezzanine between the atomic and the bulk. Somewhere on this level, situated on firm ground and in well mapped territory, are the stable molecules; somewhere else, less travelled, is the shifting desert of the clusters. In a third region, beyond the limits of the stable molecules and some distance from the clusters is the suburbia of the semistable molecules. These are unlike stable molecules in that they are difficult to isolate and put in a bottle, and unlike most clusters in that they are made up of a mix of elements.

These were what Harold Kroto set out in search of at the University of Sussex near Brighton in England. A microwave spectroscopist at heart, Kroto wished to establish the identity of his new molecules by recording the microwave spectra of the radiation they emitted as they fell, in the moments between their creation and inevitable decomposition into more stable species, from higher quantum states of rotation to lower ones.

Kroto's approach had a pedigree logic. Dmitri Mendeleev, a contemporary of Kekulé's, sorted the chemical elements into sequence according to the number of protons in their nuclei, a quantity later termed the atomic number. By breaking this sequence in the right places and rearranging it in rows and columns, Mendeleev revealed for the first time the underlying order that explained why certain elements have very similar properties even though they are not sequential in atomic number. While the rows maintained the atomic number sequence, the columns of what became known as the periodic table revealed these groupings at a glance. Mendeleev's table even had gaps from which it was possible to predict the existence of still unknown elements such as gallium and germanium some years before they were isolated. It became a powerful tool for predicting the properties of any element from its row and column and remains on every school and laboratory wall to this day.

Kroto began his search for semistable molecules by constructing a similar table not of elements but of very simple compounds.[27] This table was three-dimensional, like a Rubik cube. The three axes of the cube represented related properties – similar molecules with the place of one atom taken in turn by other elements in the same column of the periodic table vertically; compounds with different constituent atoms but with the same number of electrons available for bonding along one horizontal direction; and compounds with more complex substituents that nevertheless remained homologous with the root compound along the third direction at right angles to both of the others. Stable small molecules already filled some positions. Others were pencilled in, awaiting ingenious synthesis and challenging spectroscopy.

Consider the table that has formaldehyde as its cornerstone. Ethylene, with the same number of electrons as formaldehyde, lies along one horizontal axis; acetaldehyde, analogous with formaldehyde but with one of its hydrogens displaced by a methyl group, lies along the other. Both these compounds exist. Directly below formaldehyde in the table is thioformaldehyde in which a sulphur atom replaces the oxygen atom in the formaldehyde carbon-oxygen double bond. This one did not exist. Kroto made it one of his targets.

This was chemistry as a board game. What moves could be made? Which squares could be filled? The principle was simple: to use heat or light of an appropriate frequency to bring about decomposition of a stable precursor that would yield the desired product, all the while maintaining experimental conditions that would limit the likelihood of further reactions and thereby prolong its existence so that it could be examined spectroscopically.[28]

The instability of some of these species arises from the ill-matched sizes of the atomic orbitals of neighbouring atoms such as the small carbon and the larger sulphur in thioformaldehyde which suggest that pi bonding is likely to be weaker than in formaldehyde itself. A similar argument explains why, for example, carbon dioxide is a gas of discrete molecules held together by two carbon-oxygen double bonds made by the overlap of orbitals of very similar size, whereas silicon dioxide, with large silicon orbitals and small oxygen orbitals, is a single-bonded crystalline solid – quartz sand. In the event, Kroto was beaten to his molecule by scientists at the United States National Bureau of Standards who used

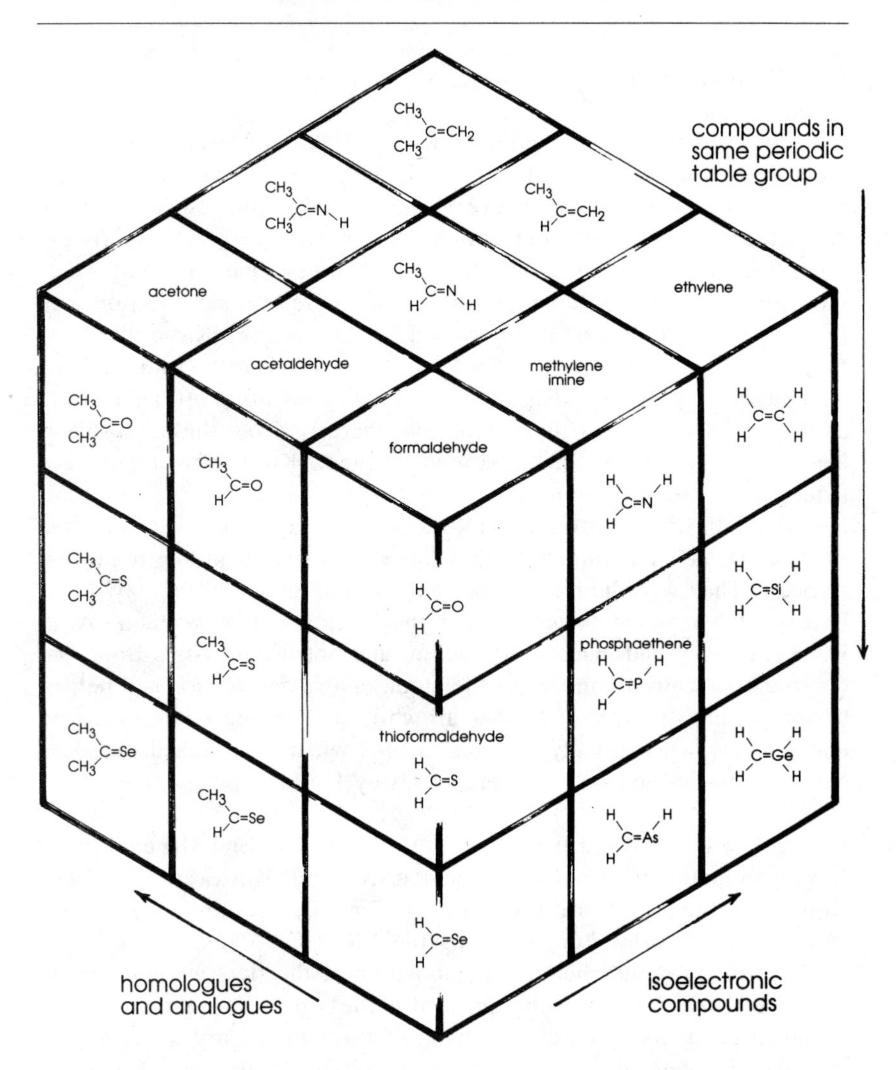

Formaldehyde, CH_2O, and related molecules, some well known, others unknown and targets for synthesis

little more than a Bunsen burner to bring about the required decomposition. To add insult to injury, they were able, from spectroscopic evidence, to confirm the presence of thioformaldehyde in space. Kroto did go on to fill other gaps in his grid, making phosphaethene, which has the same number of electrons as thioformaldehyde, and a family of related compounds centred on its carbon-phosphorus double bond. But the thioformaldehyde defeat also had a beneficial outcome in that it stimulated

Kroto's interest in astronomy, which was to contribute greatly to his later work.

Kroto conforms to some extent to that other caricature of the scientist, the man who has an insatiable interest in anything and everything, rather than the specialist who knows everything about nothing. His appearance and demeanour bolster the perception. His straggly grey hair, slightly too long, is sometimes combed, but often not. He dresses casually, with a certain personal flair. His name appears to owe more to science fiction than to science fact. (His parents shortened Krotoschiner to Kroto after settling in Britain having come from Berlin in 1937.) Harry Kroto is a man who measures progress along life's highway by counting off the avenues as they flash past never to be explored: there is Kroto the might-have-been tennis ace, Kroto the frustrated designer, Kroto who might have made a career in television....

Kroto and Smalley are quite unlike in most ways. But, in 1985 when they began their collaboration, they were very similar in two important respects. They were both ambitious and both of an age when they might be expected to have begun to fear they might not make their mark. Humphry Davy had discovered sodium and potassium, Niels Bohr had devised his electronic theory of the atom, and James Watson had helped to unravel the structure of DNA – all while still in their twenties. Kroto and Smalley were already in their thirties when they staked out their career territories and were in their forties by 1985.

It would be hard to find two sciences that appear at first glance to have less in common than chemistry and astronomy. However, they have enjoyed a long and fruitful collaboration. The rudiments of spectroscopy that became known during the 1860s facilitated astronomers' attempts to assay the chemical composition of the stars while chemists sought to hone the new technique and harness it in the search for new chemical elements on earth. During the twentieth century it has gradually been recognized that space supports a lively chemistry all its own. In the seconds after the Big Bang, the lightest atoms formed first, mostly hydrogen, rather less helium, and very little of anything else. The heavier elements are believed to have been made later in stars and therefore form a small proportion of the total matter in the universe. Even today, only one atom in a thousand, roughly one per cent by mass, is carbon.

In the 1930s, scientists began to believe that matter was not confined to stars, planets and other large objects but that there was also substantial matter between the stars. Some diatomic species such as carbon-hydrogen and carbon-nitrogen were discovered during the 1940s, but for the next twenty years nothing more was seen. Diatomic molecules were believed to be all there was.

The breakthrough came in 1968 when Charles Townes and his colleagues used a radiotelescope to detect microwave radiation at frequencies known from experiments on earth to be associated with transitions between quantum states of rotation of the ammonia molecule. Their shock announcement that ammonia was present in space was followed a year later by the still more exciting detection of water.

These were some of the molecules that had been used in a famous experiment some fifteen years earlier by Stanley Miller and Harold Urey at the University of Chicago. By passing "lightning" discharges through supposed early earth atmospheres containing mixtures of methane, ammonia, water vapour, and hydrogen, Miller and Urey found they could make a "primordial soup" containing several amino acids and simple DNA component molecules.[29] The detection of the gases in space of course suggested to some an origin of life outside our solar system. For mavericks such as the astronomer Fred Hoyle it confirmed the feasibility of life on other planets. For professional sceptics, these were just the sort of unprovable – and undisprovable – claims that gave astrophysics a bad name.

Before long, many other molecules were found in space, among them hydrogen cyanide, formaldehyde and thioformaldehyde, ethylene, acetone, and ethyl alcohol, "enough", according to Kroto, "for 10^{28} bottles of schnapps in Orion alone".[30] As the headcount of molecules increased so did the realization that there must be a rich and varied chemistry of the interstellar environment.[31]

Space between the stars is far from uniform. Much of it is very cold and very empty, with on average no more than one atom of hydrogen for every matchbox-size volume of emptiness. But there are also clouds of various types. Some are dense and dark with masses greater than a thousand suns. The black regions of space once thought to be holes in the firmament opening out onto the void are now known to be such clouds. These dark clouds are a little warmer and have a far higher density of particles. Here, some atoms combine to form molecules. Other, diffuse, clouds are positively hot by the standards of space with temperatures above –200 degrees Celsius but are less dense. Although both environments are considerably more eventful than some other parts of space, the prospects for chemistry are still rather poor. There is a Catch-22: in dark clouds, there is enough density of matter for chemical reactions to occur but little penetration by radiation that might stimulate them; in diffuse clouds, on the other hand, radiation penetrates without difficulty but there is an insufficient density of potential reactants.

Elsewhere, the environs of mature stars that are spewing out heavy elements are home to many and complex reactions. These stars are rich in carbon, and carbon-containing molecules are abundant in their effluent. Also in the region of carbon-rich stars are solid grains, containing millions of atoms but of uncertain shape and composition. These grains may facilitate the formation of larger molecules by providing a surface upon which their constituent bits and pieces coalesce. In all, stellar, circumstellar, and interstellar matter is widely varied in density, texture, and temperature, criss-crossed by rays of all frequencies, jarred by the shock waves of cosmic events, pushed and pulled hither and yon by the great forces of the cosmos.

There is still much more about the chemistry of space that we do not know. We are at least aware of our ignorance, though, because we have probes that can detect what cannot yet be identified. We cannot for the most part capture material samples so we are compelled to analyse the light that reaches us. We can do as Newton did and split the light from our own sun into its colours. We can compare the quality of light emitted by different suns and learn a little of their age and health. We can see the reflected red light that is the light not absorbed by the minerals on the surface of Mars. All this is spectroscopy of a rudimentary kind. But the naked eye quickly reaches its limit. It cannot separate the frequencies or see beyond red and violet. Modern detectors can do these things, recording peaks and troughs of light intensity – visible or invisible – created by interaction with matter. Each chemical compound has its own characteristic graphical trace or spectrum that is quite different from that of any other compound. It is these fingerprints which enable us to identify the composition of matter whether in a sample cell on earth or light-years away in space. These features reveal not only the identity of an atom or molecule but also its personal history. The height of a spectral peak, a slight shift in its position, the shape of the peak, whether it is a broad hump or a sharp spike, all say something of the conditions that prevail upon that atom or molecule.

The universe puts on a dazzling light-show. Each spectral signal of light emitted or light absorbed carries information revealing, if it can be deciphered, what lies where, how much there is of it, how hot it is, how fast it is moving. Different spectroscopic techniques reach into different regions of space, the near and the far, the hot and the cold, the dense and the sparse, gathering news of the conspicuous features such as stars and comets and of the remote limbo in between. But – and here's the rub – many of the signals we can detect do not correspond with any signals recorded from atoms and molecules monitored on earth.

Several phenomena require special mention.

Most of the absorptions observed in space are broad indistinct features, but within this blur a platoon of forty or so sharp spikes known as the *diffuse interstellar bands* stand out, contemptuous in their clarity. First observed in the 1910s in the visible region of the spectrum, they surely signify the existence of something quite distinctive. There have been many attempts to solve this long-standing mystery. Each tentative assignment to some species whose spectrum has been recorded or theoretically predicted on earth and found to produce a passable match typically stands for a few years before new evidence shows that the match was not close enough. It is still hotly debated whether these bands are better explained by small molecules or by larger grains of matter. Initially, their relatively small number suggested that one species was responsible. But astronomers have found more and more bands over the years. They now stretch into both the infrared and the ultraviolet regions of the spectrum. Is the cause a single, but probably complex, polyatomic molecule or a family of related molecules or some improbably regular grain? If the bands had different relative proportions from place to place, one could conclude that more than one species was at work. But the bands always appear together suggesting that just one object or class of objects may be responsible.

In the late 1960s, astronomers spotted a new feature, an *ultraviolet extinction* (which means that light is both absorbed and scattered) at a wavelength of approximately 220 nanometres (2200 ångströms). The absorption is a quantum phenomenon, explainable by the presence of atoms and molecules which are promoted into higher-energy states. Scattering on the other hand is largely a classical Newtonian phenomenon. It happens when light strikes much larger particles, just as when reflections come off a ballroom mirror chandelier. Atoms and molecules are smaller than the wavelength of the light associated with this feature, however, and far too small to scatter it in this way. So the fact that light is both scattered and absorbed suggests that both quantum and classical physical processes are occurring and therefore that whatever is responsible for this effect is larger than the molecules found in space so far.

The next few years also saw the detection of a number of *unidentified infrared* emission bands. Some were confidently assigned almost as soon as they had been detected. But bands are being found faster than explanations for them. In 1985, NASA scientists made the bold claim that polycyclic aromatic hydrocarbons might account for the outstanding emission bands. Having compared the astrophysical data with spectra of partially burned fuel laden with these carbon-rich molecules, they playfully subtitled their paper "Auto exhaust along the Milky Way".[32] Meanwhile, German chemists had made chain molecules with as many as thirty-three carbon atoms, and in 1977 Alec Douglas, who had shown that triatomic carbon molecules are present in comets, made the suggestion that very long chains like these might explain the diffuse interstellar bands.[33] Despite this effort, all these spectral features remain incompletely explained.

One reason for the continuing mystery is that it is hard to create space-like conditions in terrestrial laboratories. But space provides a fine laboratory in its own right. It serves, as Kroto put it, as "a colossal new spectroscopic sample cell, containing a plethora of exotic molecules".[34] Experimental findings from this vast laboratory have given astronomers new clues to the origin of the stars and planets and the universe itself. They have led chemists and physicists to explore what they might not otherwise have thought to explore, the chemists making new molecules inspired by what has been found in space, the physicists working a couple of orders of magnitude up the scale striving to create simulacra of the grains and larger particles, both seeking a match with the various unidentified bands.

One such physicist is Donald Huffman who joined the University of Arizona at Tucson around the time the ultraviolet extinction was found. His expertise was in the study of the interaction of light with small particles, and the university's astronomers latched onto him as a good candidate to undertake calculations of the likely properties of interstellar dust. At the turn of the century, Gustav Mie had found that particles scatter high-frequency radiation far more effectively than low frequencies. Mie's theory explains that the sky is red at sunset because particles in the atmosphere scatter the sun's blue and green light more efficiently than the red, and the sun's light slants through more of the atmosphere when it is low on the horizon than when it is overhead. It also explains why microwave radiation may be received from farther out in space than higher-frequency radiation thereby allowing the detection by microwave spectroscopy of molecules light-years from earth. His theory has several shortcomings of possible relevance to the study of interstellar dust. Mie assumed that his particles were spherical which, for various reasons, does not seem a reasonable assumption in this case. The theory was also bound to break down for particles so small that they approach atomic sizes. Interstellar dust, Huffman thought, might happen to lie just at that intriguing point where the quantum picture of the world takes over from the classical.

Huffman's approach was an empirical one. He would try to make the particles whatever shape they were and then measure their properties. The Arizona astronomers favoured a silicate dust. Huffman leaned towards carbon which also happened to be easier to work with. He alternated in efforts to make small particles of each material, only gaining the chance to concentrate his attention on carbon during a sabbatical in Germany in 1976. Here he set about a collaboration with Wolfgang Krätschmer at the Max Planck Institute for Nuclear Physics in Heidelberg. While Huffman had been toiling away on silicates, Krätschmer had been fulfilling a postdoctoral contract to study the tracks left by cosmic rays when they penetrate lunar or meteorite material. Cosmic rays are extremely high-energy charged particles of unknown origin, but despite this a rather unglamorous area of physics. Krätschmer was conscious of the irony that the experimental physicist who studies cosmic rays is working with particles far higher in energy than any that can be created in the most expensive accelerators but is nevertheless likely to spend his days in relative obscurity.

Upon completion of his project, Krätschmer was offered a permanent contract provided he switched his attention to interstellar dust. From the point of view of his employer, the problem was a practical one. Such dust always lies between an object and an observer and is thus something of a bugbear for astronomers. On the other hand, some information about this dust is inherent in the data accumulated by astronomers. Careful interpretation of their spectra might unearth new information about the nature of such dust and grains. The work was not as fundamentally exciting at first as the work with cosmic rays, but it did give the young Krätschmer the chance to serve a form of apprenticeship with Huffman who was regarded as a leading figure in the field which he had himself christened solid-state astrophysics.

Variously, in Arizona and at Heidelberg, Huffman and Krätschmer sought to make particles as small as they could in an attempt to understand the various spectral anomalies – notably the ultraviolet extinction. Away from routine campus pressures, Huffman pressed to redo his Tucson carbon experiments in more detail. The two physicists vaporized carbon and recorded ultraviolet spectra of the material that condensed from it, occasionally producing a feature close in profile to the 220 nanometre absorption – but not close enough.

The problem was to gain control of the condensation process. Vapour molecules cool and coalesce like steam striking a cold bathroom window and forming droplets of water. If the condensation continues unabated, the droplets merge and water begins to run down the window. Huffman and Krätschmer believed that their carbon particles were doing something rather similar, clumping together to form untidy superclusters rather than remaining isolated as they probably would in space.

They thought to slow the whole process down by trapping the first tiny molecules formed from the carbon vapour in a matrix of frozen noble gas. Then, by slowly warming the cold matrix the carbon molecules might diffuse through it and nucleate to form the desired larger particles in a controlled fashion. This too was unsuccessful.

Huffman's next opportunity to visit Heidelberg came in 1982. The two physicists immediately resumed work where they had left off. This time they managed to record better spectra of the soot condensed from the carbon vapour. The absorption at 220 nanometres reappeared, its single hump now split into two humps. They called it the camel spectrum. The new detail suggested for the first time that a single well defined molecule rather than some amorphous mush of uncertain size and shape might be the cause of the absorption.

Meanwhile in Britain, Harold Kroto was beginning a very different sequence of experiments that would also probe the zone between atoms and dust, not by examining ever smaller particles, but by working his way up, locating molecules larger than any yet seen in space.

Sussex was the first of a wave of new universities established in Britain in the 1960s. Its surprisingly fine architecture forms an idyllic campus that rides the rolling chalk downs just inland from the seaside resort of Brighton. The idealism of the time is apparent from the existence of such curricular features as an Arts/Science scheme under which all arts students were required to write a dissertation on a science topic and *vice versa*. Kroto had come back from the United States and Canada, lured by the prospect of an academic job, and wanting to catch the England of the Swinging Sixties and Prime Minister Harold Wilson's white heat of technological revolution. British scientists habitually take a perverse pride in chronicling a continual decline, so it is worth recalling that, although taking a considerable salary cut for the privilege of sharing this experience, Kroto was also returning to Britain at a time when spending on science was increasing at a record rate.[35]

The catalyst for Kroto's most spectacular molecular finds was the establishment of another course known as Chemistry by Thesis. After just two terms of course work, undergraduates pursued a research project for a full two years leading to examination on a written thesis. For such a course to be a success, research topics had to be selected with care and then tailored to balance plausibility and feasibility. They had to combine genuine new science with a goal likely to be achievable within the time frame. As if this were not enough, there was a requirement that the course be interdisciplinary. Each student was to be supervised by two faculty members with different areas of expertise. In its day, Chemistry by Thesis was arduous but rewarding for both students and staff.

> It could only have been successful in the interdisciplinary scientific research and teaching environment which the university pioneered when Sussex was founded in the "optimistic 60s". Sadly this and other courses have been "regulated" out of existence by bureaucrats who have little understanding of how student research expertise is brought to maturity and no awareness of the dire consequences for our future scientific capability.[36]

It was in this context that Kroto was to begin an enduring collaboration with another chemist, David Walton. Walton had pioneered methods of synthesizing complex acetylene molecules. Kroto was interested in these molecules not so much for their chemistry but because they represented an ideal system for spectroscopic study, a straight rod of pure carbon atoms, uncomplicated by angles and branches and foreign atoms. Under Walton and Kroto's guidance, a student, Anthony Alexander, was to synthesize molecules of this type and then record their spectra.

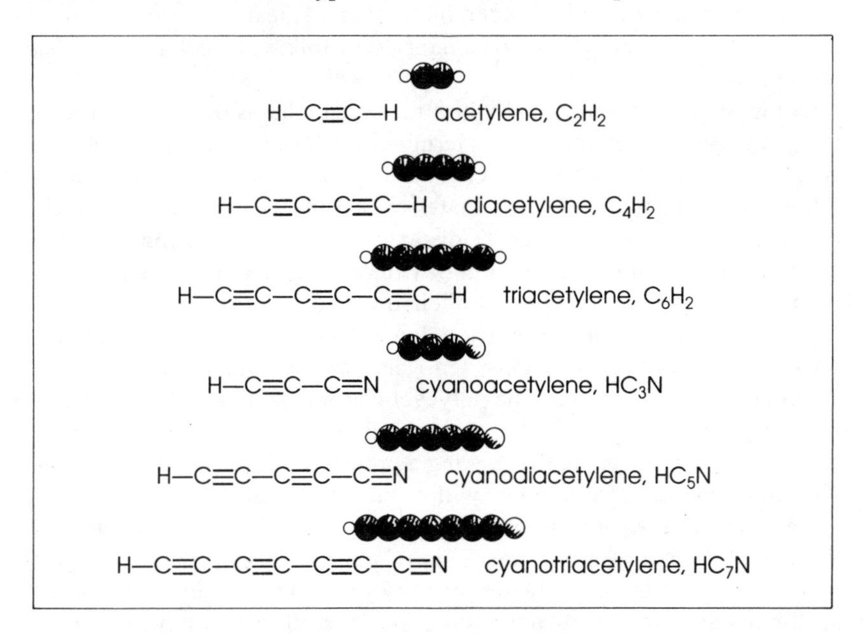

Polyÿnes and cyanopolyÿnes are chain molecules with alternating single and triple carbon-carbon bonds

In the full range of their intellectual pursuits, both Kroto and Walton take the Sussex credo of breadth of disciplinary interest considerably further than is strictly necessary. Of his many careers *manqués,* Kroto's love of design has found a practical outlet in artwork for university and faculty activities.[37] Walton's other talent is no less unexpected. In parallel with his chemistry, he has made time for ambitious projects to design a computer compatible with the three scripts in use in Sri Lanka (Roman, Sinhala, and Tamil) and to work on a dictionary for translation between the related languages, projects that exploit his interest in orthography and reflect a fondness for Sri Lanka and its people developed during a sabbatical year there helping to establish a university.

Acetylene, once the everyday stuff of bicycle lamps and street lighting, consists of two carbon atoms joined by a triple bond and capped at each end by a hydrogen atom. It burns in oxygen at a very high temperature which gives it its one remaining well known use, in oxy-acetylene welding. Acetylene has a strange and neglected chemistry. It can, for example, be persuaded to "burn" in the absence of oxygen to give a hydrocarbon residue of uncertain composition. Longer acetylenes – polyÿnes in chemical newspeak – consist of concatenated triple-bonded units and are progressively more unstable. Under the action of heat or light the triple bonds burst open and the carbon chains cross-link with such rapidity that an explosion is often the result.

In the 1950s the longest polyÿnes made had chains of seven two-carbon pairs. Bulky chemical groups terminating the chains act as protecting spacers to prevent explosive cross-linkage.[38] Ostensible reasons for interest in these molecules are that their carbon chains are electrically conducting and that they offer the distant prospect, through control of the cross-linkage, of the construction of carbon networks to predetermined designs. But they are curious in their own right. For example, successively longer chains are brightly coloured, pink for triacetylene, blue, green, and black for the eight-, ten- and twelve-carbon chains[39] – the direct consequence, as with the polycyclic aromatic hydrocarbons, of the changing spacing of their molecular orbital energy levels. Walton continued this sequence, eventually making a protected chain of thirty-two carbon atoms and the bare polyÿne with twenty-four carbon atoms.

The reason for Kroto's interest in the acetylenes was more abstract. As linear molecules essentially made up of pure strings of identical atoms (one can afford to neglect the terminating groups) they could be regarded as simple quantum-mechanical systems. In collision a polyÿne would behave like a bendy rod. Strike it amidships and it would vibrate. Strike it at one end and the impetus would cause it to begin to rotate. The homogeneous chain could be expected to exhibit vibrational and rotational behaviour simpler than that of many smaller molecules and might therefore provide a new way to study the quantum-mechanical interaction between the two.

The only snag was that polyÿnes are symmetrical, and symmetrical linear molecules do not yield rotational spectra. The nearest asymmetric polyÿne look-alike is constructed by replacing the carbon-carbon triple bond at one end of a chain with a carbon-nitrogen triple bond – a cyanide group. So it was these molecules – called cyanopolyÿnes – that Alexander, Walton and Kroto set about making.

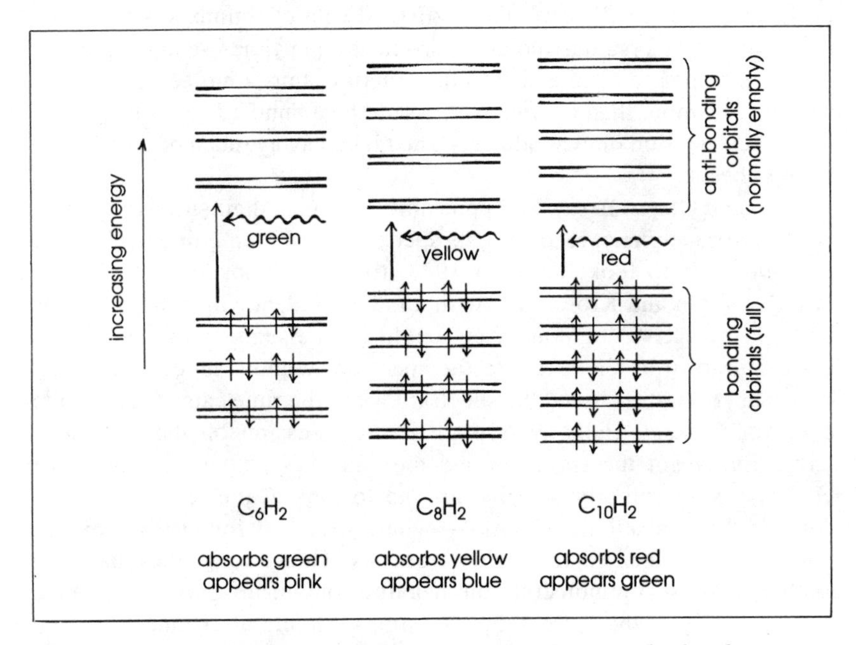

Absorption of light of different wavelengths gives the family of cyanopolyÿnes a range of distinctive colours

The simplest of them, cyanoacetylene – three carbon atoms in a chain terminated at one end by hydrogen and at the other by nitrogen – had been found among the growing clan of molecules known to be present in space. In 1974, the Sussex group made the next chain in the sequence, cyanodiacetylene, a five-carbon chain with the same hydrogen and nitrogen terminals, and duly recorded its spectrum. Stimulated by the cyanoacetylene find, Kroto contacted a former colleague at the National Research Council in Ottawa, Takeshi Oka, who readily agreed to collaborate in the interstellar search for this longer molecule.

The collaboration was the usual pragmatic affair. Oka and Kroto were both primarily spectroscopists. They teamed up with Canadian astronomers, Lorne Avery, Norm Broten, and John McLeod, who could point the 46-metre Algonquin Park radiotelescope and who knew the good places to look. They found what they sought the following year.

Cyanodiacetylene was not a molecule to seize the popular imagination like water or ammonia, but with more heavy atoms than any interstellar molecule then known it was a significant find. Based on an argument that interstellar molecules assemble through collisions with smaller molecules and ions, it was thought at this time that larger molecules would be much rarer than small ones. The statistical rule of thumb was that each new complex interstellar molecule would be ten times scarcer than the last for each heavy atom added to its structure; thus a molecule with two heavy atoms more than its precursor would be a hundred times less abundant. The detection of cyanodiacetylene kicked away much of the support for this theory.

Kroto and Oka soon saw an opportunity to repeat their success by making cyano*tri*acetylene, recording its spectrum, and then going back to the radiotelescope to look for it. In 1978, time was booked at Algonquin Park.[40] Walton and Kroto, now with another student, Colin Kirby, made the highly reactive compound, but could not persuade it to vaporize as was required in order to record the spectrum without its decomposing. They had still not recorded the spectrum when the time came for Kroto to leave for Canada. Halfway through Kroto's session on the telescope, Kirby finally got the spectrum and the data was phoned through. The telescope was promptly wheeled around to look at the region – a dark cloud in the constellation Taurus – where they had found the previous molecule and tuned to the right frequency. Hours later, they had it – another long, heavy molecule, the heaviest known in space, a molecule that, according to the current models, simply should not be there.

How far could they go? Oka now extrapolated data from the sequence of known chains – hydrogen cyanide, cyanoacetylene, cyanodiacetylene, and cyanotriacetylene – to predict the location of the spectral lines for the nine-carbon chain. This too was found to be present in space at a far higher concentration than anyone had supposed. They had broken their own record. It was not even necessary to make it in the laboratory, which was probably just as well given the fractious nature of its smaller cousin.[41]

All this while, the tally of molecules in interstellar space was increasing, and with each discovery the increased number of chemical permutations available made it likely that many more molecules were present as well. Shortly after the synthesis and detection of the cyanopolyÿnes, a number of chemists showed that the same molecules could be made in extreme conditions, for example by passing a discharge through a mixture of hydrogen cyanide and acetylene.[42] Akira Sakata of the University of Electro-Communications in Tokyo conducted a particularly striking experiment. In a parallel with Miller and Urey's attempt to simulate an early earth atmosphere, Sakata generated a plasma of hydrogen, carbon, oxygen and nitrogen that might be like that near a star. After bombarding the resulting gas mixture with acetylene, ammonia, and hydrogen molecules, Sakata found many large molecules, both chains and rings. Among the former were polyÿnes with up to ten carbon atoms and cyanopolyÿnes with up to nine carbon atoms. Among the latter were benzene, naphthalene, and other aromatic molecules.[43]

However, none of this addresses the paradox that the large molecules are found in space where they apparently cannot form. The problem with the standard ion-molecule collision theory is that for two species to combine to form a larger molecule there must be a third agent present to bear away the excess energy from the collision. Otherwise, the two colliding species will simply rebound. But triple collisions are far less likely than simple impacts. They may occur in and around stars where matter is very dense and mobile. But matter is comparatively sparsely distributed where the long chain molecules are found. Thus, either the molecules must have been transported from one type of environment to the other, or local conditions must have changed, or the wrong mechanism was being used in the attempt to explain what is happening. The surfaces of interstellar grains may play an important role, acting rather like park benches providing sites where molecules may rest, thus increasing the chance that two will meet. But, as we have seen, the nature of grains is even less well understood than the chemistry. Meanwhile, the problem simply grows more acute with the discovery of each new large molecule. A central reference on interstellar chemistry is frank in admitting the mystery of it all:

> Some families of interstellar molecules are intriguing, especially the interstellar polyynes. How are they formed? Are they present in diffuse clouds, and if so how do they arise in such objects? ... in fact, the chemical routes to most of the large molecules are quite obscure.[44]

High on the list of unanswered questions is: what molecular precursors of life are present in space? Most of Miller and Urey's "primordial soup" ingredients have been found. The detection of larger molecules suggests new routes for the potential formation of what Kroto calls "biologically emotive molecules".[45] And yet, glycine, the simplest of the amino acids that are the basis of proteins and the other molecules of life, with only five heavy atoms, far fewer than in the longer cyanopolyÿnes, has still not been spotted, although to be fair it is expected to be much harder to detect than the cyanopolyÿnes.

In Houston by this time, Rick Smalley had embellished his cluster beam apparatus by introducing a powerful laser to vaporize his samples. Now, for the first time, he could assemble clusters from materials with very high boiling points. The elements that cried out to be put in the new machine were the semiconductors, highly refractory and virtually guaranteed to yield results that would be of commercial interest. Together with his colleagues at Rice, Frank Tittel and Robert Curl, Smalley began a series of experiments to examine the clustering behaviour of silicon and germanium.

Nowhere is the no-man's-land between the atomic and the bulk scales of more pressing relevance than in semiconductor technology. Driven by the need to develop ever smaller circuit elements in order to miniaturize computer hardware, technologists know that they will one day reach an ultimate limit where quantum effects overtake the conventional electronic processes of the bulk. Knowledge of how semiconductor elements behave when present in very small quantities may shed light on this region of transition.

Like the carbon bonds in diamond, those of silicon and germanium, in the same group of the periodic table, remain tetrahedrally disposed like jacks, each bonding orbital placing itself as far from the other three as it can get. Clusters of these elements always have unsatisfied bonds dangling from them no matter how big they grow. And yet, despite this impediment, they very quickly begin to behave like the bulk material, as Smalley found.

Exxon Corporation had supplied the laser and contributed funds for the construction of the apparatus. Two of its scientists, Donald Cox and Andy Kaldor, had travelled to Houston to work on the design. Upon completion of the work one apparatus stayed at Rice University and another was shipped to Exxon's laboratories in New Jersey. With the assistance of two postdoctoral scientists, Eric Rohlfing and Rob Whetten, Cox and Kaldor used their new apparatus to look not at semiconductors but at two other hard-to-vaporize elements, iron and carbon. In July 1984, Rohlfing, Cox and Kaldor gave a paper at a conference in Berlin reporting the existence of clusters with even numbers of carbon atoms up to more than forty for the first time. At the same conference, Krätschmer and Huffman reported the results of some spectroscopy of small carbon molecules isolated in a matrix of frozen inert gas, a by-product of their failed attempts to control the condensation of carbon clusters.

Smalley, too, had coincidentally moved to look at carbon in addition to silicon and germanium. His latest coup had been to create silicon dicarbide, less like his earlier clusters, more a semistable molecule, an intriguing periodic table analogue of the three-atom carbon chain found in comets.

In the spring of that year, Kroto and fellow microwave spectroscopist

Bob Curl had attended a conference in Austin, Texas, and at Curl's suggestion, Kroto stopped by at Rice University on his way back to Britain. Kroto was excited by what he saw. With the best cluster apparatus in the business, Smalley had produced large clusters of silicon and other elements and now a mixed silicon-carbon species. There was every chance that carbon would follow suit. The hot laser vaporization region was similar enough to the conditions near stars, while the cold vacuum into which the vapour disperses was like interstellar space. Having found the carbon chains that were the longest molecules in space, Kroto now saw a way to simulate the conditions of their formation. He was convinced Smalley's machine would model the circumstellar chemistry that leads to the formation of cyanopolyÿnes and perhaps provide more definitive spectroscopic evidence of the nature of large carbon clusters than had been obtained in other experiments. Any such evidence would be one more nail in the coffin of the ion-molecule collision theory that Kroto was sure was incorrect.

Kroto and Curl talked over the possibilities, and Kroto pressed Curl to lobby Smalley to let him put carbon in the machine. Curl himself wanted to prove or disprove Douglas's conjecture that the diffuse interstellar bands might be explained by long carbon chains (he was inclined to believe that they were not) by using the machine to record real spectra for the first time. Familiar as he was with Kroto and Walton's work on the cyanopolyÿnes, Curl thought that the project would be of interest to all concerned.

Having introduced Kroto and Smalley, Curl now embarked upon the shuttle diplomacy that would eventually bring about their collaboration. It was not until a year later that everything was right. Smalley and his students had reached as much of a logical stopping point as they were likely to in their semiconductor investigations. They could afford to take a break. Curl reminded Smalley about Kroto, and after some discussion was able to go back to Kroto and offer him a choice: either they could do the carbon experiments and send him the results or he could come to Houston to do the experiments jointly. It was mid-August 1985. Kroto had to teach in October, so it was now or never. Kroto got off the phone and booked a flight. He would be in Houston within the week.

2

SEPTEMBER 1985

The story of what happened during those few, intense days in September 1985 is not an easy one to tell. This is not because the science involved is especially complicated. It is not. Enough can be deduced from the observations that were made of the physical processes that took place inside the apparatus known as AP2 in Rick Smalley's laboratory to enable the construction of a clear, if simplistic, picture of what was going on. No. The difficulty centres on the players, not the play. For it is through them and them only that we may attempt to learn what really happened. The *dramatis personae* comprises three senior scientists – Kroto, Smalley, and Curl – well known in their respective specialities, different in character and temperament; two students, Jim Heath and Sean O'Brien, just embarking upon the second year of their doctorates, also different in character and temperament; and two more students, Yuan Liu and Qingling Zhang, whose roles are less prominent. Each has a story to tell and the stories are not the same.

Were I to give them equal time, could I assemble from their separate tales an objective account of the events leading to their discovery? Of course not. So should I let you, the reader, do the work? Should I relate three or five or seven separate accounts from which you might draw your own conclusions and perhaps a truth or two? This can be a powerful technique when employed in fiction, as, for example, by Akira Kurosawa in his film (of Ryunosuke Akutagawa's stories) *Rashomon*, and by the Egyptian novelist Naguib Mahfouz in *Wedding song*. But such a device would be suspect if what results still purports to be a factual account. Both courses of action would fail. As there are different stories, so there are different tellings. The scientists do not only disagree on aspects of the story, they recount it in different ways. Some are self-effacing and underplay their part. Others hog more of the limelight than is rightfully theirs.

None of this presents an unfamiliar human situation. But it leaves us with a dilemma because science is founded on principles of objectivity. Many scientists seek to be objective in conducting their research, and claim to achieve that objectivity. Yet the process of research is certainly not objective; scientists are prey to obsession, jealousy, and prejudice like the rest of us. Neither are the observations that they make entirely objective; they are open to interpretation, criticism, analysis, "deconstruction".[1] Even the very data, some have argued,[2] cannot be objective. Notwithstanding these cavils, scientists often extend the orthodoxy of objectivity to encompass a belief that the account of their work can be objective as well. It is against this background that Jim Heath and Rick Smalley were to prepare a chronology – "the chronology" – of the events that I am about to relate.[3] They began it thus:

> The following is a running account of the activities at Rice in late summer of 1985 that led to the discovery of C_{60}.... The account is based on entries in the AP2 log book for that period, and on the raw data files that were saved. The dates and times of each data file (excepting a few as noted below) were recovered from the archival floppy disc and are guaranteed accurate, but most other comments and recollections entered here are in the category of honest but distant memories.

Where there is agreement between all those party to the events being related, then to the extent of that agreement, an account can, with reservations, be called coherent (although it still cannot be called objective if all involved share the same agenda, by virtue of that very fact). Bob Curl and Sean O'Brien (with varying degrees of satisfaction) have agreed the Heath version. But Harry Kroto has not, preferring that each member of the team should write his own, independent, account. There is a coded admission that this version has not in fact met with everybody's approval in the final sentence of the preamble: "By September 1, 1992 none of the team members had sent in any major changes to the version that appears below."

As the significance of buckminsterfullerene has come to be realized and as the potential has grown for its commercial application, so it has become important for those who made the discovery and others to put down accounts of it. But the more these are written in consciousness of the enormity of the find, the less valid they become as texts for learning about the events of 1985.

Both Kroto and Smalley wrote a number of review articles in the two or three years after the discovery was made.[4] The term review is used in scientific literature to describe an article written by a scientist that does not contain a first report of new science, but which typically places several such observations in a broader context often for a less specialist readership. The penning of reviews is one of the responsibilities that comes with success in science and part of the necessary procedure to consolidate a claim of priority.

With the passage of time and the clouding of memory, however, the two scientists have grown to disagree over some details of what is said to have happened. Journalists' accounts, for all their attempts at impartiality, are also prey to revisionism – and the one you are reading now can be no exception. Contemporary reports of the discovery appeared in the *New York Times* and in the chemistry industry's trade magazine in the United States, *Chemical and Engineering News*.[5] Somewhat later, *Discover* and *New Scientist* published more detailed features on the discovery of buckminsterfullerene and the new field of chemistry that it opened up.[6] The articles served to focus attention on the growing divergence between the two scientists' views of their individual roles in the discovery. Shortly afterwards, they each took the unusual step of publishing their own personal account of the events of 1985, Smalley in *The Sciences*,[7] the journal of the New York Academy of Sciences, and Kroto in *Angewandte Chemie*,[8] the leading journal of general chemistry in Europe. Their earlier reviews had summarized the understanding gained in the course of research using largely impersonal language. These new accounts employed vivid narrative and laid occasional emphasis on the actions of individual protagonists. They came into conflict not over questions of what and how and why, but over who and when. Who named the molecule and when? Who realized the key to its shape and when? Matters came to a head with the publication in *Science* – a learned journal that does not habitually air scientists' dirty washing – of a journalist's attempt to unravel the controversy.[9]

Even the primary sources disagree on such basic information as the dates of Kroto's arrival in and departure from Houston. It is notable, however, that the discrepancies are over minor details and that disagreement centres on memories and not on the interpretation of observations. There is no dispute between the members of the team over matters of science. (Far more compelling in this regard is the bitter dispute that was to break out between the Rice group and other researchers who did not believe their revolutionary result. This argument did centre on the science itself, and it was conducted, with all the discretion of a bar-room brawl, at conferences and in the open literature for several years after the discovery. Beneath the dry crust of critique and counter-critique of experimental method and interpretation seethed a cauldron of casuistry – belief and disbelief, faith and heresy, envy and fury – all the things that science is supposed to do without.) For now, the unresolved issues are important to those involved but comparatively trivial for science at large. They are important also to us since we are interested in the scientists as well as the science, in the myth as well as the "fact". Both Smalley's and Kroto's accounts have a highly personal tone, valid in their context, but understandably aggravating to those who, unsuspecting, find themselves playing a part in the story that is being unfolded. Written more than five years after the events which they relate, both also smack of a certain revisionism. There are episodes in each scientist's account to which the other takes the strongest exception.[10] At worst, the goal of such manoeuvring might be seen as the improvement of one scientist's claim to fame at the expense of the other's. Ultimately, it might be to render one scientist the genius of the piece, the sole architect of a new molecule. But it is obvious that both Smalley and Kroto (as well as the other members of the Rice group) contributed, albeit in different ways and at different times. The lesson that may be drawn from this controversy is then a modest one: it is not that the discovery *could not* have been made by one without the other; it is only that it *was not* made by one without the other.

I draw upon these accounts but do so in the awareness that they have been written with the benefit of hindsight and an eye on posterity. We must treat comments obtained from interviews[11] some years after the events with equal caution. What else is there? As the Rice chronology indicates, there is really only the AP2 log book.[12] This is the only contemporary document that could form an authentic record of what happened in the laboratory minute by minute and day by day. This now battered book whose chemical-stained cover bears the academic seal of Rice University is far from complete, however. Kept as it was by a rota of the students, it is scrappy and ambiguous. It is in its turn merely a commentary on the computer data files that record the experimental results in the graphical form of mass spectra. So these are perhaps the only truly impartial data available, and yet they too have been in a sense sculpted by human hands. Some files have been lost, others deleted, still others, for example when the apparatus was poorly calibrated, were never stored.

The only other record that has some claim to the objectivity we naively seek is the letter in the journal *Nature* that announced the discovery.[13] This paper is important not just because it was the first announcement to the world of this new form of carbon, but because it will remain for all time the primary published source available for independent study. If objective, then, unlike the log book, it is very consciously so. The paper also has the not inconsiderable virtue that it is the only account that bears the signature of all five principals.

I reprint this paper in its entirety at the end of this chapter. It is short, well written by the standards of scientific papers, and contains little jargon. It is the only description by the people who made it of one of the most important discoveries in chemistry this century. To read it is to share the excitement of the act of discovery in a way that cannot be re-created by an outsider. But first, we return to Houston in September of 1985.

The news of Kroto's coming was greeted with no great enthusiasm by Rick Smalley's group – his own students, Heath and O'Brien, and "the Chinese girls", Zhang and Liu, seconded from their respective advisors, Bob Curl and Frank Tittel, who with Smalley made up a triumvirate of professors working on semiconductor clusters at Rice. They had been getting some respectable results on the semiconductors, and time spent on a new project would be time out from this work. Smalley knew of the Exxon group's results with carbon and had no desire to compete with them by also looking at carbon.[14] Sean O'Brien's project was ostensibly to study the laser spectroscopy of semiconductor clusters. But his results so far had barely begun to indicate a suitable project for a doctoral degree. Kroto's arrival clearly was not going to help. He and Liu would have preferred to keep working on semiconductors.

Then again, there were some fundamental arguments that favoured at least a quick look at carbon. The element does stand at the head of the appropriate column of the periodic table. The four students had spent much of 1985 looking at the related elements, silicon and germanium. As graphite, carbon, too, behaves like a semiconductor in some respects. Yet carbon was bound to be different. Its behaviour in forming the strong chemical bonds that make possible the profusion and diversity of organic chemistry signalled that. Heath and Zhang were curious.

It is necessary at this point to describe the apparatus with which Smalley had conducted his studies of metal and semiconductor clusters and which he would now commit to looking at carbon.[15] In laboratory conversation it was called AP2, simply the second such apparatus Smalley had built during his career at Rice. But in the formal parlance of the scientific literature it was "the laser-supersonic cluster beam apparatus".

The very name is an intimidation. So many glamorous buzzwords of modern science in one title. What did it look like? The core of the apparatus was a large steel drum about a metre in diameter. It was pierced at intervals around its perimeter by several circular holes around which sturdy flanges had been fitted. Some of the flanges had polished stainless steel plates bolted to them with many bolts. Small quartz windows in the plates allowed laser light to enter the chamber. Other flanges provided a point of attachment for heavy stainless steel plumbing that led to other parts of the apparatus. The whole contraption sat on tables constructed to bear its considerable weight. AP2's attendants had to clamber up ladders and step carefully along the tabletops to make their adjustments. Around the apparatus on more tables sat a battery of lasers ready to fire through an assortment of optical devices such that their light struck through a quartz window and into the centre of the drum. Under the tables hung the apparatus's entrails – gas feed lines, vacuum pump tubes, electrical wiring and circulating coolant.

Such an apparatus does not come off a manufacturer's shelf. If one wants to do experiments that have not been done before, one of the best ways is to build an apparatus that has not been built before. Smalley designed and built AP2 initially to look at clusters of the transition metals. It was, however, a versatile tool. In principle, one could study the clustering behaviour of any element of the periodic table simply by placing it in a suitable form inside the machine.

One placed a rod or disc of the chosen substance in a holder at the centre of the steel drum where the laser light struck. A motor rotated this sample to expose a fresh part of the surface to each new laser pulse. Just to the side lay a tiny valve which could release a jet of gas. Opposite the valve – downstream from it – was a short tube forming a nozzle that led out into the high vacuum of the chamber. Protruding into the chamber from the far side and some centimetres from the nozzle was a large cone of machined metal which had been drilled down its centre. This borehole led out of the chamber to a secondary region buried within a solid steel cube. This was a busy junction. On each side of the cube was a circular faceplate that provided a perfect fit with similar plates on diagnostic equipment that could be bolted to the cube. A window on one side of the cube admitted pulses of light from a second laser. At right angles to this laser's line of fire a steel tube ran vertically upward, leading to a mass spectrometer. Other contraptions could be bolted onto remaining faces of the junction cube as required. The whole ensemble was under the control of a computer by means of which an experimenter could select pulse times and durations for both lasers, the timing of the gas pulse, and the parameters of the mass spectrometer.

In typical operation, this was what happened. A short, intense laser pulse fired and created a hot vapour of the chosen material just above the surface of the solid. The temperature of this plume of plasma was more than ten thousand degrees Celsius, hotter than the surface of most stars, and easily hot enough to vaporize any known material. From the valve at the side, a high pressure burst of gas flowed at near sonic speed across the line of the laser light, sweeping the plume away from the surface and blowing it along the downstream tube towards the vacuum chamber. The pressure pushed it onward. Within the plume, rapid collisions between the particles of the hot plasma and the cold, dense puff of gas cooled the plasma very fast to around room temperature. Clusters began to condense. All the while, the gas carrying the condensing matter was hurtling towards the vacuum chamber. As it expanded into the vastness of the chamber, the gas and the clusters it carried with it expanded and cooled

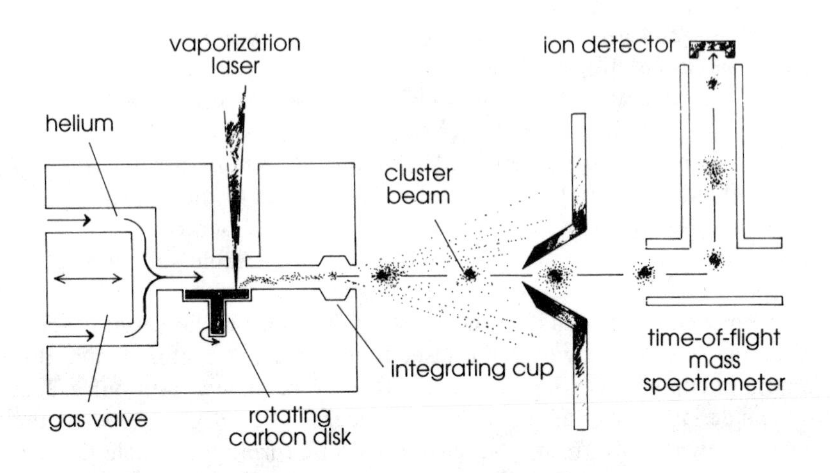

In AP2, clusters are made by vaporizing material from a disk and analysed in a mass spectrometer

once more, this time to within a fraction of a degree above absolute zero. As the gas rushed into the chamber, the clusters were caught in the wind and accelerated anew, fanning out like an aerosol spray in their pathetic attempt to fill the space. The bulk of the clusters struck the inside surface of the drum, but a small number promptly disappeared down the borehole towards the mass spectrometer. Here the second laser pulse ripped electrons from the clusters, thereby ionizing them. Electrostatically charged metal plates in this region generated a strong electric field which redirected these positively charged ions up the long tube and drove them on towards the detector. The electric field imparted the same kinetic energy to all the ions. Lighter ions travelled faster; heavier ones brought up the rear. The detector recorded the arrivals in order of size, noting the time of flight of each cluster, from which it was possible to calculate its mass.

The speed at which all this happened was breathtaking. The laser pulse lasted a matter of nanoseconds. The time between the release of the gas and the firing of the first, vaporizing, laser was around a millisecond. It took another half a millisecond for the plume to condense and the resulting clusters to traverse and exit the vacuum chamber and reach the foot of the mass spectrometer region where they suffered the blast of the second, photoionizing, laser. These times were known respectively as F2 and F3 after the labelled function buttons at the top of an IBM personal computer keyboard. Programming the times was a simple matter. Judging them was not.

Consider the sheer number of parameters to be set for each experiment.

Getting F2 and F3 right was delicate enough. The lasers had to fire at precisely the right moments, first when the gas passed over the sample surface and then when the clusters arrived in the spectrometer. Both the gust of gas and the supersonic skein of clusters flashed through the sights of their respective lasers for the briefest of moments. The density of the gas that was released could be controlled by altering the pressure of the gas reservoir dammed up behind the valve and the interval of the valve opening. The gas itself could be changed, as the sequence of experiments that will be described will amply demonstrate. For both lasers, it was possible to alter not only the duration of the light pulse, but also its energy, either by tuning the colour of the laser beam using a dye or by swapping one laser for another. Then there were banal but still critical factors, such as the pumping speed used to maintain the vacuum inside the apparatus and the cooling that maintained parts of the apparatus at required temperatures. Finally, within the mass spectrometer it was possible to vary the electrical potential of the various charged plates to accelerate and deflect the charged clusters in a great variety of ways.

Variation of these parameters altered the likelihood of cluster formation. Choosing the right combination of conditions from the permutations on offer was a largely intuitive process. It had been Smalley's skill to dare to build this apparatus and, no less, to make it work once built. Many an elaborate and expensive construction has proven to be a turkey. No amount of tweaking will turn a bad machine into a good machine. But this one worked, and worked well.

A successful apparatus, too, seldom remains for long without some embellishment. These jobs were assigned to the students. O'Brien had recently designed a new drive for the rotating platform on which the disc-shaped sample for vaporization was placed. It was Liu's job to see to the computer side of things.

Two refinements play important roles in this discovery. In addition to the easily varied conditions, it was known that one less tractable property of the apparatus affected the propensity to cluster formation. This was the shape of the channel immediately downstream from the site of the laser vaporization of the sample where the incipient cluster beam was cooling fastest. By sculpting this segment of the cluster beam route in different ways, it was possible to change the profile of the cooling process. But the relation between the shape of these expansion tubes and cones and the clusters, if any, that are produced was far from clear. Learning from their experience with metal clusters, the Rice group had tried various modifications during the course of their semiconductor studies. By the time of the carbon experiments, Heath had machined a tool kit of several nozzle extensions rather like cake icing accessories. One of these involved the addition of an "integration cup", a region of larger diameter than the preceding tube, a sort of antechamber before the beam plunges into the vast emptiness of the vacuum chamber. With a knowledge of the dynamical behaviour of flowing gases, it was possible to exercise a small degree of control over the extent to which certain atoms and ions assembled into bound clusters.

The second nuance concerned events at the other end of the apparatus. The cold, isolated clusters produced under the conditions in the drum were ideal for study by spectroscopy, a family of techniques that permits matter to be characterized according to how it affects radiation that strikes it. The clusters had so little internal energy that they did not rotate or vibrate as molecules do at room temperature. Neither were they perturbed by interaction with nearby species because, in this rarefied atmosphere, there were none. The photoionization stage, then, provides the perfect opportunity for some spectroscopy. The laser in this part of the apparatus was generally used to eject an electron from a cluster thus ionizing it. In order to do this, the energy contained in each photon of laser light had to be at least as great as the energy required to pluck the most vulnerable electron from the cluster. If the laser pulse was less energetic than this, the electron would merely undergo a transition from the ground quantum state to an excited quantum state but remain bound to the cluster. Another pulse of light, from the same laser or at a different energy and wavelength from a third laser, was then required to complete the dislodging of the electron. If one knew the energy of a transition for a cluster one suspected to be in the distribution, one could tune the photoionization laser to that energy and fire it as the cluster beam passed. If these clusters were present, they would be pumped by this laser into their excited state. The cluster excitation was then said to be resonant with the laser pulse. The subsequent laser firing then completed the ionization in order that the presence of the resonant species might be recorded further down the line. This process was known, reasonably enough, as resonant two-photon ionization or R2PI. Because the two photons in these experiments were often photons of light of visible wavelengths, Smalley's team often looked for what they called two-colour enhancement. If they found it, they could calculate the energy difference between the ground and excited states of the cluster from their knowledge of the laser energies they had been using. Alternatively, if one already knew an energy difference or had a reliable theoretical prediction for it, one could search for the requisite cluster by arranging the laser energies such that resonance would occur if it was present. The same equipment was also useful for selecting a particular cluster for study. It was possible to separate wheat from chaff like this by judging the photoionization laser energy finely so that it only just surpassed the ionization threshold of the desired cluster. The subtleties of this technique would become a point of contention later, when other researchers challenged the Rice group's remarkable discovery.

For the moment, however, it suffices to summarize the capabilities of Smalley's apparatus. It could accommodate any solid element or mix of elements and probe its behaviour in that shadowy "mesoscopic" region between the ideal of single atoms and the everyday of bulk materials. Within its halls and corridors were extremes of hot and cold and of crush and void. Running this gauntlet each time the vaporizing laser fired were several billion clusters of atoms. The mass spectrometer needed just one of these to record a signal.

At the end of the week before Kroto was due to arrive, Smalley told the students to try carbon in the apparatus to make sure that it would work smoothly for the new experiment. The Exxon group, Eric Rohlfing, Don Cox, and Andy Kaldor, had recorded a distribution for clusters larger than about forty atoms – a peak for each even-numbered cluster of carbon atoms; virtually nothing, gaps in effect, for the odd numbers. Smalley wanted to see that picket fence.

O'Brien, Liu, and Zhang put in graphite and recorded several spectra, varying the deflection voltage of the mass spectrometer to look at different ranges of cluster size. They found the even peaks for the larger clusters. Then they looked at the smaller clusters in the size range that Kroto was interested in. Switching to a different photoionizing laser with a different pulse energy, they noted clusters of two, three and four carbon atoms. Finally, they recorded a spectrum that they liked. Liu wrote in the log book that it was the "same distribution as E.A. Rohlfing".

It was anything but.

The person monitoring the spectra had the ability to select different portions for display and could magnify those portions on the computer. In this case, the intermediate carbon cluster peaks filled the screen. The peak in the spectrum for the sixty-atom carbon fragment was way off-scale, far bigger than its neighbours, and so far off in fact that even if O'Brien had rescaled the image so that the entire peak had appeared on the screen, cutting the others down to pinpricks, it would quite probably have been dismissed as a huge, if unexplained, artefact, simply too improbable to be anything else. Naive graduate students are eager to be seen to do the right thing. Even though everything they see is new to their eyes, they wish to appear sophisticated. And so they write: "same distribution as E.A. Rohlfing". Before going back to their work on silicon and germanium for what would be the last time during the final week of

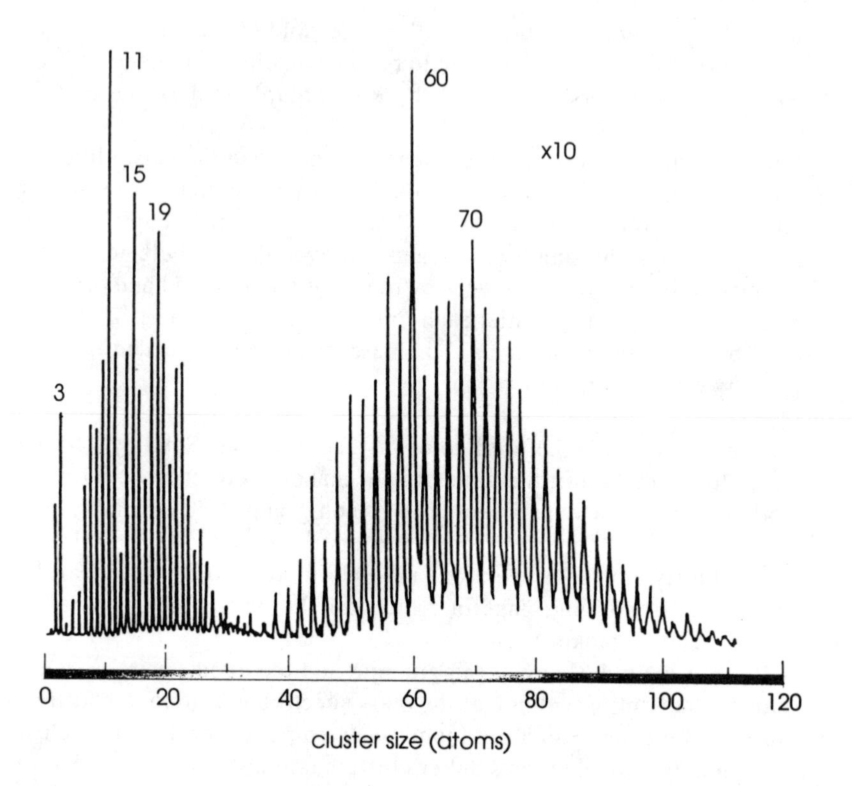

The Exxon group found small clusters with odd numbers of atoms and larger ones with even numbers

August, the students tried resonant two-photon ionization of carbon clusters. Their results suggested that the spectroscopy that interested Curl was at least a realistic prospect. Kroto's visit might not be a complete waste of time.

Harry Kroto was coming to Houston as Bob Curl's guest. So far only Curl really knew what was on Kroto's agenda. When he arrived at Rice, Kroto knew he had to do everything he could to persuade Smalley and his students that the experiments he wanted to do were exciting. Barely pausing to introduce himself, he launched into his spiel. He told them about carbon in space, about his discovery of the longest molecules in space – carbon chains capped with hydrogen at one end and nitrogen at the other – and how he was sure he could make these same molecules in Smalley's apparatus by feeding hydrogen and nitrogen gas through the hot carbon vapour created by the laser blast. He described how this carbon was generally thought to have come to be located in the interstellar medium far from any stars, and why, in his view, that hypothesis could not be the right one. He told them about the diffuse interstellar bands, the family of spectral peaks indicating the presence of some molecule or molecules unknown that absorbed infrared light at wavelengths that did not correspond with any known terrestrial spectrum. Perhaps these could be even longer carbon chains. If only he could somehow make those chains and know their spectra....

It was all new material to the students. Most of it was new to Smalley too. They lapped it up. Having built an apparatus to look at ultracold molecules, Smalley saw that here were some molecules worthy of his machine. He was not convinced that it would do the chemistry Kroto wanted, which in any case did not strike him as all that interesting. It seemed obvious that if you had a chain then it would have ends and that you could clip onto those ends pretty much what you wanted. Indeed, the Exxon group had already demonstrated as much by adding atoms of potassium to the ends of carbon chains. Heath thought rather more of the idea even though the Exxon group had done this experiment. He felt they could repeat the experiment with different elements and get a cleaner result to boot.

But the diffuse interstellar lines – little else was known about these spectra other than the positions of their lines, precisely the information needed for resonant two-photon ionization. To both Heath and Smalley, this was a problem worth studying, with the distant prospect of a dramatic *denouement* – an explanation of a long-running mystery in science. Heath saw the beauty of the problem and instantly felt he had better try to solve it. Above all, it was a puzzle, a nice piece of pure, fundamental science. For his part, O'Brien would have preferred to get back to his semiconductor clusters, but he was soon to realize that he had done the last work on them that he would ever do.

Jim Heath and Sean O'Brien prepared briefly to set aside their embryo doctoral projects and explore Kroto's sooty byway. In demeanour they could hardly be more different. Jim was precocious. He had been doing research while his peers were doing no more than they were meant to by attending the required classes. He was cocky as well, often dominating the apparatus, Smalley's apparatus, and scorning the abilities of the other students. Sean was more measured. He was not given to bragging or even sometimes to speaking up for himself when perhaps he should have. As experimentalists, the two young men were very different too. Jim had green fingers with apparatus, a quality that Rick greatly admired, and one reason why Jim appeared as Rick's favourite son in the laboratory. Jim could occasionally devise an experiment, intuit how to make it work, and get a result. He might not play with variables systematically and was as likely as not to record only his conclusion and not how he got there. Sean was more methodical, more the stereotype of a good scientist, making up in orderly procedure what he appeared to lack in creativity.

Jim was so fired up he wanted to get started right away on Harry's experiments. On the first day of September 1985, a Sunday, Jim and Harry went into the lab. As they worked, they became fast friends. Jim was excited by Harry's infectious enthusiasm. Harry talked nineteen to the dozen and not just about the problem at hand but about science in its broadest sense. Out of hours, they would play guitar, Harry playing folk, and Jim jazz.

Harry's first need was to learn how the apparatus worked. With Jim he ran through the hidden drama that was enacted with each firing of the apparatus to try to gain an appreciation of the importance of each step – what it means to fire the laser at the beginning or the middle or the tail of the gas pulse, what was happening to the vapour and the gas in the nozzle, ranking the options, weighing the implications of varying one parameter or another.

As they did this, they recorded carbon spectra. They tried a range of vaporizing laser pulse energies. With a high energy pulse and the open nozzle, there was no useful clustering. With the integrating cup in place and more modest laser pulse energies, clusters large and small checked in at the detector. These distributions were like the Exxon spectra. The sixty-carbon peak was there too, several times larger than adjacent peaks, big enough to be worth monitoring as a general measure of clustering quality, but not so remarkable as to cause any consternation on its own account. The large even-numbered clusters from about thirty atoms upward were mysteries, perhaps chains, perhaps planar sheets like slates of graphite, perhaps something altogether messier. The Exxon group had speculated that they might be bundles of carbon chains cross-linked in some unspecified manner.[16] Rick had discussed this hypothesis with his former co-worker, Andy Kaldor, but he remained sceptical. The presence of these peaks now was reassuring because they had been seen in previous work but not worthy of special note. What was important was that the peaks were present for the low-numbered chains that Harry wanted to study.

The following day work fell into the weekly routine. The mornings were largely taken up in discussions between Rick and Harry while the students were off attending classes or teaching. Harry and the students did their experiments in the afternoons and evenings. The first experiment was to try to produce the carbon chains terminated with hydrogen. The idea was simply to introduce hydrogen as the carrier gas. This would bear the carbon vapour away from the graphite sample reacting with it as it went. Then it would simply be a matter of detecting the hydrogenated carbon chains with the photoionization laser. Harry explained to the students the theory by which it was hoped they might do this. Simple acetylene has a high ionization potential; it is hard to remove an electron from the stable carbon-carbon triple bond. In the longer acetylenes with concatenated carbon single and triple bonds, the polyÿnes, the ionization potential decreases as the number of acetylene units in the chain increases. Whereas in acetylene itself the pi-bond electrons are held within the orb of just two carbon atoms, in a polyÿne they are free to travel the length of the chain, held by four or six or more carbon atoms. The most easily removable of these electrons are less tightly bound to a six-carbon chain than to a four-carbon chain, and less tightly bound to a four-carbon chain than to a two-carbon one.[17] Plotted against chain length, the successive ionization energies describe a neat curve from which it is possible to predict the ionization potential for a chain of any length. With the right laser energies, therefore, it should be possible not only to use resonant two-photon ionization unequivocally to confirm the length of carbon chains detected by the mass spectrometer, but also to record spectra for the longer chains that just might match the mystery interstellar lines.

The reality was rather different. In place of helium that the Rice group habitually used as an inert carrier gas when making semiconductor clusters, here was hydrogen, one of the most reactive gases. A plume of ten-thousand-degree carbon vapour erupting into a stream of hydrogen was opening the path not only to a tidy xylophone rack of polyÿnes but to the entire extended family of hydrocarbons. The twelve-times-table peaks for the low-mass pure carbon clusters disappeared to be replaced by a smear of oil and tar fragments. Only the large, even-numbered clusters of pure carbon seemed strangely impervious to hydrogen attack.

To try to understand the disruption that the hydrogen was causing, they turned late in the afternoon to look more systematically at the apparatus parameters under their control. It is important with any complex apparatus to have a strategy for experimentation. Time spent playing with controls and praying for results is likely to be time wasted. There are too many conditions available to be changed, and a change in any one has an effect on all the others. A better method is to play with a principal variable, the one where a small adjustment produces the largest change elsewhere. Using information obtained by changing this variable, it is then necessary to dream up a hypothesis, however far-fetched, before altering other experimental conditions to construct a test of that hypothesis. In this experiment, the principal variable was the interval between the release of the pulse of hydrogen gas and the firing of the vaporization laser. Jim, Sean, and Harry had come to regard the sixty-carbon mass spectrum peak as a convenient landmark whose coming and going could be monitored as they adjusted this interval in order to produce a better yield of small clusters. It was in this supporting role that "C_{60}" made its first entry in the log book on Monday 2 September. Having found that C_{60} became the dominant carbon cluster when there was a long delay between gas pulse and firing of the vaporization laser, Rick, Jim, and Harry concurred on the hypothesis, familiar from the semiconductor work, that the large clusters were forming slowly in the thin tail end of the gas pulse, rather than cooling fast in the denser middle of the pulse. The reason for the poor yield of small chains seemed simply to be that the large clusters were eating up most of the carbon. Either the carbon vapour condensed directly to form large clusters or the large clusters formed in some secondary process by consuming the small clusters. With this model, they had what they needed to propose more favourable conditions for the formation of clusters in any desired range.

After a lay day during which Liu and Zhang modified the computer and its software, Harry and the students returned to the apparatus on Wednesday 4 September. They replaced hydrogen with nitrogen in an attempt to produce dicyanopolyÿnes – acetylene chains with nitrogen atoms on each end instead of hydrogen atoms. They tried several runs over a period of an hour but without success.

Two brutal truths now confronted them. They could reproduce the Exxon results with an inert helium carrier. But with the gases for Harry's chemistry, both hydrogen and now nitrogen, they could only get garbage. Logic dictated a middle course: start with the helium carrier and gradually bleed in the new gases. The first run, at eleven minutes past six in the afternoon, was with pure helium. The mass spectrometer was set as it had been for the carbon test runs before Harry's arrival. The spectrum showed the expected carbon chains at low mass numbers. By now in the habit of monitoring the prominent sixty-carbon peak, everybody was amazed to see not the picket fence of even-numbered high-mass clusters with the sixty-carbon peak slightly taller than the others, but a completely new distribution. The peak for C_{60} was huge. Nearby was another large peak, for C_{70}. The result was so astonishing that they repeated the experiment as fast as they could. Two minutes later they got precisely the same result. This was even more astonishing. Repeating an experiment is standard practice used to reveal whether an odd result is genuine or a freak. This was genuine. Harry, Jim, Sean, and Yuan felt the first thrill of discovery. Yuan Liu wrote in the log book that "C_{60} and C_{70} are very strong". She added an exclamation mark and underlined the remark.

What was going on? The question was of little immediate concern to the students. They logged the data and pressed on. The reactions on the small chains were still not working. Their spectra showed nitrogen atoms, nitrogen molecules and odd-numbered carbon chains but no signs of carbon and nitrogen in the same cluster. As they worked, they now routinely looked in the high-mass region of the spectra, religiously monitoring the presence of C_{60}, but the focus of their effort remained on the small clusters. For the next couple of hours, Yuan, Jim and Sean performed a sequence of experiments in which they progressively increased the amount of nitrogen in the helium carrier from five per cent to ninety per cent. As the evening wore on, they began to notice a strange thing. Although nitrogen was not giving them the cyanoacetylene products they were hoping for, it did diminish the presence of the carbon clusters. But through it all, the sixty-carbon peak stood out bold as brass, even at the highest concentrations of nitrogen. Forget the picket fence. This was a flagpole.

Rick and Harry, meanwhile, began to discuss what "the kids" were seeing. Many ideas were batted around. Almost equally many were rapidly discarded. It seemed to come down to a flat graphite-like sheet like a piece of chicken wire or something three-dimensional but of otherwise uncertain morphology, what Harry described in British slang as a "wodge". The term was new to Rick. He liked it. With C_{70} prominent as well, there were perhaps many wodges. Rick christened C_{60} the mother wodge; Harry called it the god wodge. However, it was not clear why either of these basic shapes, the two-dimensional sheet or the three-dimensional wodge, would stabilize at sixty atoms or how it could meet the seemingly impossible requirement to have no edge or surface with unsatisfied chemical bonds that would insulate it from the ravages of the carrier gases and other reactive carbon fragments. If resolution of this problem was not the highest concern at this time, then that was understandable. The exact disposition of atoms and bonds in the larger metal and semiconductor clusters was far from certain and had always taken second place to examination of their electronic properties.

It was not until the following day, Thursday, that speculation began in earnest. Jim showed the best of the previous day's data to Rick and Bob Curl. The C_{60} peak bulked nearly thirty times larger than its neighbours. The two Rice University professors and Harry Kroto took stock. None of the proposed structures answered the questions that needed to be answered. They did not tie up the dangling bonds. There was nothing sufficiently special about any sixty-atom wodge. The best candidate was a layered structure of graphite hexagons that Harry drew up: a four-decker sandwich with a hexagonal ring of six carbon atoms top and bottom clamping two larger chicken-wire plates each of twenty-four carbon atoms. The atoms added up – six, twenty-four, twenty-four and six – to sixty. But there were dangling bonds all over the place – six on each hexagon and twelve on each of the larger layers, thirty-six in all. This represented an embarrassing amount of bond energy. These were not missing pi bonds that could perhaps be explained away by some hand-waving bluster about electron clouds, but thoroughgoing sigma bonds, the bonds that give diamond its hardness. In theory the half-bonds bristling from the edge of one plate needed to be married to those on its neighbours. Thanks to the symmetry of the hypothetical sandwich, they did at least correspond in position. But they stuck out parallel from their respective carbon planes. What could help them join up? Bending through 180 degrees in the length of a molecular bond was clearly impossible. But perhaps if the graphite plane were to bend as well?... No one had a mechanism for curvature. And it was still not clear why sixty (and seventy) were so special. But confronted with all those dangling bonds, perhaps it had to be.

Jim and Sean finally found conditions under which nitrogen began to react to form dicyanopolyÿnes. The ratio of the C_{60} peak to the adjacent carbon peaks remained high all the time. Now Jim began to think how strange this persistent feature really was. After dinner that evening, he returned to the apparatus to do the first experiments specifically intended to manipulate the mystery peak.

His first thought was the one they had on every occasion when the apparatus produced one very large cluster signal and nothing else. Perhaps the photoionization laser energy happened to coincide with the ionization energy of this cluster, making it the principal one to reach the mass spectrometer detector. For such a large specimen to exhibit a strong resonance was unlikely, so Jim proceeded to run a check. He swapped the comparatively low-energy laser they had been using for one of a higher energy. If the first laser produced this resonance, then the second was sure not to. He put the second laser into place, took all the precautions he could think of, and ran the experiment. Up came the spectrum on the computer screen. C_{60} was still dominant. It was not an ionization artefact. It had to be real. It had to be new.

On Friday afternoon, 6 September, a council of war was held. Harry was due to fly back to England early the following week. Bob felt that this was the right time to review the progress made on the carbon chain experiments to decide what still remained to be done so that Harry could depart with as complete a set of data as possible.

One of Bob's main roles in professional discussions was to be the devil's advocate. He was the one who made objections, and what's worse, they were usually good ones. He had a reputation as a worrier. Even his eyes set deep behind a weathered brow seemed to convey this trait. If Rick and Harry and Sean and Jim were the cowboys on the apparatus, Bob was the one with the reins yelling Whoa! One could paint Bob as something of a wet blanket, but this would be very wrong. Bob's cautious attitude, borne of the spectroscopist's professional good practice not to assign molecules to spectra without corroborative evidence, was an insurance policy. If Bob thought you were right, you probably were right. If he thought you might not be, he would oblige you to support your wilder claims. If they survived, they were the more compelling for it. They had passed the Bob Curl Test.

It was agreed that more needed to be done on the chains. Then Bob moved on to C_{60}. They still knew too little about it to publish a good paper. A paper with a title along the lines of: "C_{60} flagpole: some unexplained cluster" was not going to do.

One thought was that the sixty-carbon cluster might not form directly from the impact of the vaporization laser on the graphite disc but from vaporization of other wodges of graphite already loosened from the disc by a previous laser pulse. Such a mechanism might explain why C_{60} survived in conditions where the other clusters disappeared. Bob thought someone should do an experiment to check whether this was the case. Depending on the results of this experiment, the next priority was to attempt a more detailed study of how the formation of C_{60} varied with laser vaporization energy. And with that, the meeting adjourned. Rick and Bob went home. Harry took Bob's second car and set off to spend the weekend in Dallas.

After the meeting, it was Sean who took up Bob's challenge. He took a fresh graphite disc, loaded it into the apparatus, introduced the helium carrier gas, and fired successive laser pulses at the disc, rotating it a little each time. He put the disc through one full rotation. Then, rather than resetting the apparatus so that the laser blasted out a new circle of pits on the disc, he rotated it again, firing the laser at the first ring of pits rather than at smooth, new graphite. The C_{60} peak was large the first time around and remained large the second and third times around. So C_{60} was not forming from any inadvertently prepared carbon surface. However, Sean did notice that the presence of C_{60} seemed to vary according to the vaporization laser energy. C_{60} was thirty times more abundant than its neighbour, C_{58}, with a laser pulse energy of forty millijoules but only twice as abundant at a hundred millijoules.

On Saturday, Jim took over the weekend watch. He began by trying to repeat Sean's result with the laser energy dependence. Among other things, the laser energy changes the behaviour of the passing carrier gas. Jim believed that Sean had failed to allow for this when measuring the cluster populations as a function of the laser energy. Perhaps the meaning of Sean's observation was that the formation of C_{60} was a function of the carrier gas pressure, not of the laser power. To confirm his hunch, Jim set about varying the pressure of helium passing over the graphite disc. During the course of a firing, the carrier gas passing over the graphite reached a maximum density and then fell to a minimum again in half a millisecond. Jim needed a way of simulating this process in slow motion so he could see the change in the cluster peaks as he watched. He decided to do this by simply turning off the helium cylinder and monitoring successive mass spectra as the gas thinned out. He found C_{60} thirty times more prominent than its neighbouring clusters when the helium pressure was high. At lower pressures, in the tail end of the gas stream, C_{60} fell back in line with the other clusters. Here was the first systematic data on the new cluster.

Jim's intuition was that some sort of cooking process was going on. The evidence suggested that C_{60} was not forming in a single physical process as the laser pulse struck the graphite surface. In his experiments he had employed a short nozzle extension that somewhat delayed the emergence of the cluster beam into the cold vacuum region of the apparatus. There must be chemistry going on in the hot carbon vapour during this time before the cluster beam products are frozen and shot across to the mass spectrometer. To test this hypothesis, Jim removed the extension. He wanted the shortest nozzle possible in the hope that it would inhibit the clustering process. It did. The C_{60} peak was now only twice as big as its neighbour. The following day, Sunday, Jim checked his experiment by reversing the process. He put the nozzle extension back on. The C_{60} peak duly bounced back into place. Satisfied with this, he then turned to logging the results of reactions both of C_{60} and the smaller carbon chains. The first set of reactions he recorded was with oxygen. He found to his surprise that an oxygen atom added not only to C_{60} but also to the other large even-numbered clusters. He then swapped the short extension for the integrating cup and re-ran the failed nitrogen and hydrogen experiments from earlier in the week and began at last to get some reaction products here too.

On Monday, Jim presented the results of the weekend's experiments to the assembled group, including Sean's news that the C_{60} peak could be made thirty times as large as C_{58}. The new findings added substantially to the body of knowledge about C_{60}. The group now knew that it was chemically virtually inert, they knew that it was photochemically stable and that you could not fragment it under laser light, and they knew it was no artefact. Jim described the "cooking" hypothesis. Harry, in particular, saw the appeal of this. The low number of interstellar spectral absorption bands suggested that they might arise from a very few species, perhaps just one. Here was a model in which out of the chaos of the hot carbon vapour, one species emerged triumphant.

The need to come up with some explanation for all this was now overwhelming.

Everything was coming good at once. With the reactions between the gases and the small carbon chains working well, Jim and Sean were dispatched back to the lab to get more data. But Jim's first action of the day was once again to upstage Sean. With cooking time added, with the integrating cup in place of the short nozzle extension, Jim found that he could now make C_{60} a full fifty times more prominent than the other large clusters. Sean, meanwhile, was growing understandably disgruntled. Just as things were getting interesting, he was fed up with carbon. Jim worked on alone on the apparatus, this time reacting water with the carbon vapour and producing polyÿnes of different lengths.

In Rick's office next to the lab the senior scientists once again sat down to weigh their results. Rick was forced to agree that Harry's reaction studies, about which he had expressed such reservations before Harry's arrival, had turned out rather well. They had a paper to write on those. The results were those of a chemistry experiment and the obvious place to publish was in the *Journal of the American Chemical Society*. At this point, Bob made one of his characteristic interjections. He reminded his colleagues that although they had made polyÿnes by adding hydrogen atoms and dicyanopolyÿnes by adding nitrogen atoms to both ends of the carbon chains, they had not made the particular monocyanopolyÿnes, with hydrogen on one end and nitrogen on the other, that Harry had detected in space. Rick and Harry thought that Bob was being pedantic but they conceded his point.[18]

Then there was the flagpole, now taller than ever. The right journal seemed to be *Chemical Physics Letters*, a publication intended for the announcement of brief but significant results in the physics of molecules. Rick was on the editorial advisory board of this journal. It was his view that any referee worth his salt would read the paper and come back with a report asking the authors to include some conjecture as to what they thought their flagpole might be.

Harry was still advocating his wodge of graphite layers but without conviction. There was no other model worth even talking about. But there was the cooking hypothesis. This suggested that whatever C_{60} was it did not come ready formed from the graphite or the carbon vapour but assembled itself over a period of time. The problem was to imagine a mechanism in which it would take time for C_{60} to form. Perhaps a single large sheet of graphite might be blasted free from the carbon disc and then – somehow – curl around to form – to form what? What would curling chicken wire lead to? It would lead to – a geodesic dome!

How do you construct a geodesic dome? Both Rick and Harry thought that geodesic domes had hexagons on their surface. Harry remembered a "stardome" that he had once made for his children by folding polygonal facets to form a polyhedral globe. He thought that it might have had pentagonal faces. But their memories were too vague. They needed something else. Rick crossed the few yards of the Rice campus to the library and sought out a book that might show such a thing and resolve their uncertainty. At length, he found R.W. Marks's *The dymaxion world of Buckminster Fuller*. He wanted to see that it was possible to fold a sheet of hexagons around on itself without putting in pentagons. It was ridiculous enough to him that you would have to fold it around at all. If you had to use pentagons as well as hexagons to do it, it seemed to him to fail Occam's razor several times over. Rick found what he was looking for in a photograph of a railroad rolling-stock shed designed by Fuller for the Union Tank Car Company in Baton Rouge, Louisiana. The dome hugged the ground with a span of more than one hundred metres but a maximum height of perhaps twenty metres, like a giant upturned wok. The structural grid on the surface of the dome appeared to be made up entirely of hexagons. He checked out the book and took it back to his office. Bob had reappeared to take Harry to lunch. Rick showed them the picture and made the point about the hexagons. "Well, maybe," they said, and went off to eat. Harry talked of the stardome over lunch. But Bob was his usual sceptical self and thought that the chance was low that the object would have had exactly the sixty vertices required to correspond with their carbon cluster.

Towards the end of the day, having spent some time on the apparatus getting the carbon chains to react better than ever, Jim met up with Harry and Rick for dinner at the open-air Mexican restaurant that had become one of their regular haunts. Jim's wife, Carmen, a medical student, joined them. They had all the results they needed and more, even if they did not understand them all. In a sense, the pressure was off. They were in festive mood. Over dinner, the ideas flew back and forth. Rick mulled over ideas based on the dome in Baton Rouge. Harry put forward the four-decker sandwich, but without much conviction. He restated his memory of the stardome and relayed Bob's scepticism about it. Did it have sixty vertices? Did it have pentagonal faces or was it entirely made up of hexagons as the Baton Rouge dome appeared to be?

As dinner broke up, they resolved not to let it rest.

Rick could not see how a sheet of hexagons might close or why it should do so with sixty atoms. Perhaps the only way forward was to try it.[19] Jim, too, was bothered by this. Harry wanted to have another look at *The dymaxion world.* He went back to the lab to try and find it but could not. He returned to Bob's and deliberated whether to call home, wake up his wife, Marg, and have her find the stardome. But it was by then the middle of the night in Britain and Bob talked him out of it.

Jim and Carmen also went back to the lab, to turn off the apparatus which had been cooling down through the dinner. Jim wanted to build a model, but this was not the sort of lab to have a ball-and-stick molecular model kit lying about (clusters typically do not have structures that can be accurately represented by such models which are used principally in organic chemistry). So they paused to buy a packet of Gummy Bears on their way home. With the Gummy Bears as carbon atoms and toothpicks representing the bonds between them, Jim began to construct a curved hexagonal net. He quickly found himself getting into difficulties. As fast as he added one bond another came unstuck. Gummy Bears were clearly not the best way of doing this, but the exercise did at least seem to demonstrate the impossibility of forming a closed hexagonal net that would tie up all the dangling bonds.

Back at his home, Rick was trying the same thing by different means. He sat down at his computer determined to program a closed hexagonal net into existence. Rick had written much of the code on the laboratory computers, but was also something of a hobbyist. He had the latest, most powerful version of the IBM personal computer at home, and had been exploring computer graphics, then in its infancy on such machines. By programming the requirement for sixty vertices perhaps the computer could visualize what was special about the number. But after a couple of hours, Rick lost patience. If the high-tech route would not work, he would have to go low-tech. He started cutting out paper hexagons. When he had a stack of them, he began to tape them together along their sides to make a net. He knew he had to cheat if the net were not to come out flat. But, he thought, better a little cheating than that forest of dangling bonds. In terms of chemical structure, his cheating would come down to a matter of energy. There might be a small energy cost, perhaps a less stable bond, but if he could just finish the model maybe there would be an energy benefit at the end that would excuse his sleight of hand and explain the stability of the sixty-atom cluster. The hexagon-covered

Baton Rouge dome seemed to support the notion. But as he added hexagons around the central hexagon, like petals around the stamen of a flower, he found he had to cheat more each time. By the time he came to add the sixth petal to complete the flower there was no way it would go.

By now it was midnight. Rick had finally convinced himself that the hexagons weren't going to work, but he was sure that he was close to a solution. Reluctantly, he began to reconsider the pentagons that Harry had mooted as being integral to his stardome. C_{60} was already odd enough. Pentagons would just make it odder. But in desperation he began to draw his first pentagon. As soon as this was cut out, he placed five of the hexagons around it. It formed a shallow dish.[20] The structure now curved with no cheating at all. The pentagon was the way to curvature, and it was immediately clear that more pentagons would extend the curvature. Rick threw caution to the winds and quickly cut some out. He now had a choice as he stuck down each new facet of the structure. He decided to take the simplest route: each pentagon would have the same surroundings as the first one. Off he went. The flat flower with the hexagonal centre had become a dish with a pentagonal base. Rick then added a second pentagon, sticking two of its sides to the sides of two abutting hexagons on the rim of the dish. In all there were five sites for further pentagons. He built up the next layer with hexagons, ten of them to surround the five new pentagons in the same manner as the first. He now had something resembling a hemisphere. It was time to inventory the carbon atoms. The total came to forty. But some of the facets he had just added crossed the equator. If he allowed for these, there were thirty carbon atoms in the hemispherical structure. His heart leaped. There would be sixty in a sphere. He quickly finished the paper model.

Here at last was strong evidence to explain the specialness of sixty. Whether or not the molecule was aromatic or behaved in chemically familiar ways was another question, but at least they now had something to say in their second paper. It was well into the small hours. Rick was tempted to call Curl's house and get Harry out of bed and tell him the news. But he held back. Instead, he disciplined himself to go to sleep. And, strangely, he slept.

Rick awoke early on Tuesday. As he drove in to the laboratory in his black Audi, he picked up the car phone and called Bob's office. Bob was not in yet so Rick dictated a message to his answering machine. The message was that the closed solution worked, and that it was an object with twenty hexagonal faces, twelve pentagonal faces and sixty vertices, and that each vertex was equivalent to every other lying at the corner of one pentagon and two hexagons, and that he ought to call the guys together in Rick's office, and that he would be there immediately with a model of it.

The group duly assembled in Rick's office. Rick came in and tossed his paper model onto the table. They had all felt the solution, whatever it was, would be novel. But this was beyond their wildest expectations.

Bob kept his cool. He inspected the model. Then he said that he would believe it only if the bonds worked out in the conventional fashion of other aromatic molecules and had alternating single and double bonds. If the bonds did not work out, it would have severe implications for the molecule's likely stability. Rick had called it a day in the early hours without going this final distance. He had done enough origami for one night. In the event, it was readily shown with more sticky tape that the double and single bonds went around the sphere in an alternating pattern. Now Bob had to admit it. It passed the Curl Test. It really was special.

The soccer ball structure of buckminsterfullerene with the alternating pattern of double bonds demanded by Curl

Rick, meanwhile, called Bill Veech, the chairman of the Rice mathematics department whom he felt might know more about this elegant structure. Veech put it to his students and said he would call back. Rick's next call was to the biochemistry department. Having failed to coax his personal computer into producing an image of the molecule, he now set off to use the more sophisticated graphics hardware of the biochemists to produce the first picture of C_{60}. While he was gone, Veech called back with his answer to Rick's question, telling Bob: "I could explain this to you in a number of ways, but what you've got there, boys, is a soccer ball."[21] He added its technical description – a truncated icosahedron. A soccer ball! Of course! By the time Rick returned to his own lab, Sean had bought the campus bookstore's entire supply of molecular models and had built a proper rolling model of C_{60}. Jim had left for the nearest sporting-goods store to get a real soccer ball.

More serious things also remained to be done. Jim returned to the apparatus using first the short nozzle extension and then the integrating cup to produce more evidence that the formation of C_{60} was determined by chemical reaction. The integrating cup series of results produced the strongest signal yet, the one that would be used in the paper. Sean's interest in carbon had been rekindled by what he had seen that morning. Bob suggested he try to perform the Hückel calculations in order to show whether the electrons in the double bonds on the surface of the C_{60} sphere occupied stable, bonding orbitals, or whether some spilled over into unstable, anti-bonding ones. In general, the more atoms in a molecule, the more complex such a calculation becomes. On the other hand, symmetry simplifies the situation. C_{60} was large but highly symmetric. Anything could happen, but it was certainly worth trying.

The professors sat down to write the paper. Rick and Bob had earlier discussed the matter of authorship. Although his rule normally was to lead off with the name of the student who had been most associated with the experimental work, Rick argued that Harry should be first author in this case. After all, they would not have done the experiment without him.

Then there was the matter of where to submit. This was no longer something for *Chemical Physics Letters*. This was big news. Now that they had a model, the discovery was sure to be of interest not only to chemical physicists or even only to chemists but to scientists of many disciplines. The British journal *Nature* or the American *Science* were the obvious candidates for an important announcement of general scientific interest. Although dominated by the biological sciences, *Nature* was the choice. This was where Watson and Crick had published the double helix structure of DNA and where news of many of the century's great discoveries had broken.

With the elation of that day came a certain fear. This discovery was so large, someone else was sure to stumble upon it very soon. Perhaps they already had. The Exxon paper had been written seventeen months earlier. What had they been doing since? It had been published in October 1984, available for all to see – and to follow up – for almost a year. They knew that the Bell Laboratories group, Rick's former graduate student Mike Geusic among them, were on the same track. Their paper, which also found the sixty-atom ion prominent in the mass spectrometer, was in press with *Chemical Physics Letters. Nature* was not only the place where they would make the biggest splash; *Nature* was the place for rapid publication. But they had to write fast. They wrote in relays, Harry and Bob at the word processor and Rick at his draughting table drawing a diagram of the apparatus, then Harry and Rick writing. The completed manuscript showed some signs of this haste. In some parts, it benefited. There was an undeniable air of excitement in the paper as the authors strove to squeeze all their observations, conclusions, and speculation into the thousand words allowed by *Nature*'s authors' guidelines. In other parts, it suffered. A more thorough literature search would have turned up several previsions of the new molecule.[22] Time taken in discussion with organic chemists would have given them the dish-shaped corannulene precursor. But then they could have stopped to compare notes with so many people. So many sciences had an angle to offer.

The title of the paper was to be simply the chemical formula, C_{60}, together with the name they had chosen for their molecule: *buckminsterfullerene.* Whether the name was thought up as they sat down to write the paper, as Harry maintains, or the week before, as Rick has argued,[23] what is certain is that Harry was its biggest fan. He thought it tripped off the tongue. The -ene ending was euphonious as well as chemically correct. To anyone who had ever heard of Buckminster Fuller, it was in fact a remarkably economical descriptive term. Given the formula and the name, one could hardly fail to deduce the structure. The students, Sean and Jim, liked it too. Rick was less convinced. Bob, of course, had misgivings, but misgivings over nomenclature were not the same as misgivings over science. He let this one go.

By Wednesday, as the editing of the paper proceeded in Rick's office, they were already eager to perform the first ever experiments on buckminsterfullerene. Where to begin? A whole new world of chemistry confronted them, yet they had no actual stuff. All they had were a clutch of a few billion molecules that endured perhaps a couple of milliseconds in the dark heart of AP2.

The first idea that springs to mind when one sees an empty cage is what could, or should, or once did go inside it. By assuming the lengths of the bonds on the surface of the molecule were the normal ones for carbon, it was quickly calculated that buckminsterfullerene provided an empty sphere about seven ångströms (0.7 nanometres, or 7×10^{-10} metres) in diameter. An atom of any element would fit. In principle, the whole periodic table lay before them, but the strongest candidates seemed to be the transition metal ions which form stable complexes with organic compounds. The choice was still wide. Everybody but Jim wanted to try iron. Iron had blazed the way to the new field of organometallic chemistry in the 1950s with the synthesis of the first "sandwich" compound, ferrocene, in which the metal ion is the filling and two pentagonal carbon rings the bread. Here was a strong possibility of something equally fascinating. Iron is also abundant in the cosmos, and one of the suggestions for the diffuse interstellar bands was they might be due to spectral transitions of iron atoms held in some kind of special matrix. For obscure reasons of his own, Jim was keen to try another element, lanthanum. They voted to try iron first. His enthusiasm truly ignited for the first time, Sean worked with Qingling Zhang to coat one of the graphite discs in ferric chloride and ran the experiment but with negative results.

Meanwhile, final arrangements were being made preparatory to Harry's departure. A souvenir photograph was taken of the authors as a five-a-side football team, Sean, Rick, Harry and Jim kneeling on the grass outside the Space Sciences building with large framework models of their discovery, Bob standing behind them holding the soccer ball that Jim had bought. Material was gathered for the package to *Nature* – the manuscript, Jim's best mass spectrum, Rick's diagram, a Polaroid photograph of the soccer ball, and the computer graphics rendition of buckminsterfullerene from Biochemistry. Rick wrote the covering letter requesting "rapid publication as a letter in NATURE so that it reaches the widest possible scientific audience" and adding that they might like to use the computer graphics image on the cover of the journal.[24] The paper was sent off on Thursday afternoon by Federal Express. Rick took Harry to the airport to catch his evening flight home. As he was driving back, he got a call from Jim. The lanthanum had worked. This little nugget of new science could be slipped into the *Nature* paper during editing. With a little more work, there would be material for a second paper – the first chemistry of buckminsterfullerene.[25]

The paper arrived at *Nature*'s editorial office in Washington DC the next day.

The *Nature* editors sent the paper to be refereed anonymously as is standard practice. The two referees were positive if slightly nonplussed at what they read.[26] The paper was published five weeks later.[27]

To read the paper is a satisfaction in itself, but there is much more to

Buckminsterfullerene as a solid model looks very different from the skeleton that shows only its bonds

be gleaned from it than might at first be assumed. Human experience is passed on through storytelling. The penning of an account of a scientific experiment is no different from the recording of an epic journey or an historic battle. It is the first step in a myth-making process. Myth arises whether the authors wish it or not. There can be no objective account of scientific endeavour however devoutly the scientists might wish, hope, or pretend that there should be one. It happens despite the scientific style guides that admonish not only against first-person braggadocio but also against literary tropes, against entertainment or diversion of any kind, against the indulgence of inserting a phrase of the subjective critique of peers' work, and even against historical truth where that might include an account of long, fruitless periods of exploration, or the admission of wrong turnings, moments of serendipity, or lucky accidents. "Whatever the scientists' feelings, or style, while working, these are purged from the final work", as Lewis Wolpert put it.[28] It often seems that even the content is to be so disguised. Popularizing journalists scour the scientific literature in search of "stories" in a mood of fond despair, scanning each paper in turn, shaking their heads as they quickly flip to the end, to the "discussion" and "conclusion" where they know they will find the news that should, for them, have been in the very first paragraph. Such is the dull decorum of scientific writing. And yet from the straitjacket of conventional "objective" writing, sentiments, emotions, feelings cannot help but erupt. The Nobel chemistry laureate, Roald Hoffmann, imagined the thoughts of a humanist upon first reading a scientific paper:[29]

> She notes a ritual form.... There is general use of the third person and a passive voice. She finds few overtly expressed personal motivations, and few accounts of historical development. Here and there in the neutered language she glimpses stated claims of achievement or priority....

Hoffmann contended that the real content of a paper is the product of a struggle between what should be said in order to preserve the convention of the medium and what must be said in order to persuade readers of the argument and achievement of the authors. "That struggle endows the most innocent-looking article with a lot of suppressed tension."

This is true in varying degrees of all the thousands of papers brought forth by the world's scientists to form the body of information known as "the scientific literature", or, more often within the field, just "the literature". The terminology invites consideration by the techniques of literary criticism. For all its denial of "literary" pretensions, the scientific paper is a text like any other.

> The chemical article is an artistic creation.... It is not a laboratory notebook, and one knows that that notebook in turn is only a partially reliable guide to what took place. It is a more or less (one wishes more) carefully constructed, man- or woman-made text.... If one is lucky, it creates an emotional or aesthetic response in its readers. [30]

While acknowledging the attempt by scientists to use language in a neutral manner, literary theorists find no grounds to exempt such texts from critical investigation.[31] By using words such as "constructed" and "text", Hoffmann welcomes their enquiry.

Literary theorists playfully liken the arcana of their field to nuclear physics or other branches of science, but although this may be the most "scientific" technique available to the literary critic, it is hardly so in the sense that a fundamental physical scientist would respect. Fundamentalists who disown such shadowy sciences as economics or psychoanalysis will find no difficulty in rubbishing structuralism. At the very least, it does not lead to testable hypotheses. I therefore use it not rigorously in the sense that physical science uses the term, but indiscriminately, for what it is worth. And it is worth much. A close, analytical reading yields far more information than a superficial reading even of what science assumes is a dispassionate text. Some of what is revealed is no more than what a perceptive scientist would take as obvious on a quick reading but which might pass Hoffmann's humanist by. Some is inside information that would otherwise be appreciated only by a small group of researchers.

But some content can be teased out that might surprise this group. Some might surprise even the authors themselves.

We begin, naturally, with the title:

C_{60}: Buckminsterfullerene

This is a clever formulation. Despite its length, the name demonstrates a certain economy here. It says all that needs to be said and yet says it in such an intriguing way that the casual reader – and, importantly in a general science journal such as *Nature*, the non-chemist – is drawn in, wishing to learn more. Even the non-scientist finds something in this title.

H.W. Kroto*, J.R. Heath, S.C. O'Brien, R.F. Curl & R.E. Smalley

I have described above how the authors came to be listed in this order. The addresses follow the conventional format with the location at which the experiment was carried out given directly after the author line:

Rice Quantum Institute and Departments of Chemistry and Electrical Engineering, Rice University, Houston, Texas 77251, USA

and Kroto, as a visiting scientist, is accredited at the foot of the first column:

*Permanent address: School of Chemistry and Molecular Sciences, University of Sussex, Brighton BN1 9QJ, UK.

We proceed to the abstract (printed here, as in *Nature*, in bold type), the very first sentence of which makes up in completeness what it lacks in elegance:

During experiments aimed at understanding the mechanisms by which long-chain carbon molecules are formed in interstellar space and circumstellar shells[1], graphite has been vaporized by laser irradiation, producing a remarkably stable cluster consisting of 60 carbon atoms.

This brutally concise summary of two weeks of experimentation is significant for its suggestion of serendipity at the very outset of the paper: we were doing A, when we noticed B....

Concerning the question of what kind of 60-carbon atom structure might give rise to a superstable species, we suggest a truncated icosahedron, a polygon with 60 vertices and 32 faces, 12 of which are pentagonal and 20 hexagonal.

The use of the active – *we suggest* – is comparatively rare in scientific literature and frowned upon by some stylists. It provides some indication of the excitement of the discovery that is shortly to be described and of the pride felt by the discoverers. Already, words such as *superstable* and, in the previous sentence, *remarkably stable*, are in use where "stable" might have been thought to suffice. Note the use of *polygon* where *polyhedron* is meant, perhaps an error made in haste.

> **This object is commonly encountered as the football shown in Fig. 1.**

A gratuitous element of non-science, showing a buoyant sense of fun and anticipating popular coverage of the discovery. The authors quickly reassert the language of chemistry:

> **The C_{60} molecule which results when a carbon atom is placed at each vertex of this structure has all valences satisfied by two single bonds and one double bond, has many resonance structures, and appears to be aromatic.**

In other words, C_{60} meets certain basic requirements for existence as a stable molecule. Early steps are thus taken to allay disbelief that such a unicorn among molecules might be made to exist. The authors are careful not to ascribe properties to C_{60} that it is not yet known to possess. By adopting the construction *The C_{60} molecule which results when* ... they leave it open that the C_{60} molecule observed might not in fact be the one described. The body of the paper begins:

> The technique used to produce and detect this unusual molecule involves the vaporization of carbon species from the surface of a solid disk of graphite into a high-density helium flow, using a focused pulsed laser.

There is meiosis in the phrase *this unusual molecule*. C_{60} is of course unique, not just unusual. The rest of the description of the experimental method is conventional, often in the passive voice, but is no less effective for that:

> The vaporization laser was the second harmonic of Q-switched Nd:YAG producing pulse energies of ~30 mJ. The resulting carbon clusters were expanded in a supersonic molecular beam, photoionized using an excimer laser, and detected by time-of-flight mass spectrometry. The vaporization chamber is shown in Fig. 2. In the experiment the pulsed valve was opened first and then the vaporization laser was fired after a precisely controlled delay. Carbon species were vaporized into the helium stream, cooled and partially equilibrated in the expansion, and travelled in the resulting molecular beam to the ionization region. The clusters were ionized by direct one-photon excitation with a carefully synchronized excimer laser pulse. The apparatus has been fully described previously[2–5].

Short, quick-fire sentences laden with high-tech jargon build a sense of excitement. The carbon clusters *were expanded ... photoionized ... and detected.* The ball was thrown, struck, and caught. Zip! zap! splat! Adjectival phrases *precisely controlled* and *carefully synchronized* continue to indicate that something out of the ordinary is taking place.

> The vaporization of carbon has been studied previously in a very similar apparatus[6].

There is an implicit rivalry in this citation of previous work on *a very similar apparatus.*

> In that work clusters of up to 190 carbon atoms were observed and it was noted that for clusters of more than 40 atoms, only those containing an even number of atoms were observed. In the mass spectra displayed in ref. 6, the C_{60} peak is the largest for cluster sizes of >40 atoms, but it is not completely dominant.

Following this *précis* of the Exxon research, the authors describe the published Exxon mass spectra in such a way as to make the key observation sound a statement of the obvious. The authors' point here is that the Exxon group did not make this statement.

> We have recently re-examined this system and found that under certain clustering conditions the C_{60} peak can be made about 40 times larger than neighbouring clusters.

This single sentence makes the re-examination sound as easy as falling off a log, accentuating Exxon's failure to investigate. The inconsistent tense, *We have ... re-examined*, but *We ... found*, is revealing. If the authors had written *and have found*, the inference would be that they had found something long sought. But the simple past, *and found*, is more immediate and more surprised in tone.

> Figure 3 shows a series of cluster distributions resulting from variations in the vaporization conditions evolving from a cluster distribution similar to that observed in ref. 3, to one in which C_{60} is totally dominant.

The re-examination is again made to look easy in a sequence of spectra showing the peak for the sixty-carbon atom mass. Despite the logic of the narrative, it is the mass spectrum with the greatest relative C_{60} peak height that is numbered 3*a* and the most Exxon-like spectrum, where C_{60} barely rises above its neighbours, that is numbered 3*c*. (It is the latter which is termed *gaussian* below, this being the bell-shaped normal distribution that describes the probability of a certain event subject to random influences.) The effect on the page is to make the C_{60} peak preternaturally prominent while the three sets of conditions are described in the text in the reverse order of an awards ceremony:

> In Fig. 3*c*, where firing of the vaporization laser was delayed until most of the He pulse had passed, a roughly gaussian distribution of large, even-numbered clusters with 38-120 atoms resulted. The C_{60} peak was largest but not dominant. In Fig. 3*b*, the vaporization laser was fired at the time of maximum helium density; the C_{60} peak grew into a feature perhaps five times stronger than its neighbours, with the exception of C_{70}. In Fig. 3*a*, the conditions were similar to those in Fig. 3*b* but in addition the integrating cup depicted in Fig. 2 was added to increase the time between vaporization and expansion. The resulting cluster distribution is completely dominated by C_{60}, in fact more than 50% of the total large cluster abundance is accounted for by C_{60}; the C_{70} peak has diminished in relative intensity compared with C_{60}, but remains rather prominent, accounting for ~5% of the large cluster population.

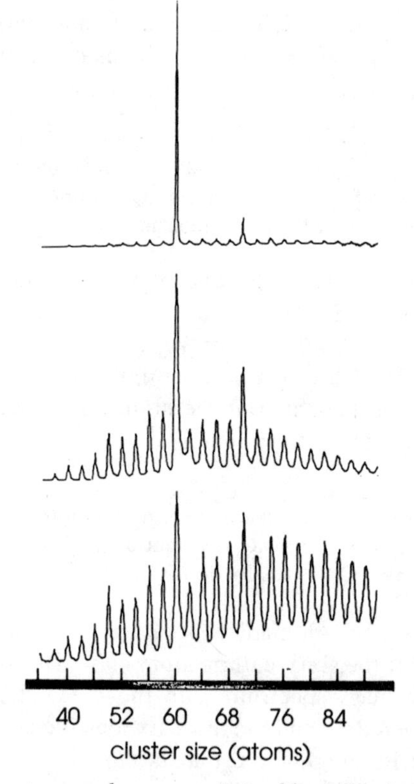

The sixty-atom peak was maximized by adjusting the flow of helium and adding an "integrating cup"

One meaning in the sentence *The C_{60} peak was largest* ... is merely to point out that one peak had to be the largest. The suggestion that *the C_{60} peak grew* begins to confer special powers on the new molecule. The Fig. 3*b* peak is *perhaps five times stronger* – exactitude is not important; better is to come. At last, the final cluster distribution is *completely dominated by C_{60}*. Authorial elation is plain in this comment and is underscored by the vernacular repetition of the fact of the C_{60} domination: *in fact more than 50%....* Thus has C_{60} been plucked from obscurity and groomed for stardom in this sequence of results. Before we proceed to the discussion of these results, note the reference to C_{70} – another unusual molecule – which will receive no further consideration in this paper.

> Our rationalization of these results is that in the laser vaporization, fragments are torn from the surface as pieces of the planar graphite fused six-membered rings structure.

The discussion proceeds unabashed that any rationalization is completely hypothetical at this stage. In the absence of sound evidence for one formation mechanism over another, the authors continue by stating their belief, but return one sentence later to statement of an observation of "fact". Observation and hypothesis are intermingled in order that some of the credibility of the former rubs off on the latter:

> We believe that the distribution in Fig. 3*c* is fairly representative of the nascent distribution of larger ring fragments. When these hot ring clusters are left in contact with high-density helium, the clusters equilibrate by two- and three-body collisions towards the most stable species, which appears to be a unique cluster containing 60 atoms.

The argument develops with an appeal to the reader to become involved: *When one thinks.... If one tries....*

> When one thinks in terms of the many fused-ring isomers with unsatisfied valences at the edges that would naturally arise from a graphite fragmentation, this result seems impossible: there is not much to choose between such isomers in terms of stability.

The elimination of one allotrope of carbon – graphite – as being able to explain the specialness of the sixty-atom fragment is followed logically by a consideration of the other known form – diamond:

> If one tries to shift to a tetrahedral diamond structure, the entire surface of the cluster will be covered with unsatisfied valences.

Both forms fail to explain it. The deduction that *this result seems impossible* reinforces the new molecule's claim to specialness. We have been led through the problem only to find it intractable. We too are still *unsatisfied.* Our expectation is building for a surprise denouement. Having begged our assistance (and we presumably having failed to divine the solution), the authors resume control of their narrative in order to enlighten us:

> Thus a search was made for some other plausible structure which would satisfy all sp^2 valences.

Logical argument is exhausted and so, without apology, the authors cast a wider net – *a search was made....* What kind of search? Any old search? Hardly. Though in the singular, the suggestion is of a search that was wide-ranging and exhaustive. The previous two logical steps are succeeded by an (intuitive?) leap to a conclusion, a sequence that also forms a satisfying one-two-three narrative structure:

> Only a spheroidal structure appears likely to satisfy this criterion,

In fact, a number of other structures, both spheroidal and otherwise, meet the observed criteria, but these are not discussed whether for the sound but unstated reasons that they are known to be unsatisfactory in some way or simply because the spheroidal solution seems supremely attractive.

> and thus Buckminster Fuller's studies were consulted (see, for example, ref. 7).

This sentence is unique in the paper. Grammatically, it comprises two independent sentences linked by a comma. The effect of this device – strictly an error – is to hurry us on faster than we might otherwise have gone. A reason for doing this is that Fuller is not the source one expects to find referenced in a scientific paper such as this. Better some standard work on polyhedra, perhaps. Nevertheless, it was Fuller's work that was consulted; the reference has the virtue of veracity and has been preserved from any sort of intellectual gentrification for publication. From here on, their bombshell dropped, the authors can relax. They do so with a freer choice of terminology and an increasingly individual tone and content:

> An unusually beautiful (and probably unique) choice is the truncated icosahedron depicted in Fig. 1.

Figure 1 in fact does not show a truncated icosahedron *per se* but a soccer ball, shot, as the caption proudly avers, "on Texas grass". The fact that they made a *beautiful choice* is an implicit admission that aesthetic intuition guided that important last leap of the imagination towards the final structure. Because such thinking is supposed to be rare in science and is, in the minds of some, suspect, the authors now move to assure their readers that chemical requirements are satisfied by their unusually beautiful, probably unique molecule:

> As mentioned above, all valences are satisfied with this structure, and the molecule appears to be aromatic.

The authors mix description of their chosen structure, an article of geometry which is incontrovertible, with what can only be conjecture about the physical actuality of the molecule that may possess this shape. This conflation is sustained while the structure and the molecule alternate as the subject of the next few sentences:

> The structure has the symmetry of the icosahedral group. The inner and outer surfaces are covered with a sea of p electrons. The diameter of this C_{60} molecule is ~7Å, providing an inner cavity which appears to be capable of holding a variety of atoms[8].

The choice of a poetic term *sea* to describe the distribution of electrons in the molecule encourages speculation on unusual properties where a more conventional scientific term might not. The paragraph concludes with an early hint of the potential utility of the new molecule. This is probably the area of greatest interest to non-expert readers. It will be explored more widely in a moment, but for these fundamental scientists the relevance of their discovery lies elsewhere:

> Assuming that our somewhat speculative structure is correct, there are a number of important ramifications arising from the existence of such a species.

The authors begin with an understatement – *somewhat speculative* – that prepares the ground for what will be a still more speculative discussion. They take possession of *our* structure at this key point, before conjecturing as to its cosmic role and earthly versatility, but they temper this move by inviting others to confirm their results. The promise of *important ramifications* heads off those who would dismiss C_{60} as a mere bauble.

> Because of its stability when formed under the most violent conditions, it may be widely distributed in the Universe.

The suggestion that this hitherto unknown and unsuspected molecule might be distributed through the cosmos bolsters the authors' claim to have made an important discovery.

> For example, it may be a major constituent of circumstellar shells with high carbon content. It is a feasible constituent of interstellar dust and a possible major site for surface-catalysed chemical processes which lead to the formation of interstellar molecules.

This discussion returns to the domain of astrophysical science evoked in the first phrase of the abstract. In fact that first phrase is scientifically gratuitous. Its use was primarily to suggest serendipity at the outset of the paper. The renewed discussion of astrophysics here is not dependent on its having been mentioned earlier. The creation and closure of this loop is part literary elegance and part a celebration of interdisciplinary science.

> Even more speculatively, C_{60} or a derivative might be the carrier of the diffuse interstellar lines[9].

This is the big one. The authors cover their bid to resolve a long-standing mystery with the phrase *or a derivative*.

> If a large-scale synthetic route to this C_{60} species can be found, the chemical and practical value of the substance may prove extremely high.

Acutely aware that they have made only a handful of molecules, the authors couch their discussion of applications in terms conditional upon finding a large-scale process. As chemists, they envisage a "synthetic" route whereby C_{60} would be assembled from component molecules step by step rather than in a physical process like laser vaporization. This done, they can hold back no longer:

> One can readily conceive of C_{60} derivatives of many kinds – such as C_{60} transition metal compounds, for example, $C_{60}Fe$ or halogenated species like $C_{60}F_{60}$ which might be a super-lubricant.

The emphasis is on chemical versatility rather than serious guesswork at applications. Nevertheless, the scientists know that applications for funding are favoured if they demonstrate "relevance". Hence one easily imagined application. Readers of *Nature* will know that Teflon is a polymer of carbon and fluorine and thus that $C_{60}F_{60}$ is a good candidate for a non-stick "ball-bearing".

> We also have evidence that an atom (such as lanthanum[8] and oxygen[1]) can be placed in the interior, producing molecules which may exhibit unusual properties.

The authors consolidate their claim to title on C_{60} adducts.

> For example, the chemical shift in the NMR of the central atom should be remarkable because of the ring currents.

This statement supports the claim for *unusual properties*. Currents are expected to flow in rings within the "sea" of electrons. The metaphor is happily sustained although the jargon is conventional on this occasion.

> If stable in macroscopic, condensed phases, this C60 species would provide a topologically novel aromatic nucleus for new branches of organic and inorganic chemistry.

The authors position their molecule not within organic chemistry or inorganic chemistry. To experiment with C_{60}, they say in effect, is to participate in a rapprochement between needlessly divorced fields of chemistry.

> Finally, this especially stable and symmetrical carbon structure provides a possible catalyst and/or intermediate to be considered in modelling prebiotic chemistry.

Having already used the adjectives "beautiful" and "unique", the molecule is this time described as *especially stable* rather than, for example, "very stable" or "extremely stable", while *symmetrical* is used in its colloquial rather than scientific sense to mean possessing highly ordered beauty. The phrase *prebiotic chemistry* is a posh way of saying "the origin of life", no less; the jargon serves to excuse the audacity of the proposal. As if embarrassed, the authors return abruptly to mundane issues:

> We are disturbed at the number of letters and syllables in the rather fanciful but highly appropriate name we have chosen in the title to refer to this C_{60} species.

Not so *disturbed*, however, that they don't go ahead and name it buckminsterfullerene anyway!

> For such a unique and centrally important molecular structure, a more concise name would be useful.

Claims for uniqueness return, no longer qualified as speculation or assumption but stated as fact. C_{60} is now a *centrally important molecular structure*. The conflation of the molecule phenomenon and the structural idea is complete. This slips by because we are no longer discussing matters of science such as chemistry and structure but merely nomenclature.[32] There is, of course, a more concise name, at least for conversational use, in the very title of the paper: C_{60}. Nevertheless, the authors proceed with their game:

> A number of alternatives come to mind (for example, ballene, spherene, soccerene, carbosoccer), but we prefer to let this issue of nomenclature be settled by consensus.

The authors protest too much. Of course, they wish their name to prevail. Indeed, this is one reason for placing it in the title of the paper. They finish with acknowledgements phrased in the usual way. Notice, however, that the apparatus receives support independent of that received for the research or by the researchers. It is considered in effect a co-star in the above production and receives its own recognition as the credits roll:

> We thank Frank Tittel, Y. Liu and Q. Zhang for helpful discussions, encouragement and technical support. This research was supported by the Army Research Office and the Robert A. Welch Foundation, and used a laser and molecular beam apparatus supported by the NSF and the US Department of Energy. H.W.K. acknowledges travel support provided by SERC, UK. J.R.H. and S.C.O'B. are Robert A. Welch Predoctoral Fellows.
>
> Received 13 September; accepted 18 October 1985.

Nature prints the date of receipt and acceptance after each paper. The importance of this information will become apparent as the C_{60} literature grows.

1. Heath, J.R. *et al. Astrophys. J.* (submitted).
2. Dietz, T.G., Duncan, M.A., Powers, D.E. & Smalley, R.E. *J. chem. Phys.* **74,** 6511-6512 (1981).
3. Powers, D.E. *et al. J. chem. Phys.* **86,** 2556-2560 (1982).
4. Hopkins, J.B., Langridge-Smith, P.R.R., Morse, M.D. & Smalley, R.E. *J. chem. Phys.* **78,** 1627-1637 (1983).
5. O'Brien, S.C. *et al. J. chem. Phys.* (submitted).
6. Rohlfing, E.A., Cox, D.M. & Kaldor, A.*J. chem. Phys.* **81,** 3322-3330 (1984).
7. Marks, R.W. *The Dymaxion World of Buckminster Fuller* (Reinhold, New York, 1960).
8. Heath, J.R. *et al. J. Am. chem. Soc.* (in the press).
9. Herbig, E. *Astrophys. J.* **196,** 129-160 (1975).

The references tell their own story. Only one (ref. 6) cites previous related work. Three (2–4) refer to the Rice group's earlier work for a description of the apparatus, while another three provide ample indication that this astonishing discovery will shortly be followed by further revelatory new science.

3

THE SEARCH FOR THE YELLOW VIAL

It is one thing to produce a handful of molecules in a cluster beam, quite another to amass enough of them to be able to do much else beyond simply proclaiming their existence. The usual techniques for the laboratory analysis of a freshly made compound – that array of probes that falls under the general label of spectroscopy – demand macroscopic samples that can be manipulated without undue cunning. Buckminsterfullerene cried out for such analysis, which would establish unequivocally its structure, but it could only be made in the most meagre quantities.

This was the predicament of the scientific community at the end of 1985 as chemists and physicists around the world read for the first time of the novel sixty-atom carbon molecule and pondered the likelihood of its really being as described. Of course, the dilemma was felt most acutely by the members of the Rice group themselves, for the onus lay more heavily upon them to prove the structure they had put forward than it did upon anybody else either to prove or disprove it.[1]

In the event, it was to be five years before anyone found a way to scale up the procedure for making buckminsterfullerene sufficiently to do the desired experiments. Rick Smalley and his team could not make enough of it by laser vaporization – it would take years of continuous operation to accumulate even nanograms of the stuff by this means. Nor could they think up any other angle of attack that offered more prospect of success. Certainly, there was no obvious way to set about synthesizing it. As long ago as 1980, Orville Chapman at the University of California at Los Angeles had tried, solely for the intellectual thrill of it, to make C_{60} and other hypothetical polyhedral pure carbon molecules by organic synthesis. He had failed. Leo Paquette's successful synthesis of dodecahedrane, with a mere twenty carbon atoms, had been no picnic. His route to what he calls the "chemical transliteration of Plato's universe" takes 23 steps.[2] The quality of thought that had gone into the design of the synthetic route is apparent in his paper, where the rationale for the chosen path is set out with the glee of a Holmes explaining to his Watson the finer points of some masterpiece of detection. The yield for each step ranges from a minimum of 36 per cent to a maximum of 99 per cent conversion of reactants into the desired product. But multiply these percentages up for all the 23 steps, and one finds that the total yield of dodecahedrane is a mere 1.5 per cent. Add to this the fact that the starting point of this synthetic route is not some commonplace set of laboratory reagents but an esoteric precursor with eighteen of the necessary carbon atoms already in place. It is an odd sort of elegance to some eyes that produces even a Platonic beauty amid such waste. And yet the Rice group could only look with envy upon a success rate of 1.5 per cent. Although buckminsterfullerene was by far the dominant cluster, their apparatus probably only transformed a tiny fraction of one per cent of its vaporized graphite into clusters within the overall range of sizes they were monitoring.

Scanning through the papers they published during this period, one can sense their growing frustration. By the summer of 1987, one paper referred to "the interim" before the structure of buckminsterfullerene was determined as if some bureaucrat had imposed a curfew on all research in this field.[3] Their subsequent paper led off with feeling: "*For some time* we have been searching for a spectral probe of the special cluster, C_{60} ..." (italics added).[4]

They could not do the spectroscopy they wanted. But this is not to say there was nothing to be done. The fact that papers were published at all tells us that much. In the event, the Rice group were to embark upon an ambitious series of experiments which, taken in their entirety, would provide abundant circumstantial evidence in support of the spherical structure they had proposed (and none to cast doubt upon it). Alongside their efforts, there were theoretical investigations into the sort of properties that buckminsterfullerene and related molecules might possess. The theoreticians had no need of physical samples, only a good excuse, and this they had in a highly symmetrical molecule ideally suspended in the limbo between the imaginary and real worlds. Although unavailable for experiment, the sixty-atom carbon molecule was no longer an entirely hypothetical bauble. The probability of its existence, however reluctant, lent a *frisson* to the theoreticians' work, bringing with it the prospect that predictions might one day be confirmed by experiment. As in a competition to guess the weight of a cake at a village fête, there was sure to be a prize of some kind for the closest estimate of the molecule's vital statistics.

The theoreticians' predictions of some unusual properties in turn sharpened the appetite of the experimental chemists for getting on and producing the stuff. One of these predictions was that buckminsterfullerene would absorb light strongly in the violet part of the spectrum. It might be not a sooty black powder like graphite or carbon black but a soft yellow solid, like butter. Thus began the search for the yellow vial.

The Rice group had set something of a time-bomb under themselves. In their announcement of the discovery of a sixty-atom cluster of carbon atoms, they had said more than they needed to so far as the cluster scientists were concerned. These specialists wondered why Smalley and his co-workers had stuck their necks out quite so far. Previous novel clusters, including those made by Smalley and his colleagues, had never been delineated in such graphic detail. Some regular chemists were equally loath to give credence to such a fancy structure on the basis of a mass spectrum but no other spectroscopy. (The difficulty with spectroscopy is that, whereas the smaller metal clusters on which Smalley had worked happily re-emit absorbed electromagnetic radiation at characteristic frequencies to enable their identities to be determined spectroscopically, large molecules such as buckminsterfullerene are able to dissipate the absorbed energy in interatomic vibrations without necessarily re-radiating it.) The attitude of those scientists who specialize in carbon was still more equivocal as will be seen in a later chapter.

The sceptical reaction from these several quarters was taken as hostility by the discoverers. The members of the Rice group desperately wanted their proposed structure to be right and for it to be recognized as the miraculous discovery they were convinced it was. Rational scientific debate quickly grew impossible between factions polarized into believers and unbelievers. And yet, with the benefit of hindsight, it is easy to see both points of view. We have seen how the members of the Rice group wrestled with the anomalous sixty-atom peak in their mass spectra, until finally they were forced to try and explain it rather than explain it away. We have seen how, fortified with further information from experiments aimed at learning how and why the sixty-atom peak came and went, they overcame their reluctance to challenge the conventional wisdom that carbon tends to form flat sheets. Even to the Rice group, this state of affairs was not ideal. They would have preferred to be able to present more and better evidence for buckminsterfullerene, just as others would have preferred to read such evidence.

To those who had not lived through this series of experiments, the apparent case for such a revolutionary structure was much less convincing. All that readers of the *Nature* paper had by way of evidence was the infamous peak and an attenuated account of how it arose. The world scientific community was being asked to believe that carbon spontaneously forms the most regular molecule ever seen out of the chaos of a hot turbulent vapour on the basis of very little more than a mass spectrum and a fuzzy photograph of a soccer ball.

Perforce, the paper did not provide an extended discussion of the series of experiments during which Smalley's apparatus was tuned to chemically "cook" the carbon vapour in order to maximize the proportion of sixty-atom clusters, nor of the thought processes of the scientists during this time or as they finally came to consider viable structures. The authors did not even stop to point out why the structure could not be something more obvious, such as a sixty-atom ring. They merely ruled out a diamond structure because of the dangling bonds. Given another week, Kroto and Smalley and Bob Curl might have done all this – and unearthed the mathematical laws behind the specialness of the number sixty, learned that viruses have similar symmetry, and made various other connections, all of which together would have built a more watertight case. As it was, no one reading the paper could deduce all that lay behind it. They could only do what scientists habitually do – wonder at the conclusion, pick at the scant experimental details for what clues they could find as to what was really going on, and then, for perhaps the best reassurance that what they had just read was not a fiction, check the authorship, the institutions involved, and the calibre of the journal chosen for the publication.

The stage was thus set for anyone to attempt to refute Kroto and Smalley's astonishing result. In practice, the potential to advance the debate was limited by the arcane nature of the apparatus that had been needed to make buckminsterfullerene – people who did not have similar equipment could not play. Even the Rice group were limited in their scope for the exploration of their claimed new molecule to experiments that could be performed within the confines of the cluster beam regime. The only other research group in possession of equipment similar to AP2, and therefore able to repeat the Rice group's experiment and comment knowledgeably upon it, was the cluster team at Exxon. It was they who became the antagonists in a scientific battle that would rage with increasing bitterness during the years of the fullerene famine. The hostilities were viewed with some dismay by outsiders who saw an exciting fledgling subject bring itself low as the two groups best able to do the relevant experiments wasted their energies in attacking each other.

The Exxon scientists found fault with the Rice group's experimental procedure. As the Rice group amassed what they considered to be further evidence for the specialness of C_{60}, the Exxon group responded with further critical comment. As the stakes grew higher, so the debate grew shriller, eventually spilling out from the scientific literature onto the conference floor.

Science thrives on theories and the conflict of theories. One theory makes for very dull science – or worse, something like the unquestioning faith of religion. Yet there are elements of the religious, too, while more than one theory persists. Each school of thought has its orthodoxy, demanding belief that a certain set of results provides evidence for a particular hypothesis and faith that one day the hypothesis will be proven correct – and meanwhile viewing all challenges to that hypothesis as heresy. And like a holy war, the struggle for dominance between rival theories is often ugly. It is, after all, a struggle to the death. Two theories cannot be allowed to stand for ever in explanation of a single phenomenon. One must prevail. Nor is the fight is always solely mental, as has been pointed out with reference to Thomas Kuhn's famous work on the nature of discovery, *The structure of scientific revolutions*:

> Since people become so attached to their paradigms, Kuhn claims that scientific revolutions involve bloodshed on the same order of magnitude as that commonly seen in political revolutions, the only difference being that the blood is now intellectual rather than liquid – but no less real! In both cases the argument is that the underlying issues are not rational but emotional, and are settled not by logic, syllogisms, and appeals to reason, but by irrational factors like group affiliation and majority or "mob" rule.[5]

In just such a fetid atmosphere, Smalley and his team would develop their "party line" on how buckminsterfullerene comes into being. Anything that detracted from its specialness was to be declared heretical. The party line did change from time to time in response to new evidence and sound scientific argument, but always there was the refrain that the C_{60} soccer ball was the most stable molecule in the universe. Anything that appeared to contradict this axiom simply had to be disproven.

This debate raged on the very brink of good scientific practice. A heavy blow dealt by one party might send the other over the edge into professional oblivion. Kroto and Smalley, and Bob Curl, Jim Heath, and Sean O'Brien, believed that buckminsterfullerene was a super-stable molecular sphere and they asked that others believe it too. They knew belief was no lasting substitute for proof. But the leap of faith did serve to raise them onto a platform from where they could launch new experiments. There is a shadowy interlude in the history of some scientific discoveries when sheer conviction is the catalyst that leads to corroboration of a hypothesis. One must state one's theory and then try one's utmost to prove it correct. It is never right, during this period, to close one's eyes to a disproof that might be offered even of the most cherished theory, but equally it is to be expected that the plausibility of such a disproof will be carefully weighed against the strength of the conviction that the theory is right.

EVIDENCE FOR THE STRUCTURE OF BUCKMINSTERFULLERENE (AND THE ADVERSE REACTION IT OCCASIONED)

EXHIBIT A: TRAPPING

We are already acquainted with the first piece of evidence that the Rice group brought forward in support of their soccer ball molecule – the lanthanum experiment that Jim Heath had done immediately the original buckminsterfullerene paper had been dispatched.[6] The paper describing this was sent off to the *Journal of the American Chemical Society* less than two weeks later, and before they knew when the buckminsterfullerene paper might appear in *Nature*. All reference to the name they had coined was carefully omitted lest this second paper be published sooner than the original announcement.

Heath and his co-authors showed a mass spectrum for the complex of C_{60} with lanthanum. This merely proved that a lanthanum atom and an associated assemblage of sixty carbon atoms survived the rigours of the cluster beam regime. Indeed, when they increased the flux of laser light, they found that the lanthanum complex survived better than the naked fullerenes themselves. However it was attached, it seemed that the metal atom was a stabilizing force. But there was nothing to *prove* that the lanthanum atom was caged within a spherical shell of carbon. On the other hand, there was no mass spectrum peak for sixty carbon atoms in tandem with two, or three, or half a dozen lanthanum atoms, perhaps an indication that the metal atom was trapped – one atom would fit inside the carbon cage; two or more would not. Heath's hunch about trying lanthanum had been based on some undergraduate work he had done on a coordinate complex of the metal with a wrap-around, electron-rich environment similar to that which an aromatic cage might provide. And the fact that the lanthanum complex was so resilient strongly implied that the lanthanum atom was not weakly attached to the exterior of the carbon cluster. (Much later, a terminology was devised to distinguish between *endohedral*, caged, complexes and *exohedral* complexes where atoms and molecular groups were attached to the outside of fullerene spheres. The @ sign became the charming signifier of the enclosure of an atom. Thus, Heath succeeded in making $La@C_{60}$. LaC_{60} would be much duller stuff.) As the paper was being written, and undaunted by their earlier failure to persuade iron to form a similar complex, Heath tried the same experiment with a number of other likely metals. By the time they had the draft, they were able to add a footnote to the effect that calcium, strontium, and barium also formed these complexes.

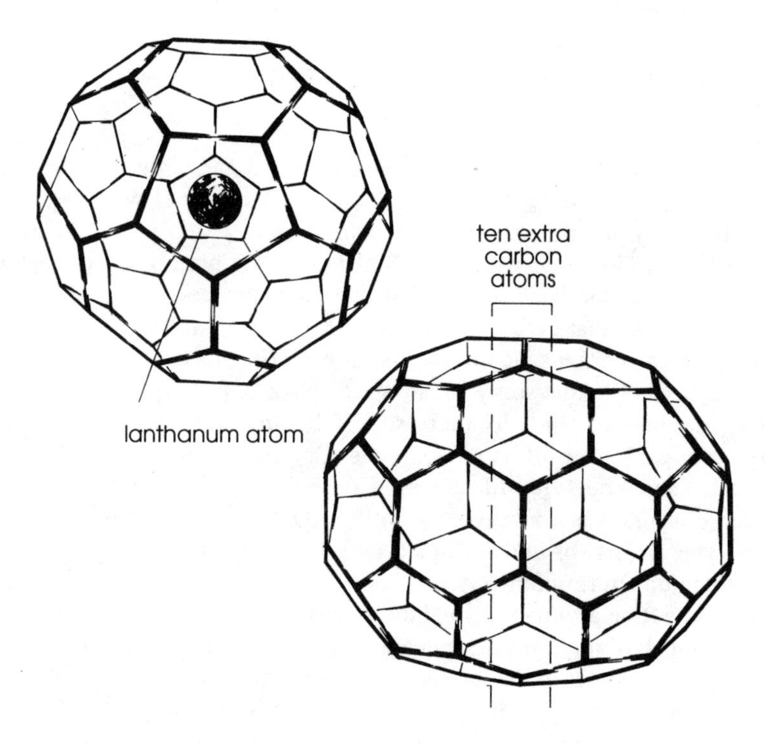

The first fullerene cage compound with a central atom of lanthanum and the second fullerene, C_{70}

(Almost in passing, the lanthanum paper announced the debut of the spheroidal C_{70} molecule, whose smaller peak in the mass spectrum had been clearly visible in the earlier paper in *Nature*, but which had been entirely ignored for want of time to think about it or of a theory to explain it that was as attractive as the icosahedron for C_{60}. Proposed as egg-shaped, a rugby ball or American football to place alongside the soccer ball, C_{70} may be visualized by pulling the C_{60} sphere apart into two hemispheres, inserting a ring of ten carbon atoms, and then putting the hemispheres back together.)

If one atom of lanthanum could attach itself to the exterior of the spherical carbon shell, there was no obvious reason why several could not. The fact that they did not strongly suggested that there was just one stable binding site per C_{60} molecule. If one accepted the molecule was a hollow sphere, its symmetry immediately indicated the existence of a unique site – its interior. However, for those not already inclined to believe that the original molecule was spheroidal, there was little here to persuade them.

Donald Cox and Andrew Kaldor at the Exxon Research and Engineering Company in Annandale, New Jersey, did not believe. It was no sin of theirs that they had not singled out the sixty-atom peak for special attention in their 1984 paper with Eric Rohlfing on the production of carbon clusters.[7] There had been news aplenty in its sequence of mass spectrum peaks indicating that carbon can exist in the form of stable clusters made up of dozens, perhaps hundreds, of atoms. The mass spectra showed series of peaks like the spiny fin along the back of a scorpion fish. There was no obvious reason why there should be anything special about the sixty-atom peak over all its companions. One peak had to be the largest, after all. The conjecture of Rohlfing, Cox, and Kaldor as to the structure of such clusters was consequently entirely generic – they were made up of polyÿne chains (of undetermined length and number) cross-linked in some (also undetermined) way.

When the Rice group published their confident claim that sixty was not just special, but very special indeed, and then promptly backed up this shocking proposal with the lanthanum paper which underlined this claim to specialness,[8] Cox and his colleagues fought back. They took the soccer ball simile and ran with it, penning a critical paper loaded with psychological overtones: "$C_{60}La$: a deflated soccer ball?"[9] The title implied deflation not only of the molecule but also of the Rice group's whole theory. Could the Exxon group burst the buckminsterfullerene bubble? They wrote of experimental evidence they had which "seriously challenges this enticing conclusion".

By repeating Heath's experiment using laser light of a variety of frequencies and intensities to ionize the clusters, Cox and his co-workers found that while the positively charged lanthanum complex ($C_{60}La^+$), the species detected by the mass spectrometer, was relatively abundant under some conditions, under others it was not. They concluded that this abundance was largely a product of the efficiency with which the laser produces these ions, and that the neutral lanthanum complex ($C_{60}La$) might not be especially abundant at all. "Attempting to infer structural or stability information about neutral clusters simply from an … ion signal … is frought [sic] with complications which can lead one astray", they admonished darkly.

Furthermore, they reasoned that if the apparent presence of the C_{60}-

lanthanum ion was exaggerated in some way, then there might also be circumstances under which clusters of C_{60} with more than one lanthanum atom, presumed by the Rice group not to exist, might become abundant. The fact that they had so far gone undetected did not necessarily mean they did not exist. If a two-lanthanum complex were detected, it would cast serious doubt on the spherical cage model since both atoms could not fit inside. The Exxon scientists stated their preference for carbon clusters like flakes of graphite whose capacity for forming complexes would not be limited to one metal atom.

Particularly offensive to the Rice group was a footnote in Cox's paper which appeared to imply that the C_{60} ion detected in the mass spectrometer was also essentially an artefact; that it was, like the C_{60}-lanthanum ion, the result of a coincidence of particular physical conditions within the cluster beam apparatus. (In the eyes of Heath and the other members of the Rice team, the likelihood of these conditions' being coincident was so low that it made the soccer ball postulate look mundane by comparison.) In their earlier paper, the Exxon group had suggested that the ionization of carbon clusters with fifty or sixty atoms was a single-photon process using a particular laser (that is, the energy of one of the laser's photons is more than enough to ionize a cluster), a notion not inconsistent with the clusters' being cross-linked polyÿnes which are expected to be fairly easy to ionize.[10] In this paper, however, while taking the Rice group to task for their supposed peccadilloes, Cox and his colleagues now implied that carbon clusters of between 40 and 100 atoms would be harder to ionize – a two-photon process – contradicting their own previous argument, and without even admitting as much.

Cox intended his paper as a warning shot across the bows of those who would sail the scientific seas blown by winds of faith more than reason. But while the Rice group admittedly could not prove that the lanthanum atom became trapped inside a carbon cage, the Exxon group had not proved that it didn't, and had not put forward any more compelling alternative to explain what both groups had now observed in their respective experiments. Rather than explain the novel phenomenon, Cox and his team had shown merely that it was possible to produce conditions under which the phenomenon would go away.

The members of the Rice group were less than thrilled to read that they might have been "enticed" into drawing a false conclusion. They reviewed what they had done and found they could not agree with Cox's criticisms. They argued the matter through with the Exxon team, but reached no consensus. The two teams, former collaborators in the supersonic cluster beam business, agreed to go their separate ways.

EXHIBIT B: INERTNESS

In their paper for *Nature*, Kroto, Smalley, and Curl had gaily speculated that if buckminsterfullerene was able to survive the harsh conditions inside AP2, it might be widely distributed across the universe. But these experimental conditions provided only a limited test of the molecule's supposed stability. The only other species present in AP2 were the carrier gases and other fragments of carbon which hardly constituted a true test of chemical reactivity. So exactly how unreactive was buckminsterfullerene?

Smalley and his group sought a vivid demonstration of the molecule's lack of reactivity by exposing the sixty-atom cluster to a battery of reactive compounds in a series of experiments like those in which they had made Kroto and Walton's carbon chains terminated with nitrogen or hydrogen simply by mixing those gases with the helium pulse. In those experiments, the gases had been mixed with the carbon vapour. This time, the idea was to make the carbon clusters first and then introduce the gases. AP2 was modified so that it became essentially a fast-flowing reactor (something Smalley had done before in order to make the first metal clusters with small chemical compounds adhering to them). They could now add all sorts of reactive gases to the mature clusters.

Sean O'Brien fed in quantities of some of the most reactive gases – hydrogen, oxygen, ammonia, sulphur dioxide, carbon monoxide, and nitric oxide. Faithful to the party line that buckminsterfullerene was super-stable, Smalley was confident that the new mass spectra would leave only the C_{60} peak showing. He was irritated when O'Brien's results showed that most of the other even-numbered carbon clusters also survived the attack. There were minor variations from gas to gas, but the picket fence stood firm.

As they were trying to comprehend these unexpected results, the Rice group received an important fillip, a preprint of a theoretical paper that dispelled a number of mysteries. In due course, many theoretical studies would be stimulated by the discovery made at Rice. But this one was simply a fortuitous coincidence. Anthony Haymet, a theoretical chemist at the University of California at Berkeley, had been working independently and quite by chance at predicting the properties of what he took to be an entirely hypothetical class of polyhedral carbon molecules while the original buckminsterfullerene paper was in press. His work provided theoretical evidence for the stability of the truncated icosahedron just when it was needed most.[11]

In describing what he called "footballene", Haymet filled in a number of significant gaps in the Rice group's knowledge. He drew their attention to the dish-shaped molecule, corannulene, whose twenty-atom carbon framework of a pentagon surrounded by five hexagons can be mapped onto part of the surface of a buckminsterfullerene sphere. The fact of its existence indicated that buckminsterfullerene might not be too strained a structure to be stable too. He did the Hückel calculation, predicting that the pi-bond electrons occupy completely filled bonding molecular orbitals, another useful indicator of stability.[12] He invoked Euler's theorem, known to schoolchildren as "faces plus corners equals edges plus two", and combined it with the supposition that each carbon atom has three bonds radiating from it to confirm that polyhedral aromatic shells must have an even number of carbon atoms. He then added the condition that the molecular shells might be made up entirely of hexagons and pentagons. Solving the simultaneous equations expressing these various conditions, Haymet showed that closed shells of spheroidal carbon may have various numbers of hexagons provided they have exactly twelve pentagons. This was the simple mathematical proof that Kroto and Smalley had needed as they wondered how to close the hexagonal graphite net. Their solution – to seek out Buckminster Fuller's domes – had been a poor guide because there are so many hexagons on the surface of such enormous polyhedra that the pentagons become hard to detect, and because the example they chose, the Baton Rouge dome, was so shallow, representing only a small part of a complete sphere, that it did not in any case sport its full complement of twelve pentagons. Amazement has since been expressed in the general science literature that chemists could be so ignorant of such a basic area of mathematics.[13]

Haymet's communication together with the growing stack of unexpected mass spectra from the reactivity experiment forced a new realization. All the carbon clusters were chemically inert! It was obvious – they were all closed spheroidal cages. With Haymet's help, the Rice group were suddenly able to provide a complete explanation for the persistence of all the even-numbered mass spectrum peaks.[14] The idea that carbon clusters of many sizes might be closed shells had quite simply not occurred to them before. Even with the working out of a plausible rugby-ball structure for C_{70}, the penny had not dropped. But now there was a rationale for a whole family of fullerenes. Knowledge of the requirement for twelve pentagons immediately gave the dodecahedral C_{20} ("dodecahedrene", a hypothetical dehydrogenated version of Paquette's molecule) as the smallest fullerene. Between C_{20} and C_{60}, and from C_{60} out to much larger molecules, there lay numerous possibilities whose likelihood could be judged according to the arrangement of pentagons and hexagons on the molecular surface. (Euler merely demands that there be twelve pentagons to form a closed polyhedron; he says nothing about where they

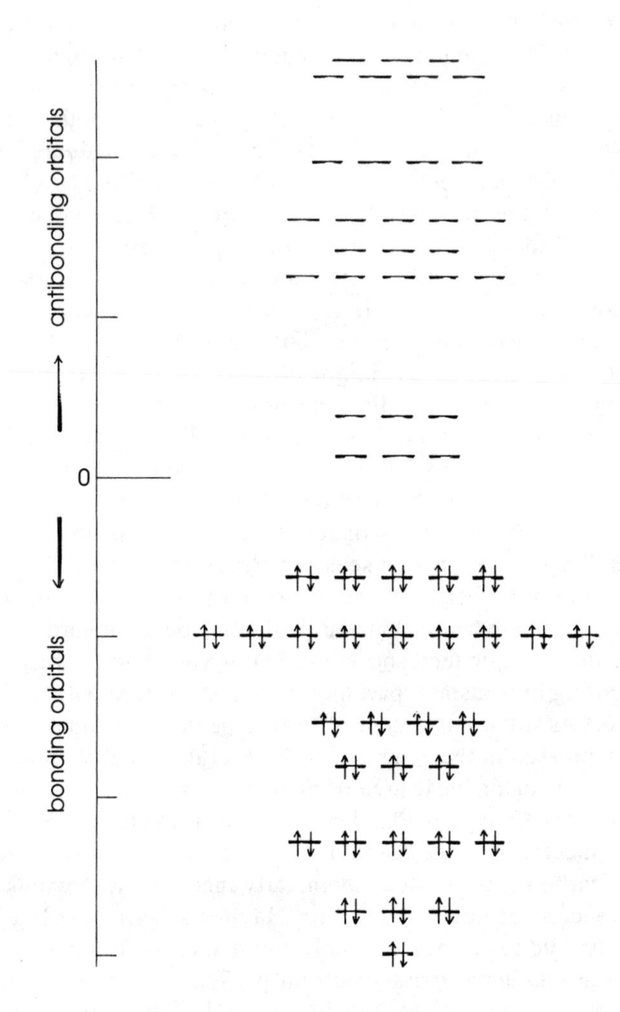

Hückel diagram for C_{60}. The picture is simpler than for a typical sixty-atom system because of symmetry

may lie. However, for chemists, their position and juxtaposition have a critical bearing on stability and reactivity.)

Don Cox and Andy Kaldor were still unimpressed. They responded – at leisure this time, in a paper submitted at the end of 1986 – to the new suggestion that all the large even-numbered clusters of carbon were spheroidal with an exhaustive series of experiments in which the full range of clusters and the conditions used to produce them were "systematically investigated".[15] Reading between the lines, there is again the coded language that apparently seeks to undermine the whole thrust of the Houston research.[16] The title openly questions the special behaviour of C_{60} and other large carbon clusters by placing the adjective "special" in inverted commas. The pages of the paper itself practically drip with scepticism: Smalley's group worked "under special experimental conditions" to show that the C_{60} ion "can be" dominant; their subsequent experiments "were taken as" strong support for their proposal. There was even the slur that the truncated icosahedral symmetry would merely yield a carbon shell that was "nearly spherical". (The paper was not published for more than a year, by which time its damaging effect was much diminished; Cox reckons it was refereed by a "believer".) At the root of the Exxon work was the insistence, once again, that the experimental procedures of the Rice group might account for the prominence of the C_{60} ion, and thus that this strong ion signal might not be an indication of the existence of a stable neutral C_{60} species in the cluster beam before it reaches the ionizing laser used for the mass spectrometry.

Where the Exxon group had been within the scientific pale in making their previous suggestion that the lanthanum complex might not be a cage (and thus that other possibilities should be admitted for consideration), it now appeared to some that they were dismissing the spherical structure – and in the face of growing evidence for it from many sources – for no better reason than that they simply did not like it.

EXHIBIT C: NEGATIVE FULLERENE IONS

The crux of Exxon's scientific argument was that the strong mass spectrum signal for the positive single ion of the sixty-atom carbon cluster, C_{60}^{+}, could not be taken as a reliable indication of the abundance of a super-stable neutral molecule, C_{60}, because it just might be that the apparatus was functioning in such a way as to ionize this species preferentially, thereby exaggerating its importance.

The believers in the super-stable molecule needed to find a new indicator. Mass spectrometry generally relies on the separation by mass of positively charged molecule fragments. From this separation it is possible to deduce the composition of the fragments and hence of the original molecule from which they sprang. The Rice group decided to reverse the voltage in the spectrometer and run an experiment to see if the buckminsterfullerene *negative* ion – C_{60} with an extra electron attached to it – was stable. If mass spectrometry produced a strong ion signal for the negative ion of the sixty-atom cluster as well as for the positive ion it would be persuasive evidence that it was the cluster itself that was the notable phenomenon and not the process used to ionize it.

The experiment worked. Pointedly, they titled their report: "Negative carbon cluster ion beams: new evidence for the special nature of C_{60}".[17] It is worth quoting their last paragraph in its entirety:

> Thus C_{60} and other neutral even clusters of similar size which we detect appear to be *survivors* of the processes taking place before expansion with C_{60} being the most inert. The fact that the negative ion distribution produced by ionization in front of the flight tube is so similar to the positive ion distribution produced by laser photoionization of the beam a meter downstream is convincing evidence that the [negative and positive] distributions shown ... reflect the neutral cluster distribution and are *not* an artifact of the ionization process. This neutral distribution, in turn, is strong evidence supporting the spheroidal carbon shell model for even carbon clusters in this size range in general and the truncated icosahedron (soccer ball) model for C_{60} in particular.

This result, too, drew flak, not this time from Exxon, but from Robert Whetten, a young chemist at the University of California at Los Angeles who had formerly worked with Cox and Kaldor. Whetten's paper was emollient in tone, at least in comparison with the barbed language adopted by the Rice and Exxon groups.[18] Whetten had spoken with Cox who had stated his unrepentant view that it could still be that it was special experimental circumstances that were generating not only the charge but also the clusters. This time, Cox's candidate for gremlin was the electron attachment process by which the negative ions were created. Whetten identified the need to find a way to ionize the clusters by means that did not involve excessive energies that might also be favourable to cluster formation. Only in this way would a true picture begin to emerge of what the "magic numbers" were for carbon clusters. He reported results of his own experiments avoiding such irradiation which showed that while the positive ion signal for C_{60} remained large the negative ion signal did not. Whetten concluded that geometry alone was insufficient to account for the observed ion abundances.

Smalley felt that Whetten's critique deserved a reply. (He had never explicitly replied in the scientific literature to Cox's attacks.) The paper, written by the five authors of the original letter to *Nature*, O'Brien, Heath, Kroto, Curl, and Smalley, discussed how the sort of idiosyncrasies proposed by Cox could be discounted and restated that the evidence stood firm for the special stability of the neutral C_{60} molecule.[19] But this time they concluded almost gingerly, standing by their experimental data, but distancing themselves ever so slightly from the soccer ball model:

> All the evidence points towards a single isomeric structure for C_{60}. A truncated icosahedron, which has fully satisfied carbon valences with many Kekulé structures, and geodesically distributes its strain over the surface of a sphere, *is a unique structure* meeting these requirements [italics added].

A year on from the discovery, this was perhaps the low point for the five researchers. They were not exactly undergoing a crisis of faith, but – to remain with the religious analogy – they could use a sign. That sign could only be one thing: more evidence, better evidence, different and varied evidence.

EXHIBIT D: PHOTOFRAGMENTATION

It was some time before the Rice group were back in the fray. The motivation for their new work was not, as had been the case during much of 1986, to rebut criticism, although the underlying hope, as ever, was to produce additional evidence for the soccer ball structure. Smalley's strategy was to return to first principles and mount a detailed exploration of the photophysics of carbon clusters in general. It was an experiment they might have done in 1985 had they not been sidetracked by the dominance of the ion signal for the sixty-atom cluster. It was also a logical extension of Exxon's pioneering work from the year before.

Forget, for a moment, the soccer ball hypothesis; you have a picket fence of mass spectrum peaks for even-numbered carbon clusters. Are they all similar? If so, what is the nature of their similarity? And of their difference? These were the new questions.

The Rice group produced carbon clusters in AP2 using the vaporization laser as before. But they modified the apparatus to do two new things. In the first arrangement, cluster ions of the size to be studied were selected by mass (by accelerating them through the appropriate part of the time-of-flight mass spectrometer). These clusters did not simply pass on to strike a detector in order to build up a mass spectrum, however. Three microseconds later, a second laser fired with a beam powerful enough to blast them into fragments. It was these fragments that then passed on into a second mass spectrometer for analysis. In the second set-up, the fragmentation laser was not used, but the clusters were held for 120 microseconds before analysis in order to probe their tendency to fragment "naturally".

Since the clusters in question carry a positive charge throughout this procedure, there was no need for an ionization laser – and so, it was hoped, no room for arguments about the photoionization artefacts such an arrangement might produce. O'Brien and Curl found that while small carbon clusters shed fragments of three carbon atoms (as was already known), the larger ones, those with 34 or more atoms, appeared to lose atoms two at a time. Taken on its own, the result was a surprise; it was known that the neutral three-atom carbon fragment is more stable than the two-atom fragment. All other things being equal, it should be the three-atom cluster that breaks away. Clearly, all other things were not equal. The result suggested that while it is the stability of the breakaway fragment that is crucial in the fragmentation of small carbon clusters which are thought to be mainly rings and chains, in the case of the large clusters it must be the stability of the remnant cluster that determines the course of events.

Fragmentation does not stop once two carbon atoms have been ejected. It keeps on going, with further losses leading to the formation of smaller and smaller fullerenes. Two things might be happening: either the loss of carbon pairs is repetitive, or there must be ways to remove larger even-numbered fragments. The buckminsterfullerene ion, for example, either sheds a two-atom unit to become C_{58}^+ which then sheds another to become C_{56}^+ and so on, or it sheds C_2, C_4, C_6... each in a distinct way to leave remnants of C_{58}^+, C_{56}^+, C_{54}^+.... It was impossible to distinguish between these alternatives by experiment. The fragmentation only ceases with C_{32}^+. When this ion feels the impact of the fragmentation laser light, it simply shatters.

Of course, nothing would dictate the expulsion of two-atom fragments like the compulsion at all costs to maintain an even number of atoms in the cluster. The removal of two contiguous atoms from the carbon shell appears simple – the four bonds that must be broken in the process lie in close proximity and can readily recombine to form two complete new bonds. (There is a small snag with this mechanism in that this rearrangement of bonds is of a kind that is usually impossible in conventional organic chemistry. Under present circumstances, however, there is sufficient energy around to make the rearrangement possible.) Loss of four-atom and six-atom linear units is hardly more complicated, involving the creation of three and four new bonds respectively across the vacancy left by the departing atoms like stitches across a wound.[20]

Looking at the larger carbon clusters with between sixty and eighty carbon atoms, O'Brien and his colleagues found that their fragmentation proceeded in the same way, but they observed a "bottleneck" once the remnant fragment was down to sixty (and to a lesser extent seventy and fifty) atoms. In the fragmentation of C_{80}^+, for example, the keenness to cascade down to seventy and even sixty was readily apparent in the relative heights of the mass spectrum peaks. For C_{74}^+, C_{60}^+ is the favoured

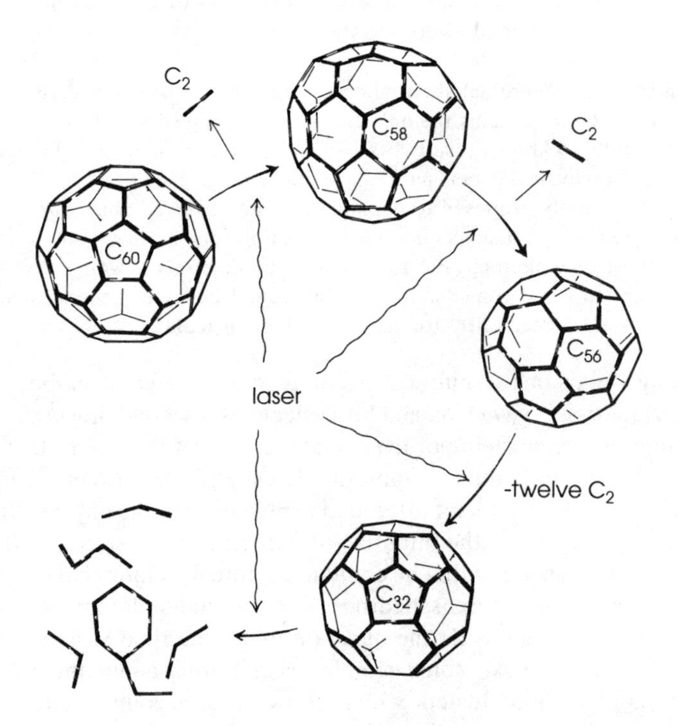

Buckminsterfullerene disintegrates in stages, losing two atoms at a time, until at C_{32} it blows apart

fragment from the word go. The fragmentation sequence may be imagined as being like marbles cascading down a flight of stairs with landings. The marbles bounce on each stair, but at any given moment there are more marbles on the wider landings. The "magic number" cluster ions with fifty, sixty, and seventy carbon atoms, evident in the mass spectra from the fast experiment with the dissociation laser, were still more prominent in the slow experiment designed to allow the clusters to decay naturally. It was yet more evidence for the special nature of the buckminsterfullerene ion which was showing itself by now as by far the most robust cluster ion in all Smalley's considerable experience. With this victory, the Rice group appeared to recover their confidence. The last paragraph of this paper read as follows:

> Although the spheroidal shell model for the large clusters of carbon is as yet unproven, it remains the only model yet advanced which is fully consistent with all known results. Its ability to explain the new, highly structured photophysical data is particularly impressive when one considers that it was originally proposed to explain only the observed abundance of the neutral clusters in a supersonic beam. Even in the absence of the original data, these new photophysics results taken by themselves would have been sufficient to force consideration of spheroidal cages in general, and the icosahedral "soccer ball" structure of C60 in particular.

The long delay in the publication of Exxon's study of carbon cluster experimentation[21] gave Cox and his colleagues a second bite at the cherry – or rather a second swing of the axe at the base of the cherry tree – with the opportunity to comment upon this latest Rice experiment. Unmoved by the logic of the argument offered in explanation of the new results, they homed in yet again on the minutiae of experimental procedure. By now, Cox was forced to admit that he could not refute the claim that C_{60} has the shape of a truncated icosahedron. But his methodological criticism remained: "If we lay aside the question of geometrical structure for the moment, can we make some simple assumptions about the electronic properties of carbon clusters which allow us to account for the experimental observations? If so, then what can be inferred about the geometrical structure of the clusters?" In other words, science should proceed by the following steps taken in sequence: observation, analysis, hypothesis, and only then visualization (not: observation, frenzied excitement, wild guess, pretty model, lame evidence, desperate justification). Throughout, Cox and Kaldor seemed oblivious of the *fait accompli* that the geometrical hypothesis had been made and that it did and does explain all.[22]

In 1988, more than two years after they had claimed to have trapped a lanthanum atom inside the buckminsterfullerene shell, Smalley and his collaborators were finally able to offer some proof, although still circumstantial, that this was indeed what they had done. The members of the Rice group themselves had been doubtful whether the metal-containing fullerenes might not be more complex than the simplistic terminology of the carbon "cage" implied, but they had been unable to perform the experiments that would tell them more.

The breakthrough came with yet another modification to AP2, the addition of an "ion cyclotron resonance" chamber which used a strong magnetic field to trap ions in a sort of holding pattern where they could be probed at more leisure than before. This enabled O'Brien, Curl, and Smalley to combine several previous experiments and explore the propensity for the metal-containing fullerenes to fragment under the impact of laser light. They found that they did so by the same process – the loss of two-carbon units – and in the same stages as the fullerenes themselves. But where the fullerenes remained as spheroidal clusters until there were just 32 atoms present in the shell, the metallofullerenes blew apart earlier. The smallest ion detected in association with a lanthanum atom was $C_{44}La^+$, with potassium it was $C_{44}K^+$, and with caesium, $C_{48}Cs^+$.[23] The highly energetic C_2 units could be stripped off as before, but the metal atoms remained apparently untouched by the ravages of the laser blast, even though the strength of a carbon-carbon bond is far greater than that of a carbon-metal bond. The final sizes of the clusters were in proportion to the radius of the metal ion associated with them. Both sets of data strongly supported the idea that the metal ions were indeed protected within the carbon cage. Successive blasts of the laser in effect "shrink-wrapped" fullerene shells to fit around their trapped metal ion. They had the best physical evidence they could hope for in support of the closed cage model. But still it was not what they really wanted.

EXHIBIT E: FIRST SPECTROSCOPY

The frustration of Smalley and his fellow chemists at not being able to summon up any spectroscopic evidence for the novel structure of buckminsterfullerene was exacerbated by the repeated attacks of the Exxon group on the quality of their other – circumstantial – evidence. It was, of course, a primary ambition to show whether or not C_{60} was responsible for any of the oddities of the various interstellar spectra. Having carried out some spectral scans in the region of some of the strongest diffuse interstellar bands, the Rice group had come to regard the chance of producing a match between the appropriate spectra of neutral buckminsterfullerene and these and other interstellar bands as extremely low, but still much higher than for any other single candidate they could think of. The odds on finding a fit with the spectra of a C_{60} ion or a simple chemical derivative were perhaps higher – if the spectra could be obtained at all. They laboured hard to produce a spectrum with the minuscule quantities available to them. A spectral absorption in the ultraviolet that Jim Heath

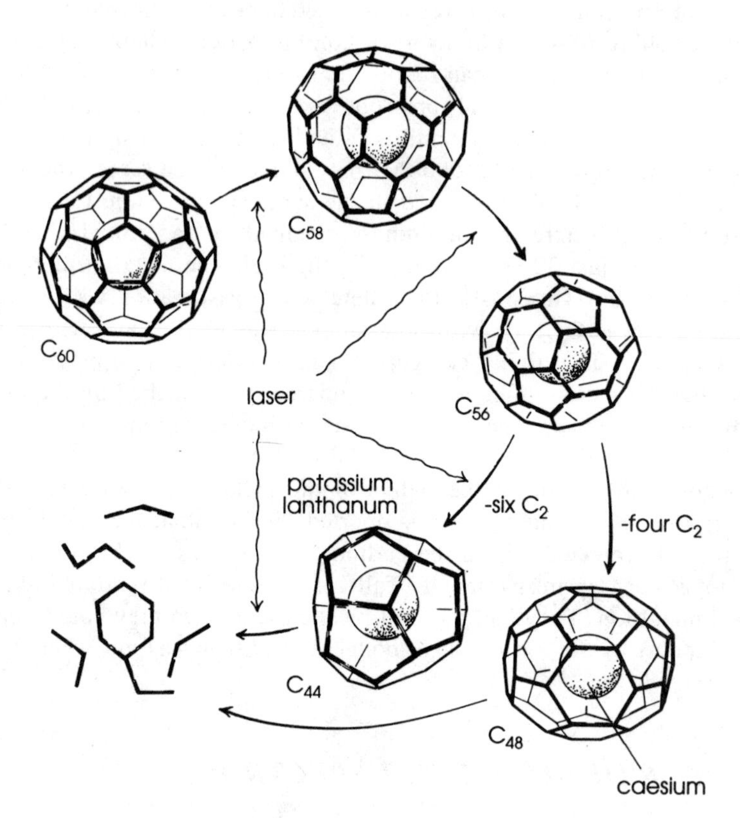

In the photofragmentation of metallofullerenes, the cage bursts according to the size of the trapped atom

assigned to the transition of an electron (from the highest occupied molecular orbital to one of the unoccupied orbitals) of buckminsterfullerene was later shown to be an incorrect assignment although still an accurate spectroscopic reading and was corrected by the Rice group, but not before it had sowed the seeds for considerable confusion among other researchers.[24]

When they did finally achieve the first real spectroscopic result, it was not by the ideal means that would yield a "fingerprint" signal to establish once and for all the molecule's supposed sphericity. The frustration, both at Cox and Kaldor's continuing scepticism and at their own inability to produce a spectrum by the preferred means, is evident in Smalley's preamble to the report:

> Although to many the current experimental evidence for such spheroidal carbon shell structures is quite compelling, this evidence still falls somewhat short of a clear proof. To most, such a proof would be spectral, involving high-resolution IR or UV spectroscopy, or (ideally) NMR spectroscopy such as that which so definitively established the structure of dodecahedrane, $C_{20}H_{20}$, the largest previously known polyhedral form of carbon. However, it will be a while yet before enough C60 is collected in an NMR sample tube for such an experiment to be feasible.[25]

The second-rate means of spectroscopy by which they did succeed is called ultraviolet photoelectron spectroscopy. It monitors the energy of successive electrons expelled from a molecule's orbitals by bursts of ultraviolet laser light. Although it has the considerable merit of being applicable to small numbers of atoms or molecules as well as to bulk materials, it is, in comparison with NMR spectroscopy, a blunt instrument indeed. Spectra display broad humps rather than sharp peaks at the energy levels of the affected electrons. The size of the humps is in proportion to the number of electrons at each level. No information may be gleaned directly from these spectra to show what atoms lie in proximity to what others; it is not possible to infer from them particular chemical compositions or molecular geometries. But what the technique can do for some organic molecules is to provide a direct visual correspondence with Hückel molecular orbital diagrams and the theoretical calculations of energy levels which they represent. The Hückel diagram is a schematic way of showing the energy levels of the pi-bonding electrons in certain molecules. It is these electrons that are most readily ejected when the molecule is energized with ultraviolet light. Thus, the intensity and frequency of each hump in the ultraviolet spectrum is simply a record of the number of pi-bonding electrons and their energies. For many molecules this type of spectrum would be a mess, with many humps overlapping each other. But molecules with simple geometric structures tend to have correspondingly simple electronic structures which in turn yield photoelectron spectra with relatively few, distinct peaks that can be used if not to derive the molecular geometry, then at least as a check on Hückel energy levels calculated on the basis of a conjectured geometry.

The ultraviolet photoelectron spectrum of buckminsterfullerene corre-

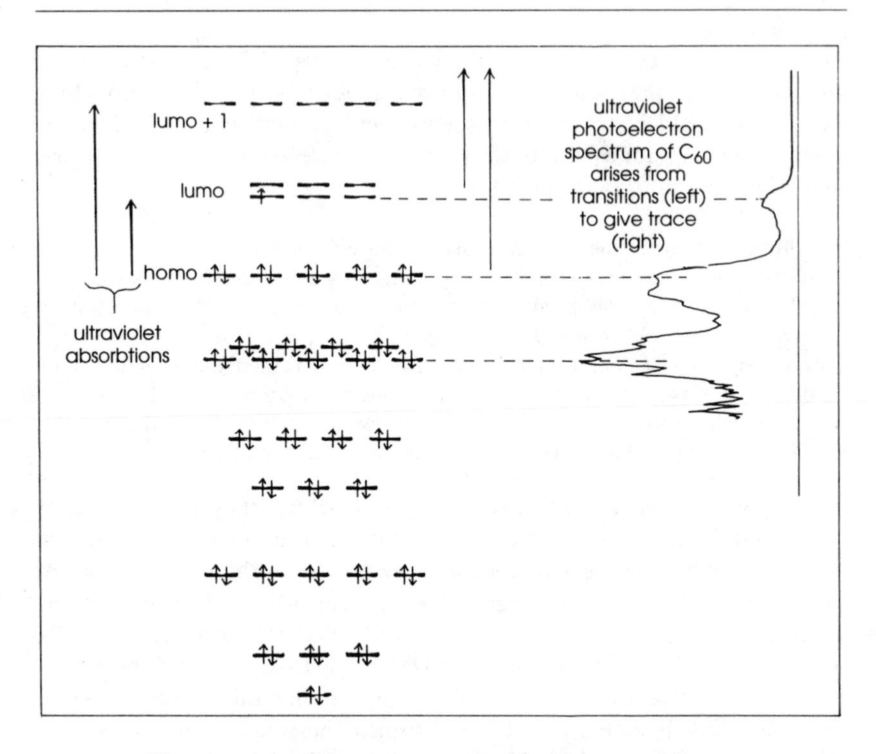

The energies of electrons measured by spectroscopy were in good agreement with the Hückel diagram

sponded well with the Hückel diagrams and with more sophisticated calculations that had been published by various theoreticians.

ON GROWTH AND FORM

Satisfaction in science stems not so much from making some new observation as from being able to explain it. To detect buckminsterfullerene and then to aver that it has the form of a truncated icosahedron with each of its carbon atoms located in an equivalent position at each of sixty vertices is all very well. But it is not to offer a complete explanation of anything. Although it may be shown that this is a more favoured structure for sixty carbon atoms than conventional candidates – most obviously a single flat sheet of graphite with dangling bonds along its edges – it does not explain how and why the shape comes about.

Kroto and Smalley and their co-workers did not forbear to speculate. As soon as they had established the fact that buckminsterfullerene is impervious to attack by a range of reactive small molecules, they were keen to use this information to comment upon the likely morphology and means of formation of common soot.[26] Their theorizing was stimulated in part by Smalley's playing with graphics of curved nets on his computer and in part by Kroto, Heath, and O'Brien's scouring of the Rice University library for literature on the science of combustion and much else besides.

In the paper on reactivity, they did something quite curious. Instead of crediting Haymet, whose paper they had been made aware of, with drawing their attention to the twelve-pentagon requirement, or even invoking Leonhard Euler, the eighteenth-century father of the theorem, they referenced one D'Arcy Wentworth Thompson, a Scottish zoologist active in the early part of this century.

Thompson's claim to fame rests largely on one extended and luminous essay, *On growth and form*; it is this work that is cited by the Rice group. The contemporary palaeontologist and author Stephen Jay Gould has called Thompson "perhaps the greatest polymath of our century" and his essay "one of the great lights of science and of English prose, the greatest work of prose in twentieth-century science".[27] Thompson's aim was to show that the shape of living things has a mathematical basis (and hence has no need for reliance on supernatural or teleological explanation). His argument is completely general. It applies to plants and animals, to airborne, waterborne, and land creatures of all sizes. He notes that the Eiffel Tower and John Smeaton's design for the Eddystone lighthouse both take the form of the trunk of an oak tree. It would be easy to conclude that nature inspired man to create these shapes. But it would be more perceptive to note, as Thompson did, that both man and nature take the most economical course of action prescribed by physical laws.

Having conjectured that the size and shape of all things is determined by the action of physical forces, it was a small, but revolutionary, step to explore the relation between different-looking creatures from this standpoint. This Thompson did in a remarkable series of graphical studies. By drawing, say, a fish on a rectangular grid on a sheet of rubber, and then distorting the grid by stretching the rubber in different ways, Thompson could produce warped new shapes that surprisingly corresponded to the outlines of quite different species. He applied the same technique to everything from leaves to skulls, finding simple mathematical relations between species often only distantly related according to Darwin's theory.

Polymath though he was, D'Arcy Thompson had little to say about chemistry. Nevertheless, he was to serve Kroto and Smalley's purpose as rather more than just an erudite ambassador for Euler's theorem. His

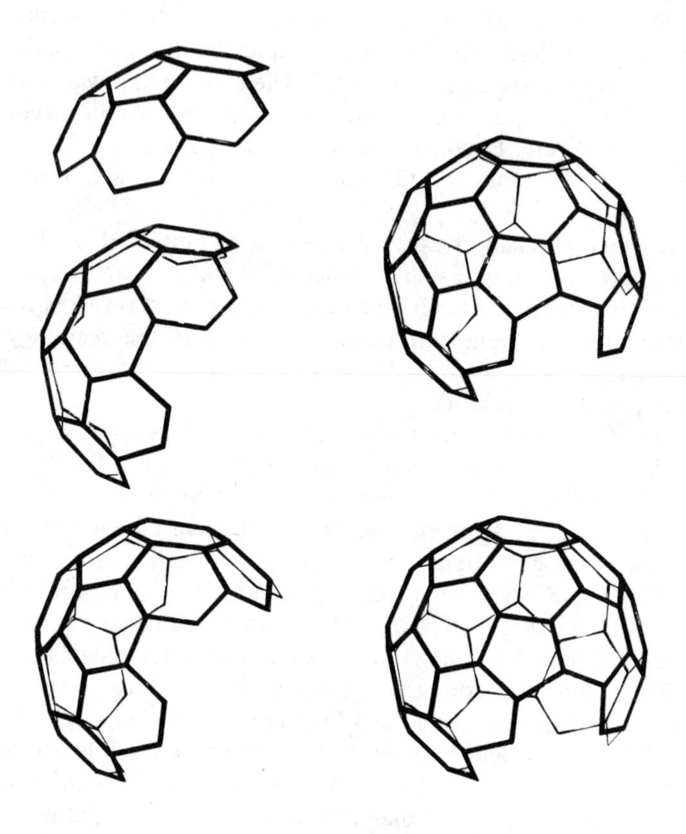

Possible mechanism for soot particle growth based upon nucleation by a central C_{60} molecule

genius can be seen as an inspiration behind a beautiful diagram in the paper on the reactivity of the fullerenes which shows a buckminsterfullerene molecule encased inside a larger spheroidal carbon framework which in turn is beginning to be enclosed by a third shell. The whole spiral scheme bears a remarkable resemblance to the spiral pattern of growth adopted by some plants and animals that are illustrated in Thompson's book. (The resemblance is in fact a little misleading. As Thompson points out, nature favours the "equiangular or logarithmic" spiral in which a radius drawn from the centre of the spiral to its leading edge increases in geometric progression – that is, by a constant *factor* – as successive orbits are scribed out. This is the mathematical relationship followed by the Nautilus sea-shell and many other gastropods. In the

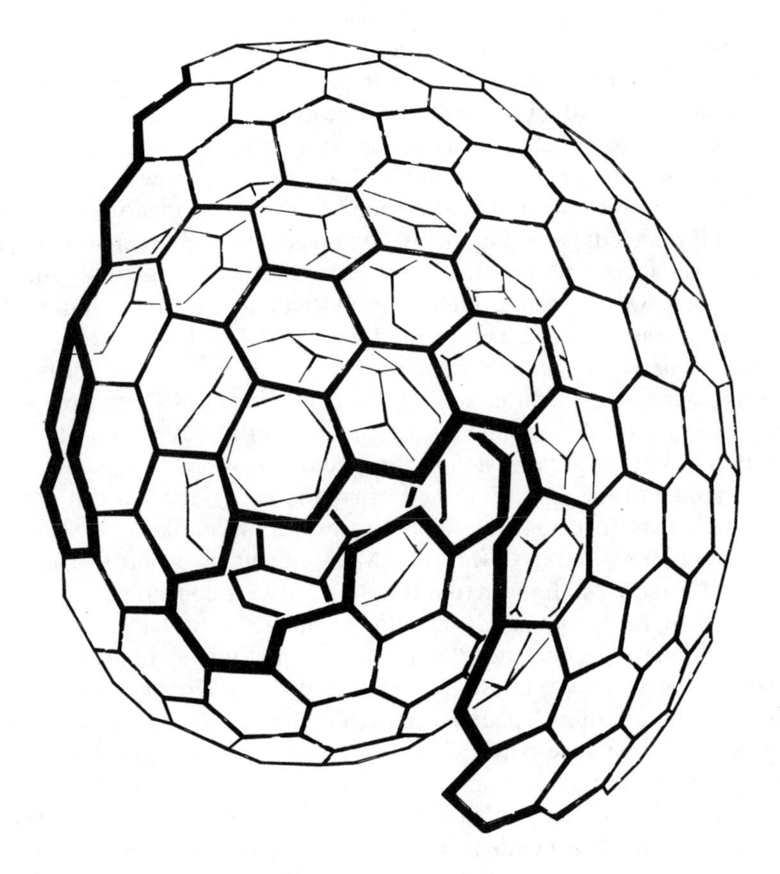

alternative, the "equable" or Archimedean spiral, this radius increases in roughly arithmetic progression – that is, by a constant *increment* – generating a spiral like that of a Swiss roll or a coiled rope. It is this latter model that lies at the heart of the proposed mechanism for the growth of soot particles, the spacing of successive layers being not the thickness of a piece of sponge cake or the diameter of a rope but the familiar van der Waals distance between layers of graphite.)

This figure stands entirely apart from the main text of the paper. It is left to the reader to make a connection between the caption describing this "one possible model of a growing soot particle" and the observation in the text that "soot ... is almost always found in the form of small spheres, which grow very rapidly to a diameter between 100 and 500 Å before they coalesce into strands of beads or more complicated structures." Could these spheres be spirals or nests of fullerene spheres? Were their nuclei molecules of buckminsterfullerene?

This tenuous thesis almost seems calculated to infuriate the sceptics. It had, however, passed muster with the Rice group's in-house sceptic, Bob Curl. The Curl Test applies to deductions from experimental observations more than to this sort of conjecture. Curl did worry, however, about how the combustion scientists would respond to the novel idea. His worries were well founded. On top of an already inadequately proven notion that there exists a bizarre, beautiful – and essentially unforeseen – form of carbon, here was a theory, utterly unsubstantiated, setting itself up to explain the entire mechanism of the formation of carbon soot, a theory apparently based on little more than a pretty picture. Needless to say, the Exxon group took exception to what was being proposed. They were supported in this by the community of combustion scientists who largely rebuffed this eccentric invasion of their field.

The model of the spiralling shell of soot is pretty, indeed. But prettiness is no reason to regard a model as especially unlikely to offer an explanation for what is going on. Many scientists would argue the reverse. Certainly, so far as Kroto and Smalley were concerned, it was no mere passing fancy, but an idea that they embellished in later work.[28]

Let us review the evidence that led Kroto and Smalley to propose this controversial idea: We know (or believe) that sixty carbon atoms can close to form a hollow, spherical molecule. We know that this molecule and others like it lose pairs of carbon atoms when energized by laser light, forming successively smaller molecules.

We know that the sixty-atom molecule (and to a lesser extent those stable species with other numbers of atoms) is in some way a survivor of the violent processes going on within AP2. By making small adjustments to the nozzle conditions, one could look at hot, young clusters right through the range to old, cold clusters. It was possible to place these snapshots in sequence in one's head to construct an animated view of the evolution of carbon clusters, from small to medium to large, from chains to rings to fullerenes. The reactivity experiments showed how impervious buckminsterfullerene was to chemical attack, but they also forced discussion of the nature of the vast bulk of the carbon pouring through the chamber that did not close up in this unique way. It seemed that there must be some kind of growing process taking place, one that continued even beyond the fullerene stage to consume the ever larger agglomerations of carbon in the cooling vapour, leaving behind only the tiny fraction of the carbon that had locked itself up as buckminsterfullerene as immune to further change.

Finally, we know, from experiments stretching back over decades, that sooting flames produce polycyclic aromatic hydrocarbon molecules in abundance. Given the existence of laboratory-made versions of such molecules with five-membered rings of carbon as well as the more usual six-membered rings, it seems likely that some of the molecules, whether made in a flame or by laser vaporization, have pentagons in among their hexagons. Such pentagons would, as we have seen, begin to induce curvature in a growing sheet of carbon.

The key to a rigorous model for the growth of soot particles now lay in understanding the unwritten rules that dictated not only why buckminsterfullerene was supremely stable but, more particularly, why fullerenes of some sizes were stable while others were not. The members of the Rice group understood – intuitively or aesthetically if not fundamentally – that buckminsterfullerene was stable because the twelve pentagons required for closure are held apart by hexagons. The puzzle was that the next largest fullerene "magic number" appeared to be seventy, a number with few factors and not traditionally associated with special numerical or geometric significance. Specifically, there was no obvious reason why seventy possessed any more claim to fame than numbers in the sixties which looked to be next in the fullerene family sequence.

It fell to Thomas Schmalz and Doug Klein, theoretical chemists at Texas A&M University in Galveston on the shore of the Gulf of Mexico, not far from Houston, to resolve the conundrum. Captivated from the start by Kroto and Smalley's discovery,[29] they had followed developments closely, at intervals offering valuable theoretical evidence for the molecule's proposed geometric and electronic structure. Schmalz and his co-workers confirmed the specialness of seventy in this context with the surprising announcement that C_{70} was indeed the first fullerene larger than buckminsterfullerene for which it is possible to construct a closed shell of hexagons and pentagons without bringing pentagons into direct contact with each other.[30] The result was highly counter-intuitive. One does not expect, having achieved mutual separation of the twelve pentagons in buckminsterfullerene, that the only way to make the larger molecules, C_{62}, C_{64}, C_{66}, and C_{68}, is by bringing those pentagons back into contact again. Science is frequently counter-intuitive, however. Fortunately, we have tools to help us counter our intuition. Schmalz and Klein had used mathematical methods in order to reach their result. At around the same time, Kroto, on one of his regular visits to Rice University, had worked with models to reach the same conclusion.

This was the beginning of what came to be known as the pentagon isolation rule – which states that when constructing fullerene structures one should try to minimize the contact between the twelve pentagons in the spheroidal framework. The knowledge that buckminsterfullerene was uniquely symmetric was thrilling, but its very uniqueness left scientists with a problem because it shed no guiding light on how related molecules might form. The knowledge that both C_{60} and C_{70}, the other prominent species in so many of the mass spectra, lived by the same rule, on the other hand, immediately lit the way forward.

Kroto started to make models of the smaller fullerenes. In continuous structures of sp^2 carbon atoms bonded to three other similar atoms, only a flat (infinite) graphite sheet is entirely without strain. Any curvature, whether the simple curvature created by rolling the sheet or the double curvature that comes from inserting pentagons into its hexagonal matrix, introduces strain. Buckminsterfullerene is stable because it is worth more to tie up the dangling bonds than it costs in straining the structure to do so. As the fullerenes become smaller, it is still worth closing, but the strain involved in doing so increases. Whereas the strain is evenly spread across the symmetrical surface of buckminsterfullerene, it now begins to concentrate at particular regions of the spheroidal surfaces as the pentagons are driven into closer proximity. A vertex where two pentagons and one hexagon meet is strained further away from flatness than a buckminsterfullerene vertex where there is one pentagon and two hexagons. Worst of all is a vertex where all three converging faces are pentagons.

The special stability of buckminsterfullerene arises because the pentagons are held apart by the hexagonal matrix. At the other end of the scale is the dodecahedral framework of the smallest (theoretically possible, but in practice probably too strained to exist) twenty-atom fullerene with only the twelve pentagons and no hexagons remaining; here, of course, three pentagons converge at every vertex. But what of those polyhedra in between these two symmetrical extremes? Working down from buckminsterfullerene, all the small fullerenes have surfaces where pentagons must come together. C_{50}, a relatively stable structure, is the smallest possible fullerene before it is necessary that more than two pentagons abut. Any smaller fullerene demands either a linear sequence of fused pentagons marching across its surface or the painful conjunction of three pentagons at a single vertex. The more stable small fullerenes are the last after which one of these geometrical sacrifices must be made. This pentagon rule, for which Schmalz had produced a proof and which Kroto quickly extended, generated magic numbers in good agreement with observed mass spectrum peaks.[31]

The pentagon isolation rule fitted neatly into an overall scheme by which carbon vapour condenses to form gradually larger particles. The standard picture at the time was that single-figure clusters of carbon atoms form chains, that clusters of between ten and about twenty atoms

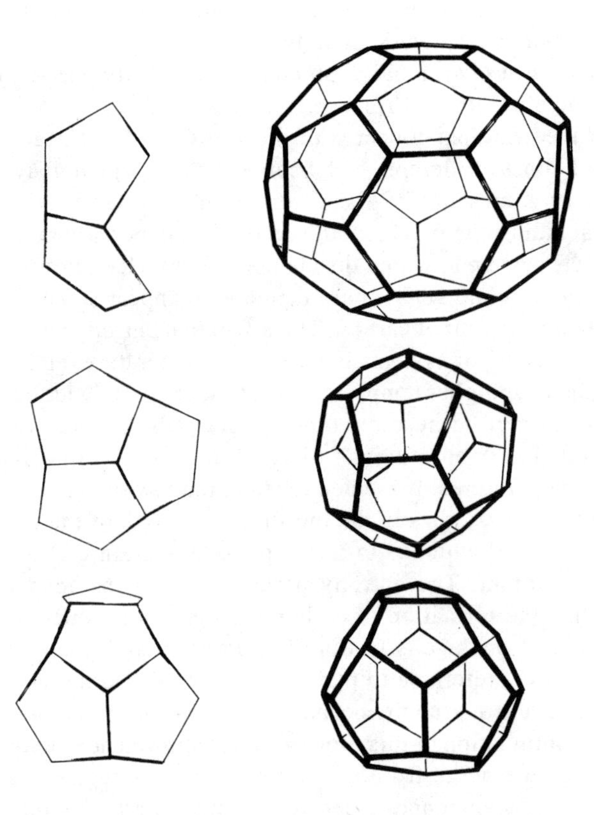

The stability of small fullerenes is governed by how many pentagons they can avoid bringing into contact

are single rings, and that larger clusters are largely based on polycyclic hexagonal frameworks. The new supposition was simply that at some point pentagons also begin to play a part in building the hexagonal arrays and that as a consequence the larger clusters curve and occasionally close.

Dangling bonds along the leading edge of growing soot particles consume individual carbon atoms and reactive groups of atoms (called radicals) as they advance adding them to the network. The occasional joining of two dangling bonds to each other to incorporate a pentagon introduces some curvature. In the spiral model, the edges continue to grow like wood shavings, curling round and round upon themselves until eventually they resemble the soot particles that the Rice group had seen in the carbon literature. In general, these edges would overshoot the previously formed part of the network as they continued to grow in an unending curled sheet. But occasionally, an edge curls tightly enough to catch its own tail and close up to form a self-contained molecule.

One of the principal concerns of those involved in the experiments to make buckminsterfullerene had been entropy (which may be loosely described as a measure of the disorder of a system). All irreversible change, including chemical reaction, requires an increase in the entropy of the complete system in which the change occurs. Here, however, the most highly symmetric molecule could be made to appear spontaneously from out of a chaotic vapour of carbon. The effect that entropy has on the energy and hence stability of systems is related to temperature; entropy becomes a bigger player at higher temperatures. Thus, crystalline ice can form from fluid water in part because the temperature falls. In the process of laser vaporization, however, ordered fullerene molecules were being created at very high temperatures. It seemed to stand the law of entropy on its head. The key to the puzzle lies in the inordinate strength of the carbon-carbon bond, the effect of which is to push up the temperature at which entropy becomes important. The entropy paradox was no paradox once it was accepted that the formation of carbon-carbon bonds is paramount. Thus, the insertion of pentagons leading to curvature and possible closure of carbon sheets is preferable to their remaining as disordered, high-energy amorphous sheets of graphite because it creates carbon-carbon bonds. The indefinite continuation of this process at last showed how a highly ordered molecule could emerge from an environment that began as one form of chaos (carbon vapour) and ended for the most part as another (assorted soot particles). In one fell swoop, the spiral growth model explained the seemingly contradictory properties of resilience and scarcity of the buckminsterfullerene molecules.[32] Whether or not it also represented a realistic mechanism for the formation of carbon soot remained to be seen.

With Kenneth McKay, a graduate student at Sussex University, Kroto continued to work away at the model-making which had proved so fruitful. Turning their attention to models of very large fullerenes, Kroto and McKay were surprised to find that these did not simply magnify the spherical shape of buckminsterfullerene. Instead, C_{240} and C_{540}, the first two large fullerenes with exactly the same symmetry as buckminsterfullerene (and respectively with twice and three times its diameter), were polyhedral in appearance. In place of the constant smooth curve of the hollow sphere, they found distinct vertices developing at each pentagonal site on the surface of the molecules.[33]

What had begun as a routine exercise simply to construct the models had unearthed new knowledge. Once again, one of the most humble – but time-honoured – techniques of science had shown its worth where more sophisticated tools had failed to deliver. First in the working out of the structure of buckminsterfullerene itself, and now in helping to determine the rules of fullerene formation and perhaps in providing clues to the ways larger carbon structures grow, models had carried the day. In the days when Watson and Crick had used models to elucidate the structure of DNA, computers were in their infancy, quite incapable of undertaking exercises of this complexity. For simple model-making to have demonstrated its worth in a similar manner more than thirty years later, when it might have been supposed that graphics computers had rendered such a primitive aid obsolete, gave some considerable satisfaction.

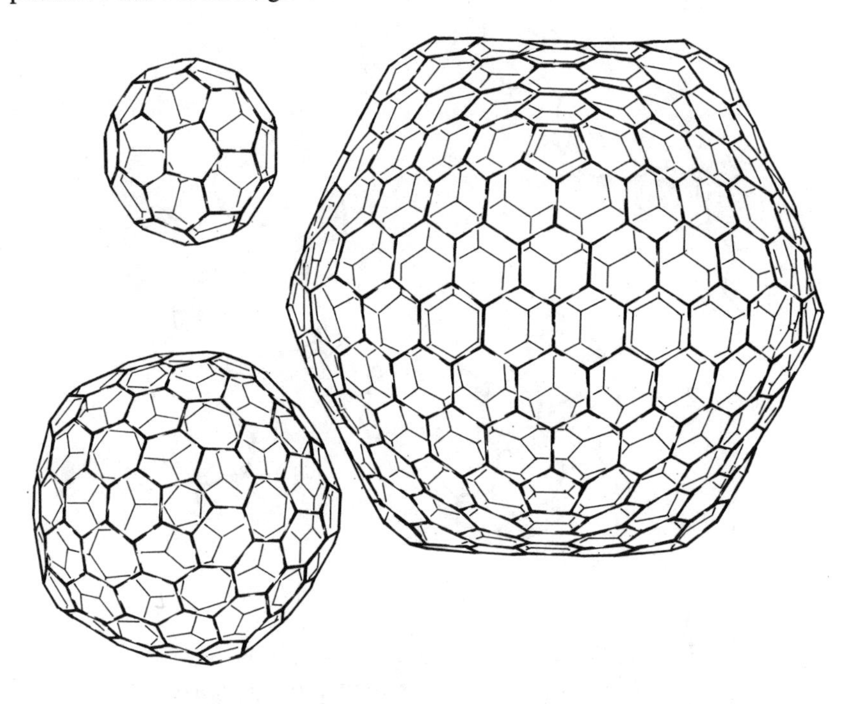

Simple model-making led to the surprising discovery that larger fullerenes would not be perfectly spherical

This further discovery necessitated a refinement of the spiral growth scheme. Kroto and Smalley's early drawings had shown complete buckminsterfullerene molecules lying at the centre of the growing spirals like the piece of grit at the centre of a pearl. The assumption was that the larger shells would adopt the same smooth curvature. Kroto now believed that as the shell continued on its spiral course, the pentagons on the outer layer would tend to align with the ones in the layers below, ensuring the preservation of the icosahedral shape. Spiral growth could take place based upon an icosahedral template. Kroto coined the word "icospiral" to describe this happy combination of circumstances. Furthermore, he believed that it offered an explanation for the formation of polyhedral graphite crystallites. (When they are not spheroidal, soot particles sometimes take polyhedral forms.) These shapes have conventionally been explained as arrangements of disjunct planes of graphite. Might they in fact be massive nested polyhedra?

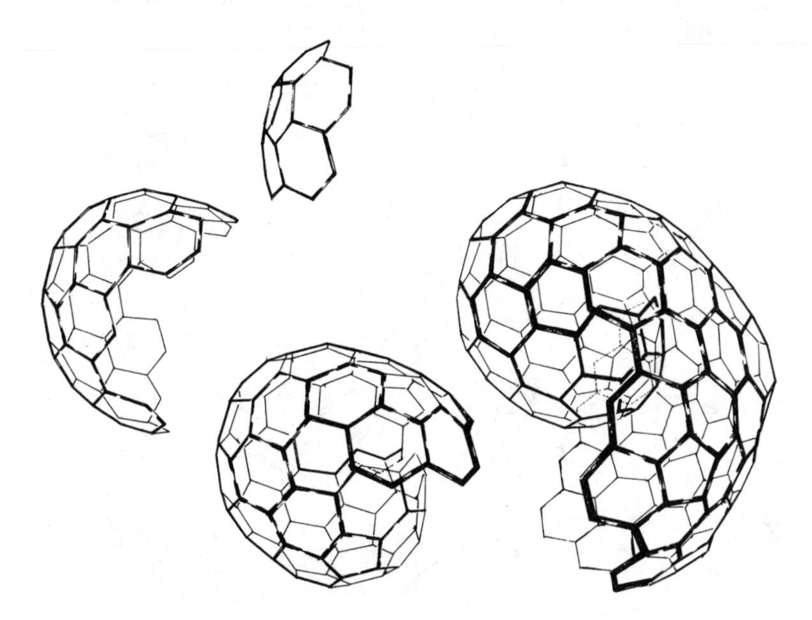

Kroto's "icospiral" shell growth mechanism dispenses with the C_{60} molecule as an initiator of growth

This closer exploration of the spiral mechanism for soot formation brought to light a new motif in addition to the spiral and the icosahedron – the helix. As the shell screws its way around and outward, looping over itself again and again, its leading edge falls into place along the line of a helix that lances the growing particle like a twisted kebab skewer. The helix is a prominent feature in nature. As Kroto wrote: "It is intriguing that the icospiral shell ... combines the two most bioemotive shapes, the helix and the icosahedron."[34]

These conjunctions – the icosahedron, the spiral, the helix – were merits of the model, but also, to the sceptics, its undoing. For them, this talk was romantic nonsense. Cox felt, not for the first time, that the simple and beautiful model was being favoured for its simplicity and beauty, not because there was hard evidence for it.

The model left plenty unaccounted for at the level of detail, too. It was not clear from the pretty picture how the few spiral shells that close in on themselves to form fullerene molecules achieve this trick. (It is topologically impossible to close such a growing spiral into a complete spheroid without breaking a lot of bonds.) Nor was it clear what happened to the dangling bonds trapped on the inside edge of the particle as it grew. Were these spheroidal soot particles of nested curling graphite sheets supposed to be especially stable like the perfect fullerenes? If so, why should they go on, as Kroto and Smalley had acknowledged they do, to form "more complicated structures"? Were the soot particles spheroidal as in the early model or icosahedral as in Kroto's revised scheme? And did either of these ideals bear much relation to real soot? Seen from the perspective of the specialist in carbon and its combustion, Kroto and Smalley's hypothesis was, to say the least, ill-informed. All in all, it smacked of too much D'Arcy Thompson and too little solid data or reading in the field.

While buckminsterfullerene and the other fullerenes remained esoterica created in an unusual way and on a minute scale, the combustion scientists could afford to snub their work as irrelevant to their field. But the omens were already against them. For in mid 1987, Klaus Homann and his colleagues at the Technische Hochschule in Darmstadt reported that they had detected buckminsterfullerene in sooting flames of acetylene and benzene.[35] It was the first sighting of the molecule (or rather its ions) outside of Smalley's apparatus. For Kroto and Smalley, Homann's flame was a beacon of hope that they would yet be able to prove the ubiquity of their special molecule.

We resume our story in Chapter Seven to consider how it was that the community of physical scientists who devote themselves to the study of carbon failed to make the premier fundamental discovery in their field during this century, before, in Chapter Eight, describing how the "yellow vial" was finally found after five years of searching. We shall occupy ourselves during "the interim" of which Smalley and his colleagues complained by acquainting ourselves with three tangential aspects of this story: the first excursus is on number and symmetry, the second on Buckminster Fuller, and the last on the art of spectroscopy.

4

ON SYMMETRY AND THE SEXAGESIMAL

It is in their search for order in the universe that science and art coincide. Scientists have always been guided in their work by the fact that nature appears to prefer processes that favour the simple over the complex. Nature's simplicity manifests itself in the frequent occurrence of objects with regular shapes, high symmetry, and "magic numbers" (although we humans have an endearing ability to overlook nature's untidy habits when it suits us). For their part, artists have sought to interpret nature in ways that accentuate or counterpoint this simplicity.

In his book, *Entropy and art*, Rudolf Arnheim remarks on the order that is explicit in both animate and inanimate nature from crystals to flowers, and asks:

> What shall we make of this similarity of organic and inorganic striving? Is it by mere coincidence that order, developing everywhere in organic evolution as a condition of survival and realized by man in his mental and physical activities, is also striven for by inanimate nature, which knows no purpose?[1]

In an essay entitled *Atomism structure and form*, the physicist Lancelot L. Whyte enlarges the question:

> What is the relation of the two cosmic tendencies: towards mechanical disorder (entropy principle) and towards geometric order (in crystals, molecules, organisms, etc.)?... Why regular forms at all, amidst so much confusion? Morphogenesis is still a mystery in nearly all realms. Only in a few restricted fields ... do we know how anything has come to have the shape it has, in terms of basic principles.[2]

One route to the answer to such questions lies in visual examination of material at hand: what shapes are present? What are the regular dimensions and formative ingredients of these shapes? What might be their causation? Thus did D'Arcy Thompson reason when he determined that the shape of a mollusc's shell is no more than the result of that creature's need to remain within a protective case while growing more or less equally and linearly in all dimensions (as most creatures strive to do with or without shells). As the animal grows in length and breadth, its shell cannot grow in this way; but by increasing in radius the shell can accommodate the expansion in two dimensions, tracing out a logarithmic spiral.

The use of visual ideas in the articulation of scientific argument is not limited to physical specimens. In the early decades of this century, the community of physicists split into two camps as they sought to understand and describe the quantum-mechanical nature of the atom. Niels Bohr headed the group of visualizers; Erwin Schrödinger and Werner Heisenberg maintained that visualization was not necessary. Yet even the non-visualizers could not help but describe the new phenomena in metaphors – spinning particles, quantum jumps, and so on. An electron may be a cloud, a wave, or a particle to suit the circumstances under discussion – and also to provide the best fit with subsequent visual episodes in a series of events (clouds merging, waves interfering, particles colliding). None of these metaphors is right, none is wrong; what is important is that they are all visualizable. On balance, the visualizers carry the day even in the invisible regions of science:

> In conclusion, the importance for creative thinking of the domain where art and science merge has been emphasized by the great philosopher-scientists of the twentieth century – Bohr, Einstein, and Poincaré. For in their research the boundaries between disciplines are often dissolved and they proceed neither deductively through logic nor inductively through the exclusive use of empirical data, but by visual thinking and aesthetics.[3]

The need – or the wish – to visualize in turn shapes the scientific agenda. As Jacob Bronowski pointed out: "This is the outlook of modern science: a search, not for numerical measurements, but for topological relations as odd as the Four Color theorem and the invented spaces of Möbius."[4]

Closely connected with the search for relations to do with shape is the broader search for beauty. Scientists do not usually confess to seeking out beauty; they refer to it surreptitiously, for example in their description of solutions to problems as elegant, but the thought is there.[5] In one of the rare unprompted speculations by a scientist on beauty, Subrahmanyan Chandrasekhar quoted Henri Poincaré:

> The scientist does not study nature because it is useful to do so. He studies it because he takes pleasure in it; and he takes pleasure in it because it is beautiful. If nature were not beautiful, it would not be worth knowing and life would not be worth living ... I mean the intimate beauty which comes from the harmonious order of its parts and which a pure intelligence can grasp.[6]

Scientists, of course, are obliged to be seekers after truth. It would be tempting to follow Keats, but in science it is not invariably the case that "Beauty is truth, truth beauty", and we saw in the previous chapter what heat can be generated by the occasional friction between the two ideals. Nor is it the case that the scientific preference is always towards truth as one might expect. The physicist Paul Dirac apparently went to the extreme of preferring mathematically beautiful theories even to ones in agreement with observations.[7] But Dirac was surely dissembling: how far would he have pushed it? For it is also the case that the beacon of beauty does not always light the right path. For example, a theory dealing with the equivalence of different groups of subatomic particles going under the name of "supersymmetry" for some time blinded theoretical physicists with its elegance. Only belatedly, guided by the brighter light of evidence, did they change the direction of their research.[8]

When pressed, scientists tend to identify beauty with simplicity. Chandrasekhar opined that "simplicity and vastness are both beautiful". Particle physics is driven by the conviction that a few simple laws of nature will be found. In the quest for a Grand Unified Theory, it is implicit that there is something grand about unification. But why should nature be simple rather than complex?

Everyday science cannot afford to become mired in such philosophical questions. Karl Popper was concerned to eliminate the aesthetic conception of simplicity. He denied any merit to simplicity when it applies merely to exposition; an elegant derivation of a proof is no better than a convoluted one. Both yield the same knowledge. So what does simplicity mean in science? Popper concluded that a simple result is no more than one that may easily be shown to be false, if false it be.[9]

By any of these criteria, buckminsterfullerene is both beautiful and simple. It is a symbol of order. As the series of experiments related in the previous chapter demonstrated, it is also now established as a truth (at least, beyond reasonable doubt). The root of its order lies in number and form.

NUMBER

There are many interesting stops on a journey along the number line. One of the oldest and grandest of these stations is the number sixty.[10]

The basis for our familiar decimal system lies in the fact that we have ten fingers for counting. Other systems, however, use base numbers which have more factors and are consequently more versatile. The duodecimal system uses base twelve, and retains its foothold in contemporary society in the way we buy eggs (by the dozen) and screws (by the gross) and measure out the hours of the day. The specialness of twelve lies in the fact that it is the lowest base for which the first four integers, 1, 2, 3, and 4, are factors. The sexagesimal system, which uses sixty as its foundation, has the same attraction. It is the lowest base to offer 1, 2, 3, 4, and 5 as its factors (and it offers 6 as a bonus).

It was the Sumerians in Mesopotamia, more than 4000 years ago, who first adopted base sixty for their calculations, although it is unlikely that cool logic entirely explained their choice.[11] When the Babylonians subsumed the city states of Sumeria, they kept the sexagesimal base. The many factors of sixty (1, 2, 3, 4, 5, 6, 10, 12, 15, 20, and 30) compared to ten (1, 2, and 5) means it is well suited to dealing in fractions. It is easier to apportion weights and measures accurately, into equal thirds or quarters, for example, by using the sexagesimal system. The number sixty continued to be of importance in Hebrew and Arab cultures and persisted in some degree until the European Renaissance, used especially in astronomical calculations up to the time of Copernicus, Kepler, and Galileo. Its principal legacy today is in the way we apportion units of time into minutes and seconds.

FORM

Regularity in three-dimensional form has two aspects. There are the continuous, repeating structures known as crystals. And there are singular regular structures which are the polyhedra. There is an infinite number of polyhedra – clearly one can keep adding facets to a many-sided object, especially if one is not concerned about maintaining its symmetry – but there are only five regular convex polyhedra, that is polyhedra in which all faces are equivalent and are themselves regular polygons. They are the tetrahedron, the cube, the octahedron, the icosahedron, and the dodecahedron. These are the objects that Plato put at the heart of his cosmology – the Platonic solids.

In some senses, however, the first polyhedron is the one that has one face and yet an infinite number of faces, the sphere. This most simple of all solid forms – simple, that is, because of the economy with which its shape may be expressed in mathematics as well as because of its beauty – is familiar in nature. The discs of the sun and the moon were suspected to be spherical long before they were shown to be. On earth, we detect the presence of the sphere, only a little distorted, in droplets and bubbles, in fruit and their seeds, in eggs and in single cells. The shape arises anywhere growth begins from a point and has no reason not to push outward from that point evenly in all directions.

In the ancient world, both the sphere and the Platonic solids were employed to represent objects in nature. They were also revered as the geometrical expression of some metaphysical Unity.[12]

> The image of the sphere ... has been used through the ages to depict physical, biological, and philosophical phenomena.... Roundness is chosen spontaneously and universally to represent something that has no shape, no definite shape, or all shapes. In this elementary sense, Parmenides represents the wholeness and completeness of the world by a sphere.... Thomas Aquinas, for example, compares God, the all-encompassing, with the boundary surface of the sphere....[13]

The telescope and the microscope uncovered new spherical forms. It was shown that the earth is roughly spherical. But the mystique of the sphere has diminished:

> As the natural sciences insist increasingly on verifying their conceptions by exact observation, the image of the sphere is limited more and more strictly to physical structures that fit it closely. However, the geometrical shape which dominated the view of nature from the beginning because of the preference of the form-seeking mind for simplicity continues to be applicable to such principal patterns of the physical world as the solar system or the atomic model. This is more than a happy coincidence.[14]

Thus, for example, it is still expedient to represent atoms as spheres in many circumstances although it is known they are not spherical in quite the same sense as billiard balls.

In two dimensions, a regular polygon is a shape with equal sides and equal angles between those sides. The three-dimensional equivalent of the regular polygon is the regular polyhedron, a faceted solid in which all the faces are equivalent and are themselves regular polygons. Just as the angles of a polygon are equivalent, so are the vertices of a polyhedron where these equivalent faces meet. Both families of geometrical forms have been studied since Pythagoras, in part because of their aesthetic appeal.[15]

There is no limit to the number of regular polygons that may be drawn in two dimensions. But there is, as already noted, an early limit on the number of regular convex polyhedra. This strange fact has contributed to the mystique of these particular forms and has served to stimulate speculation as to their role in metaphysics.

Of the five Platonic solids: the tetrahedron has four equilateral triangular sides; the cube and the octahedron are duals of one another, the former with six four-sided faces and eight vertices where three faces meet, and the latter with eight equilateral triangular faces and six vertices where four faces meet; the icosahedron and the dodecahedron are also duals, the former with twenty equilateral triangular faces and twelve vertices where five faces meet, the latter with twelve pentagonal faces and twenty vertices where three faces meet.

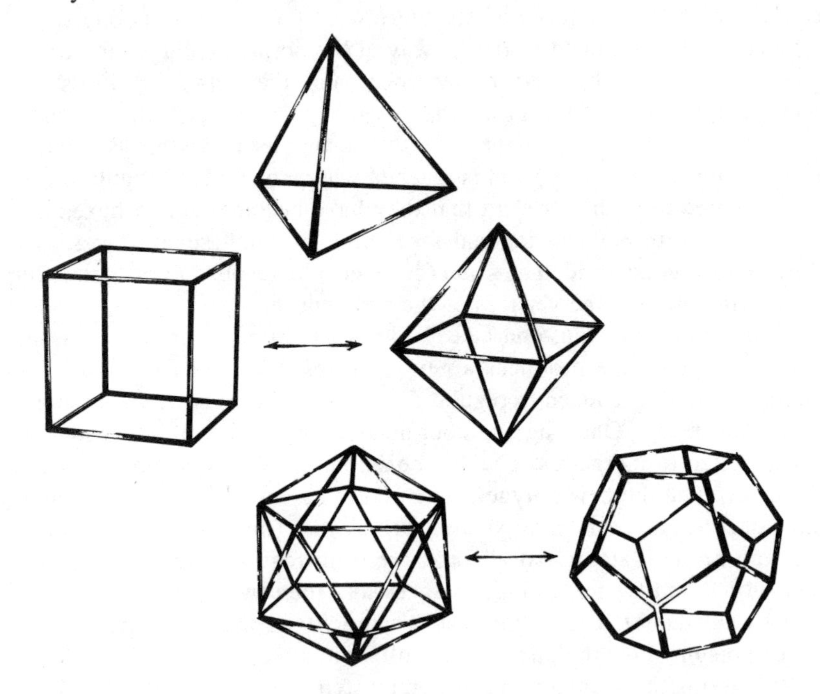

The five Platonic solids: the tetrahedron; the cube and the octahedron; the icosahedron and the dodecahedron

In his dialogue *Timaeus*, Plato outlined a cosmology in which the four ancient elements are identified with four of these solids. The earth is the cube; fire the tetrahedron; air the octahedron; and water the icosahedron. Plato singled out the dodecahedron to symbolize the "ether" that enclosed the entire universe.

In addition to the regular polyhedra that are Plato's solids, there are eight "semi-regular" solids. These are polyhedra with faces made up of a mixture of two regular polygons. Archimedes is supposed to have been the first to describe all thirteen solids, but his manuscripts perished in the fire in the library at Alexandria. It was Johannes Kepler who first published the complete list, together with a proof that there were only this many semi-regular polyhedra.

Some of the semi-regular polyhedra may be generated by demonstrating the duality of the Platonic solids. If one shaves away the corners of a cube, one eventually arrives at its dual, the octahedron. On the way, however, there are intermediate points: an object with six regular octagonal faces where there were previously the cube's square faces and eight new equilateral triangular faces where there were once corners is one semi-regular solid. Continued shaving away at the corners enlarges the triangles until eventually their points meet and the octagonal faces are reduced to squares. This is another semi-regular solid. Continue once more: the triangles begin to merge with one another while the new square faces diminish in size. A point is reached where the eight triangular faces have merged to such an extent that they have become regular hexagons. This solid, with eight hexagonal faces and six small square faces, is a third semi-regular solid. The same forms may be reached in reverse order by shaving away at the vertices of the octahedron.

If, in similar fashion, one takes a dodecahedron and shaves away its twenty vertices, one produces a new semi-regular solid with its twelve pentagonal faces reduced to regular decagons and twenty new equilateral triangular faces. Once again, continued enlargement of the triangles causes them to merge and become hexagons, and once again continued eating away at the new polygons eventually halves their number of sides and generates this time new, smaller pentagons. The next intermediate on the way to the Platonic icosahedron are semi-regular forms with twelve pentagonal and twenty triangular faces and then twelve pentagonal and twenty hexagonal faces. This last semi-regular solid, of course, is the shape presumed for the buckminsterfullerene molecule.

Witnessing this transformation step by step demonstrates that the duals share the same symmetry; that is, they can be mapped into the same set of equivalent positions by rotation and reflection. The shaving action – "truncation" – is applied equally at each vertex and thus has the same symmetry as the objects themselves. We know, then, that the icosahedron and the dodecahedron have the same symmetry, and also that this symmetry is shared by the truncated dodecahedron, with twenty triangular and twelve decagonal faces, and the truncated icosahedron, with twenty hexagonal and twelve pentagonal faces. Both truncations have sixty vertices where three edges meet. Thus, both in principle satisfy the condition for a sixty-atom carbon molecule.[16] In due course, it required sophisticated spectroscopic analysis to confirm which of these forms really represented the structure of buckminsterfullerene.

The dodecahedron and the icosahedron have been called the "most

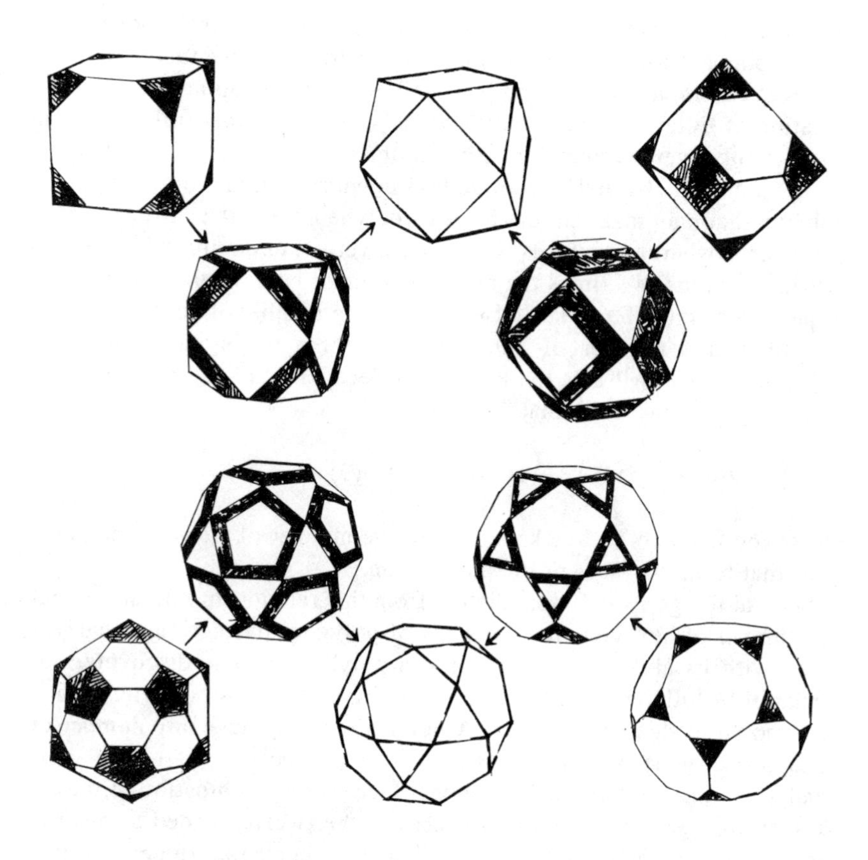

Transformation between duals of the Platonic solids generates semi-regular Archimedean solids

attractive" of the Platonic solids. It has also been noted that the truncated icosahedron is especially attractive because of the near equal size of its pentagonal and hexagonal faces, in marked contrast to most other semi-regular polyhedra and especially to the truncated dodecahedron with its large decagons and tiny triangles.[17] It is not coincidental that it is these polyhedral shapes that most closely approximate to the sphere.

So much for Platonic idealism in carbon chemistry. There is more than perfect symmetry in the form of the fullerenes. Leonhard Euler's famous theorem, "faces plus corners equals edges plus two", fits all polyhedra regardless of their regularity or symmetry. If we make it a condition that our carbon atoms are to be sp^2 hybrids, that is they are to have three bonds radiating from them, then we determine that the number of edges is simply one-and-a-half times the number of vertices (since there are three bonds per atom, and each bond is shared between two atoms). This modification of Euler's equation quickly leads to a proof that fullerenes can possess only even numbers of carbon atoms.

An extension of Euler's theorem lays down conditions for the types of polygon that may make up the faces of polyhedra. On the surface of any polyhedron where three faces meet at each vertex, three times the number of triangles plus two times the number of quadrilaterals plus the number of pentagons must be equal to twelve plus the number of heptagons plus two times the number of octagons plus three times the number of nonagons plus four times the number of decagons and so on. More economically, we may write that:

$$3(3) + 2(4) + 1(5) = 12 + 1(7) + 2(8) + 3(9) + \ldots$$

where the numbers in brackets refer to the number of sides of the polygons that form the faces of the polyhedron.

Several things spring immediately from this relationship. It shows that the number of hexagons in a network of polygons has nothing whatever to do with its ability to curl up into a polyhedron. The discoverers of buckminsterfullerene would have saved themselves a lot of grief if they had known of this law in 1985. A polyhedron may have any number of hexagons or none. It also shows that every polyhedron must have some triangular, quadrilateral, or pentagonal faces (or a combination of these). Furthermore, every pentagon in excess of the twelve needed to balance the equation can and must be "cancelled" by a heptagon. In general, it is possible to barter various polygons (pentagons demand heptagons, squares demand octagons, two pentagons demand one octagon, and so on) while maintaining a closed polyhedral structure. These properties of polyhedra are important in the consideration both of the morphology of larger fullerenes and of molecular rearrangements on their surfaces that may be key to their growth, shrinkage, and reactivity.

SYMMETRY

We crave symmetry, or to be more exact and subtle, the interplay between symmetry and asymmetry. In art and in nature, it is perhaps our prime criterion for beauty. We see it in the never quite mirrored halves of the human face, the never quite equal arms of the starfish; the symmetrical facade of the Parthenon subverted by the asymmetry of the frieze upon its pediment, and the classical paintings of the Last Supper with the symmetry of the table and the room it is in dramatically offset by the animation of the Disciples as Christ announces that one among their number will betray him.[18] István Hargittai is his letter to contributors to his vast treatise on symmetry in nature and culture called it "possibly the best bridging idea crossing various branches of sciences, the arts, and many other human activities".[19]

Symmetry stems from the physical conditions under which things take shape. Asymmetry may arise from small disruptions of these conditions. Nature exploits these imperfections to "choose" between symmetrically related effects. "Nature is never *perfectly* symmetric."[20]

When we think of symmetry in animals – the fearful symmetry of William Blake's tyger – it is usually bilateral symmetry we have in mind. Symmetry emerges during the process of cell division, but is not a direct outcome of it. In other words, living creatures are not bilateral because cells divide bilaterally. Even amphibians that grow from spherical eggs assume bilateral symmetry.[21] Newton regarded the widespread presence of bilateral symmetry in nature as evidence of the existence of a Creator. He apparently missed an explanation for this symmetry in his own laws of gravity and motion: in its three dimensions of length, breadth and height, an animal retains symmetry only across its breadth; it loses it in the vertical dimension because of the requirement to stand up against gravity, and in the longitudinal dimension in meeting the requirement for unidirectional motion. (The same constraints lead to the appearance of bilateral symmetry in wheeled means of transport of our own design.) By the same token, species which are protected from the effect of gravity by virtue of their small size or suspension in air or water, such as the radiolarians, or which are largely immobile, such as starfish or sea-anemones, are free to manifest higher symmetry.

Though it has given humankind much pause for meditation, bilateral symmetry, by which a shape is mapped onto itself by means of reflection about a single, often vertical axis, is one of the lowest forms of symmetry. Symmetry operations are actions upon an object that leave its appearance unaltered. They are classed in three types – reflection, rotation, and translation – all of which make their appearance both in nature and in art. Many things exhibit a combination of these types of symmetry. Regular brickwork, for example, has translational symmetry in the ordered repetition of brick and half-brick. The brickwork pattern may be *shifted* vertically or horizontally to a point where it maps onto itself. The brickwork also has vertical and horizontal *axes of reflection*. Reflections produce a similar mapping but by different means, *swapping* top for bottom or left for right. Many natural objects such as flowers have rotational symmetry about their centre point and axes of reflection that pass through this point.

Repeated structures such as wallpaper and crystals are comparatively limited in the number of their symmetry operations. Rotational symmetry is only possible in measures of a half, a third, a quarter and a sixth of a complete revolution, for example. Attempts to build a crystal with five-fold rotation fail because it is not possible to retain the translational symmetry while introducing the five-fold rotation. To put it another way, it is not possible to build an indefinitely repeating pattern based on five-fold symmetric units, either in two or in three dimensions.

Individual objects such as species of flora and fauna or individual molecules have no translational symmetry and as a consequence may exhibit many more reflection and rotation symmetry operations, including those that take place about the five-fold axes precluded in crystals.

The quintuplicity plentiful in nature and in icosahedral and dodecahedral forms has long held a special place in human hearts. The pentagram or five-pointed star has been believed to possess magical properties. The golden section, the generative proportion of Euclid's logarithmic spiral and of much in architecture, may be derived from the pentagon. In 1658, Sir Thomas Browne published *The garden of Cyrus*, a fantastical catalogue of the five-fold in nature, noting for example those five-leaved flowers "which concurre not to make Diameters, as in Quadrilaterall and sexangular Intersections" and also that the feet of "perfect animals" divide into no more than five toes![22]

In general, it has been determined that there are 230 symmetry operations possible about a point in three dimensions. When we speak of an object with higher or lower symmetry, we are referring to its score against this benchmark. Thus, an equilateral triangle (which may be rotated by 120 degrees and has three reflection axes) has lower symmetry than a regular hexagon (which offers rotation by 60, 120, and 180 degrees and twelve reflection axes), and, in three dimensions, a tetrahedron has lower symmetry than a cube. There are 120 symmetries present in the icosahedron and the dodecahedron and their intermediates including the truncated icosahedron. It is in this sense that buckminsterfullerene is the most symmetric molecule.[23]

The perception of symmetry was formerly an inducement to embark upon (more or less) scientific investigation; today, it is more important for the way that it can facilitate it. Recognition of symmetry enables approximations to be made, calculations to be simplified, and proofs to be constructed, nowhere more so than in the application of quantum theory. It is symmetry that determines whether a molecule may undergo one transformation or another. Knowledge of its rules forms the basis of the analytical science of spectroscopy. It is symmetry of another kind that determines the success or otherwise of many biochemical processes. The power of symmetry lies not only where it is physically manifest, but also in the symmetry of mathematical equations and reaction pathways. Symmetry is a quality of events as well as of artefacts, spanning time as well as space.

Physicists and chemists look for symmetry because they find it allows them to give expression to cause and effect in an economical way. For any other sixty-atom molecule, it would have been impossible to mount a systematic assessment of likely properties; but these properties could be predicted for buckminsterfullerene as for no other molecule in the past because of its high symmetry. It is this that distributes the strain in its bonds evenly around its surface, thereby suggesting its lack of reactivity under certain conditions. It is this that dictates the disposition of its molecular orbital energy levels, which in turn suggests the unique properties that will be discussed in later chapters.

FOLLIES IN THE GARDEN OF CYRUS

Powers of ten, a short film by the designers Charles and Ray Eames, is constructed as a continuous zoom shot. A technical *tour de force* for the time of its making, it opens with a view of the universe. The camera seamlessly closes in upon a galaxy and then a solar system and then a planet. The planet happens to be ours; as we come closer we see an aerial view of a city with people picnicking on some grass. The camera presses on, zooming in on the hand of one of the picnickers, piercing the skin, to bring us "views" of cells, molecules, atoms, and subatomic particles.[24]

One of the things that the film illustrates, in a way more powerful than words or mathematics, is the uneven grain of the universe. Matter accretes within various narrow ranges of scale. There are galaxies and stars and planets and atoms and subatomic particles but no significant clustering of matter between these things. This is not the place to explore this curiosity. I propose merely to follow a similar path in order to point out a few examples of real or imagined three-dimensional symmetry at various scales – follies, if you like, in the garden of Cyrus.

Johannes Kepler thought he had explained the structure of the solar system – and proved the existence of God to boot – when he found that the orbits of the known planets could be explained by a scheme of proportions generated by nesting the five Platonic solids.[25] Kepler was a pioneer of observation as the basis for scientific hypothesis, and his geometrical scheme was no mere fancy. "Plato's polyhedral theory of the elements is designed to deal with unquantifiable properties, whereas Kepler's polyhedral theory of the solar system is concerned with the measurable intervals between the planetary orbs."[26]

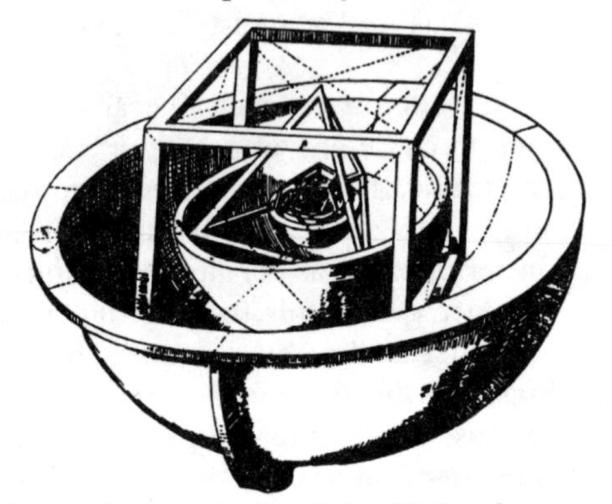

Kepler's solar system "explained" the planetary orbits by nesting the regular polyhedra in the best sequence

Kepler's scheme, published in his *Mysterium cosmographicum,* sought to explain apparently arbitrary gaps between the orbits of the planets in the Copernican heliocentric solar system by invoking the inspheres (the largest spheres that may be enclosed) and circumspheres (the smallest spheres that may enclose) of the five convex regular polyhedra. Matching the circumsphere of his innermost chosen polyhedron with the insphere of the next in sequence and so on, Kepler produced a remarkably good fit with the observed orbits. Thus, the orbit of Saturn lies within a spherical shell that touches the vertices of a cube. The orbit of Jupiter lies within the smaller spherical shell that touches the faces of that cube. This shell in its turn grips an enclosed tetrahedron. The orbit of Mars lies inside the tetrahedron and outside a dodecahedron, while that of Venus lies between the dodecahedron and an icosahedron, and Mercury's lies inside the icosahedron and outside an octahedron at whose centre is the sun.[27] The scheme appeared to explain both the number of planetary orbits and their relative diameters. The duality of the dodecahedron and the icosahedron, which Kepler characterized respectively as male and female, found added meaning in its pairing of the mythological opposites, Mars and Venus. Kepler modified his scheme by introducing more elaborate polyhedra as new experimental data became available, and not until the discovery of Uranus in 1781 was the final blow dealt to his pretty theory.

Coming down a few orders of magnitude to the scale of our own planet, the only large-scale symmetry we find on earth is that which humankind has placed upon it. The domes of Buckminster Fuller are the most notable embodiment of the icosahedral and dodecahedral forms; they are discussed in more detail in the following chapter. Among other works of architecture, the pyramids may be regarded as semi-octahedra. However, our need to build vertical load-bearing walls means that of all the polyhedra it is the cube that appears most frequently in our built environment.

Descending in scale by another factor of a million or so, we come to the manifestations of icosahedral symmetry in nature. Many creatures reveal fragments of high symmetry; the tortoise-shell, for example, has panels that are generally hexagonal, but which become pentagonal where curvature is required to form a secure carapace. The largest of nature's creations to exhibit complete three-dimensional symmetry are the radiolarians. These beautiful and fascinating marine organisms are the most highly symmetric living things. In life, they range from the poles to the tropical oceans. In death, the organic matter within their lacy cages wastes away, and only their silica skeletons remain. Like snowflakes in three dimensions, they exist in thousands of variants but always constructed around a dominating central symmetry – that of one of the five Platonic solids. The creatures are described in loving detail in D'Arcy Thompson's *magnum opus*.[28] Other radiolarians, not discussed by Thompson, have triangulated structures like Fuller's domes.[29]

Ernst Haeckel, who documented and drew hundreds of radiolarians in his accounts of the voyages of *HMS Challenger* during the 1870s, considered the sponge-like creatures "biocrystals", halfway between inorganic crystal and organic secretion. In the detail of the hexagonal skeletal network of some radiolarians we see, looking closely, pentagons, and often heptagons too – a living reminder of Euler's theorem.

Continuing on our journey, we arrive at the point where life dissolves into lifelessness. The matter of where exactly this point lies has been hotly debated. J.B.S. Haldane wrote:

> Before the size of the atom was known there was no reason to doubt that
> Big fleas have little fleas
> Upon their backs to bite 'em;
> The little ones have lesser ones,
> And so ad infinitum.
> But now we know that this is impossible. Roughly speaking, from the point of view of size, the bacillus is the flea's flea, the bacteriophage the bacillus' flea; but the bacteriophage's flea would be of the dimensions of an

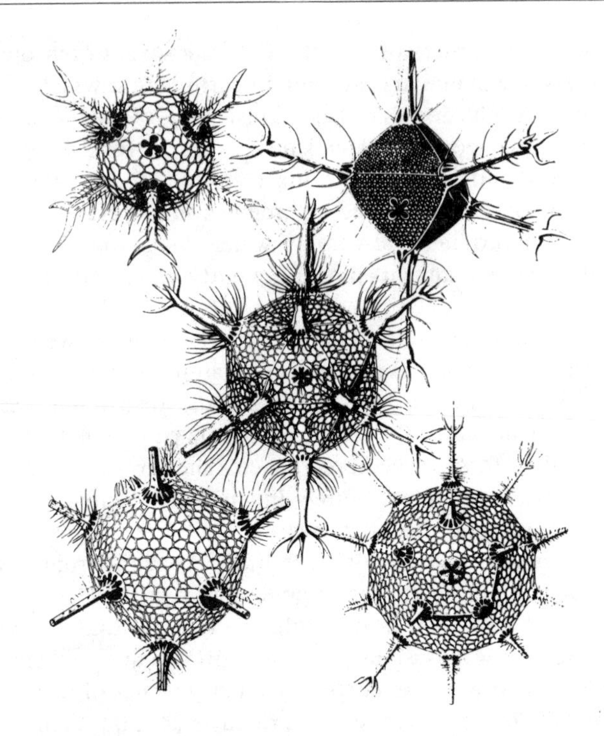

Haeckel's drawings of radiolarians showed a wide variety of forms based on the regular polyhedra

> atom, and atoms do not behave like fleas. In other words, there are only about as many atoms in a cell as cells in a man. The link between living and dead matter is therefore somewhere between a cell and an atom.[30]

Consensus today favours the virus as this intermediary. Small viruses such as the polio virus can form crystals like a mineral, but can also reproduce like a living thing.[31] Many viruses have a roughly spheroidal appearance, and the simpler ones frequently reveal icosahedral symmetry.[32] The regular polyhedral shape allows the proteins that clad the viruses to pack themselves in the most economical way around the central volume containing nucleic acid. This coating would tend to form as a flat hexagonal sheet of interlinked protein molecules, but for surface tension effects of the plasma membrane that encases the nucleic acid which force the sheet to accommodate pentagonal units thus leading to its closure into a polyhedral cage.[33]

Pressing on downward in scale, we leave behind the animal and vegetable worlds and come to the mineral kingdom. A distinction was made earlier between continuous materials such as crystals which have a lattice structure with translational symmetry and isolated objects with polygonal or polyhedral form. Before the advent of the microscope and modern X-ray diffraction techniques, gemstones and other mineral crystals were regarded as specimens of the latter rather than the former. Crystals are observed to adopt three of the five Platonic forms. They may be tetrahedral, cubic, and octahedral; but may not take dodecahedral or icosahedral forms with their forbidden five-fold axes of symmetry.[34] Because crystals secretly obey the rules of translational symmetry at the atomic level but appear to be perfect unitary objects, they have become a misleading "idol for symmetry".[35]

It is in part because five-fold symmetry is apparent in living nature but not in minerals that it has been regarded with wonder, and also why it has been presumed to be absent from inanimate nature even down to the level of individual molecules. Thus it was that one group of theoreticians predicting the spectra of buckminsterfullerene based on calculations involving symmetry could begin a paper: "In his famous treatise on group theory [of 1962], Hamermesch makes the statement that the 'icosahedral group has no physical interest and no examples of molecules with this symmetry are known'."[36]

Today, there are known to be a number of more or less stable molecules with icosahedral symmetry. One of the cage-like boron hydride ions known as boranes has twelve boron-hydrogen units disposed at the vertices of an icosahedron. Dodecahedrane has the same symmetry.[37] Buckminsterfullerene and its derivatives are new additions to a short but growing line. The geometer H.S.M. Coxeter noted in the preface of his book *Regular polytopes*, written nearly fifty years ago, that the observation in nature of crystals in the form of three of the regular polyhedra – the tetrahedron, the cube, and the octahedron – had done much to stimulate interest in these forms.[38] Could it now be that buckminsterfullerene will do the same for the higher polyhedra?

5

THE FULLER VIEW

The special geometry of buckminsterfullerene that is the key to its beauty might never have caught the imagination of the world beyond carbon chemistry had its discoverers not chosen for it such an idiosyncratic name. So what might Buckminster Fuller have made of the molecule that Harry Kroto and Rick Smalley named in his honour?

If people have heard of Fuller, it is usually as the creator of geodesic domes, in particular the giant dome that enclosed the American pavilion at the 1967 Expo in Montreal. Among architects of a certain generation, he is a cult hero. But Fuller was much more than this. His inventions spread well beyond the realm of conventional architecture. He made observations on science as well as on design. Never one to be pigeon-holed into conventional job titles, however many of them one found could apply to him, Fuller was a futurist, a humanist, and a poet as well as a technocrat and a geometer.

Richard Buckminster Fuller was born in 1895, the year of Wilhelm Röntgen's discovery of X-rays and of Lord Rayleigh and William Ramsay's reports of the isolation of the noble gases helium and argon, into a Brahmin New England family that had connections with the transcendentalists, among them Ralph Waldo Emerson and Henry Thoreau, of the nineteenth century. Their contemporary, Margaret Fuller, the proto-feminist, was his great aunt.

Even allowing for a measure of myth-making, it is clear that Fuller's interest in geometry flourished at an early age:

> In 1899, still not fitted with glasses, Bucky Fuller was manipulating by touch the toothpicks and dried peas on the kindergarten table. He had still never really seen what shape houses were, and having no idea it was orthodox to make cubes, he proceeded to discover the tetrahedron, a three-legged pyramid, and the octahedron, eight triangles born of interlacing three squares. Archimedes and Euclid had been among their earlier connoisseurs. His fingers told him these triangulations were rigid.[1]

This brief episode provides an uncanny premonition of Jim and Carmen Heath's evening in September 1985 spent toying with toothpicks and Gummy Bears in their attempt to elucidate the structure of the sixty-atom carbon molecule. There is, perhaps, more than chance behind the congruence of these two vignettes, as we shall see.

Geometrical ingenuity was only one element in Fuller's philosophy. For him, it was merely a facility that he directed towards a much larger goal which he called "design science", a technocratic vision that aimed to reveal how humankind could live better and in closer harmony with nature. But already, the paradox of Fuller begins to emerge. For while his vision was audacious, he was apparently unwilling to express his ideas in ways that might have won them – and him – more support. For "design science", Fuller gives a definition that describes the desirability of discovery that can be directed towards increasing "humanity's consciously competent participation" in local change, adding the observation that:

> Humans have thus far evolved the industrial complex designing which is only of kindergarten magnitude compared to the complexity of the biological success of our planet Earth. In its complexities of design integrity, the Universe is technology.[2]

This, it must be said, is a fragment of one of Fuller's more lucid trains of thought. Fuller's disciples and acolytes (they themselves use the terminology of religion) excuse their hero's opaque style of writing by calling it poetry. But this is not acceptable. Fuller's arguments lack detail at key points. He ignores science and fact where they stand in the way of some mystical point.

Fuller crossed the path of many professions, but he was not quickly claimed by any:

> Is Bucky an architect? No, say the architects, an engineer; no, say the engineers, a mathematician; no, say the mathematicians, but perhaps some kind of poet; no say the litterateurs, a jargon factory.[3]

Engineers regarded him condescendingly. *Time* magazine chided him for "arriving incoherently at logical conclusions".[4] The American Institute of Architects blackballed him in the 1920s, fearing the implications for their profession of his "design scientific" notion that houses should be mass-manufactured, though it and similar institutions later showered him with honours. A few mainstream architects have been more favourably disposed towards the man. Frank Lloyd Wright consulted him, and shortly before Fuller's death in 1983 at the age of 87, the British architect Norman Foster was collaborating with him on the design of a house.

Above all, Fuller was an energetic geometer. The crux of his geometry had its origin in those childhood experiments, taking the triangle, the tetrahedron, and the 60-degree angle as its basis in favour of the square and the right angle "because that is nature's way to closest-pack spheres".[5] Fuller termed this geometrical kit of parts the "isotropic vector matrix" because of its ability to spread forces evenly throughout a structure built according to its precepts.[6] Over decades, Fuller progressively refined this basic geometrical vocabulary to derive the structural system for his geodesic domes based on units of the:

> only three possible cases of fundamental omnisymmetrical, omnitriangulated, least-effort structural systems in nature: the tetrahedron, with three triangles at each vertex; the octahedron, with four triangles at each vertex; and the icosahedron, with five triangles at each vertex.[7]

Fuller employs these forms repeatedly in his projects. The icosahedron, for example, is the basis of a projection of the globe that he claimed exhibited "less visible distortion than any previously known map projection system".[8] The projection divides the earth's surfaces into 20 surface triangles and then flattens them to form the faces of an icosahedron. Despite Fuller's insistence that it is well suited to navigation, the projection has been little used. Perhaps its lasting value lies in the way that the triangles can be shuffled in order to challenge our preconceptions by showing a world in which north and south are no longer invariably associated with "top" and "bottom" and in which all the oceans may appear to join to form one ocean or in which, alternatively, the five continents may be made to appear in more intimate linkage than in familiar projections.[9]

As with others of his innovations, Fuller applied the name Dymaxion to his map of the world. In Fuller's lexicon, a Dymaxion invention is one that yields the greatest performance for the available technology. The "isotropic vector matrix" of triangulated structural members is a Dymaxion system, and his hexagonal design for a Dymaxion house of which a model was exhibited in 1929 makes use of this principle. His Dymaxion car of 1933 was aerodynamic and fuel-efficient. In 1945, he actually built a prototype house in Wichita, Kansas, that was a larger, more sophisticated successor to the Dymaxion house. The Wichita house was perhaps the supreme illustration of "design science", aiming to alleviate the shortage of housing in the United States at the end of the Second World War by adapting an aircraft factory production line to the fabrication of lightweight duralumin cladding panels.

In 1941, in an article that was rejected by one of the leading American architecture magazines, Fuller had written that:

> ...housing, as we have known it, provides no simple and practical means, economically or mechanically, for swiftly interpreting each scientific research gain in principle and precision into the everyday environment mechanics of man.[10]

It was a theme to which Fuller returned again and again:

> Shelter is by far the greatest single item among man's requirements in point of physical volume, weight, cost and longevity of tenure. Yet it is among the last to receive his scientific attention.[11]

The criticism stands today.

Fuller's preoccupation with the weight of things stemmed from his Navy days.[12] Applied to structures rooted in *terra firma*, however, it became an eccentricity. Nevertheless, Fuller was well ahead of his time in seeing a connection between the mass of an object and its implicit cost to the environment in the quantity of material consumed in its making. The Dymaxion Dwelling Machine weighed just three tonnes,[13] less than one fiftieth of a conventional house. Reducing the quantity of materials used in building was never in Fuller's mind an ideal to be pursued for its own sake. There was – and there remains – the distant gleam of a grail that everybody could understand: "Ephemeralizing weight by a factor of fifty might in long production runs ephemeralize cost by something comparable."[14]

One particularly effective way of reducing mass is to minimize the number of compression elements and maximize the number of tension elements in a structure. Compression elements such as load-bearing columns must have a certain bulk to transmit the force of their load to the ground without buckling. Tension elements, on the other hand, may be essentially one-dimensional, the barest material expression of a line of force.[15] "The last tensile wires", Fuller predicted, "will be simply the chemical bonds."[16]

Once this is realized, it follows that the design that is the lightest and most economical in its use of materials is the one where the ratio of components in tension to components in compression is high. Because all buildings rise against gravity, it is not possible to exclude the compression component entirely. However, it is possible to arrange for a greater proportion of structural components to be in tension than in compression. Consider the erection of a basic tower. This may be accomplished by stacking one floor on top of another with each floor below bearing successively greater compression loads. Such a structure would be entirely in compression. An alternative is to put up a single core from which the remainder of the structure is hung. Now, only the core is in compression like the trunk of a tree. Tensile forces hold the outer elements of the structure to this core against the downward pull of gravity.

This argument considers only the wholly vertical component of a structure's forces. There are also lateral forces that must be accommodated in order to build a conventional piece of architecture with a roof span and with habitable floors. The crypt of an old church, for example, typically presents an array of thick columns rising to vaulted arches that support the floor above. The topmost stone in each arch squashes tightly against its neighbours and the neighbours against their neighbours until the load reaches the columns that transfer it to the ground. The greater the load, the greater the compression in the arches. The stonework holds up, if it is properly designed and constructed, because there is no force pushing it out of line with the main compression force. It has nowhere to go and is strong enough not to crumble.

The intermediate floors and roofs of more modern structures, on the other hand, are sometimes constructed of regular three-dimensional assemblies of steel beams called space frames. These rigid networks are like the structure of the sides of many railway bridges, but crisscrossed over an area rather than stretched along just one linear span, and designed to withstand loads from whatever direction they are imposed. In particular, the depth of the space frame allows the vertical load of gravity in the middle of such an area to be borne without undue deflection and transferred, as in the case of the vaulted arch, to vertical supports at the edge of the frame.[17]

Around the world and down the ages, cultures have found it convenient to adopt an orthogonal basis for the construction of shelter (which is why, among other things, we call right angles right). Because of this inheritance, there is an understandable temptation to use space frames built up from cubic units. However, because squares and cubes can be skewed by shear forces into rhombuses and rhombohedra, some at least of these cubes must be strengthened with steel diagonals in order to produce a rigid frame. Buckminster Fuller's "omnitriangulated" systems, on the other hand, are inherently more efficient, if awkward to reconcile with the rectilinear conventions of building, since every unit contributes to the overall rigidity of the structure.

For Fuller, there is no conflict. A whole building can be a space frame. Fuller saw that it was possible to modify the geometry of the elements of a space frame in effect by bending it to produce not a flat plane but a curved lid that could enclose a volume over a circular floor. From there, one could increase the curvature still further – the geodesic dome was born.

The domes were Fuller's one commercial success. As well as possessing lightness and exhibiting economy in the use of materials, they achieve the more time-honoured aim of architecture of enclosing a maximum of space with minimal intrusion into that space of intermediate structural supports. In this, Buckminster Fuller joined the company of such illustrious figures as Filippo Brunelleschi, the first great

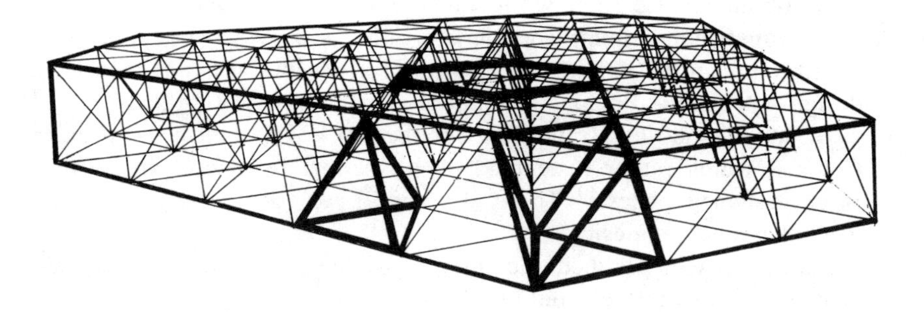

Lightweight space frames may incorporate the tetrahedral and octahedral units that Fuller favoured

Renaissance architect and creator of the *duomo* in Florence. But unlike previous domes of whatever scale, Fuller's filigree structural frames were also largely transparent, and this transparency could be retained for the finished shelters simply by covering the exterior of the curving framework with glass or clear plastic panels.

Although Fuller was awarded a patent in 1954, he was not the first to build domes based upon a structural framework and a layer of cladding. He was not even the first to build a dome by using interconnecting metal struts in a regular geometric pattern. That distinction probably falls to Walter Bauersfeld, chief designer at the Carl Zeiss optical equipment company at Jena in Germany. In 1922, Bauersfeld built a small planetarium on the roof of the factory in order to test a new projector. He covered a steel structural frame sixteen metres in diameter with a ferroconcrete shell, thereby also creating the world's first lightweight concrete dome.[18]

Long before this date, of course, the dome had been a popular feature of the grander forms of architecture. From the Renaissance through the nineteenth century, it was employed for reasons that were as much to do with symbolism as with structural efficiency. The general method of construction was to use masonry or brick in compression, like a simple arch rotated through a circle. Vast hoops of steel helped to contain the outward thrust of the lower parts of some larger domes imposed by the great weight of the spans. Giant domes have played an important part in fantasy architecture, particularly in the work of the French Neoclassical architects Claude-Nicolas Ledoux and Etienne-Louis Boullée. The most spectacular example of such a structure is surely the latter's scheme for a cenotaph for Sir Isaac Newton which would have required the construction of a colossal dome whose height can be estimated (from drawings which show a dwarfed avenue of cypress trees girding its equator) at well over 300 metres. Ed Applewhite, one of Fuller's principal collaborators, cites a nineteenth-century example of a dome with evidence of a geometric structure in the polyhedral pattern inscribed in the hemisphere of a design by Eugène-Emanuel Viollet-le-Duc and the 1914 glass-domed pavilion designed by Bruno Taut for the 1914 Werkbund exhibition in Cologne which rather resembled a pith helmet in overall shape with lozenge-shaped panes.[19] These structures were geometric to varying degrees but not geodesic.

Fuller's patent claimed the novelty of the geodesic logic that all elements of a structure should trace the shortest lines between points on the curved surface. In this respect, Fuller's domes were more rational and efficient than their antecedents in which the steel hoops of the larger domes were arranged like lines of latitude rather than along the geodesics of the spherical surface. Even the domes at Jena and Cologne, which made good use of modern materials and had fairly regular structural frames, were still not geodesic.

The simplest geodesic domes have single skins. Regular dodecahedral structures with twelve pentagonal panels or truncated icosahedral structures with twenty hexagonal and twelve pentagonal panels are adequate for small domes provided that the panels are fairly stiff. Fuller's collaborator on the Montreal Expo dome, Shoji Sadao, patented such a "Hexa-Pent" design in 1974, intended for use in low-cost housing.[20]

Larger domes demand the rigidity contributed by struts that zigzag back and forth between two closely spaced concentric spherical frames – the bent space frame. Between the outer and inner spherical frames lies a fully triangulated three-dimensional structure; every strut participates in the formation of rigid triangles. One result of this requirement is that the outer frame of the larger domes is composed of triangular panels while the inner one is composed of the hexagons and pentagons. One shell is the polyhedral dual of the other.

This was the principle that Fuller adopted in his best known domes such as the Union Tank Car Company dome in Baton Rouge, Louisiana, completed in 1958. At the time, this blister-like structure was the largest clear span of any kind anywhere in the world, enclosing twenty-three times the volume of St Peter's in Rome. Nearly 120 metres in diameter and rising to nearly 40 metres, such a dome, Fuller pointed out, could enclose Seville cathedral, the world's second largest. At 1200 tonnes, it weighs about the same as just three of that building's stone columns.[21]

The dome for the United States Pavilion at the 1967 Montreal Expo was two thirds of the spherical radius of the Baton Rouge dome. The reason why it attracted widespread admiration was that it rose higher, like a puffball, appearing almost as a complete sphere floating next to the St Lawrence river. Every facet of its surface was fitted with transparent plastic panels, adding to the magic.

In these larger domes, the underlying icosahedral symmetry of the structure is almost lost in the sea of tiny triangular cladding panels. This is because, in order to form a sound structure, the struts that carry compression loads must be kept comparatively short. This requirement together with the need to spread the loads evenly across the surface of the dome leads to a structural frame with many more struts, each short in relation to the diameter of the dome as a whole. The effect of this is to produce a dome that is more nearly spherical in appearance. The large scale structure still has overall icosahedral symmetry which brings with it the economy of scale of allowing repetition of structural units. The finer texture, however, must be judged with great precision. Although the triangles appear roughly equilateral still, they require many slightly different lengths of straight struts to produce the appearance of smooth curvature on the dome as a whole.

Imagine, first of all, an icosahedron as the basic dome. With twenty flat triangular faces it appears more faceted than spherical. Now imagine a truncated icosahedron, with its familiar hexagonal and pentagonal soccer ball facets, as the internal frame. The complementary outer structure that follows with the use of a fully triangulated space frame comprises sixty triangular units (by taking the centre of each hexagon and pentagon and drawing lines between all these points). Each of the pure icosahedron's triangular faces is replaced by a triangular unit composed now of three smaller triangles. This subframe is given a slight convexity, not by bending the struts, but by increasing their lengths slightly and unequally in order to form no longer a flat triangle, but a very low pyramid. Repeated twenty times over, the effect is to bring to polyhedral dome closer to a spherical surface. A closer approximation to a sphere comes with each further subdivision of the basic equilateral triangular faces of the icosahedral dome. But each time, the number of struts with unique lengths also multiplies. The dome that Fuller and Sadao designed for United States Pavilion at the Montreal Expo in 1967 has sixteen struts stretching between each pentagonal centre. The icosahedral supertriangle that may be drawn between neighbouring pentagon centres encloses 256 triangles, many of which are not identical. Nevertheless, the systematic nature of the components meant that they could be simply coded for assembly. No plans were needed to assemble the frame of the pavilion aside from a few hub detail drawings.[22]

Despite the fact that nature employs highly spherically symmetric forms mainly at the microscopic level, in some viruses and radiolarians,

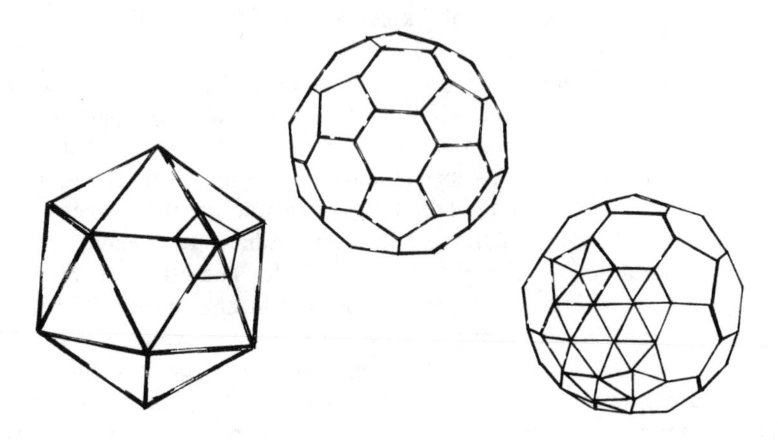

A smooth dome is built up by fragmenting the triangular faces of an icosahedron into smaller and smaller triangles

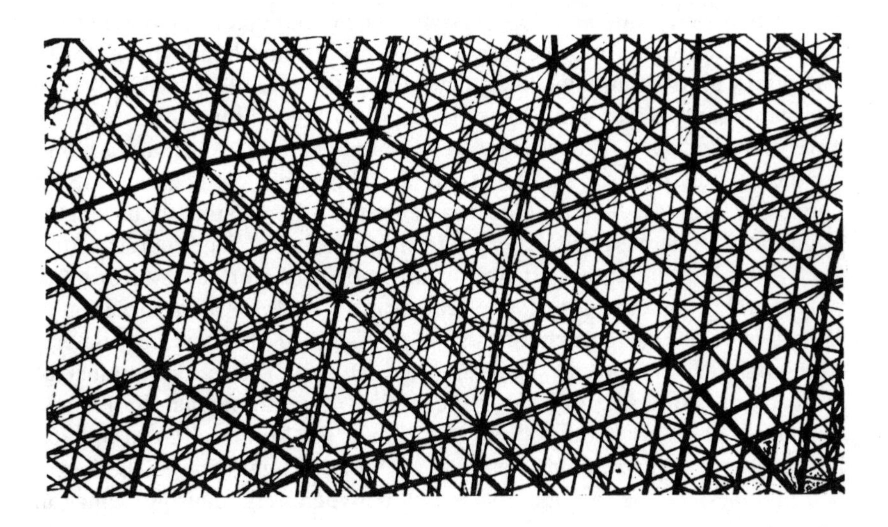

In close-up a typical Fuller dome structure shows a layer of triangles and a layer of hexagons

Fuller saw no reason why far larger structures could not be built according to nature's example. In the aluminium alloy used for these frames, he had the material to exploit this universal structural principle at seemingly any scale.[23] The published apotheosis of this train of thought – for Fuller, no doubt, it was just a step to something still grander – was his plan to cover midtown Manhattan with a dome stretching from the Hudson to the East River and comfortably enclosing the Empire State Building and other skyscrapers.

Fuller's reputation could stand proud solely on the basis of these achievements as a visionary architect. But he and his followers claimed he was much else besides, and not least that he was a scientist. Here, his reputation is far less secure. The introduction to *The dymaxion world of Buckminster Fuller*, the work that Smalley and Kroto consulted when they were looking to see whether Fuller's domes employed pentagons as well as hexagons, states:

> In times past, most pure scientists confined themselves to the physical world and its system of exact relations.... Fuller departs from this tradition in that he is equally concerned with exact and social science.[24]

This at first appears as an admirable sentiment, but in the light of Fuller's decidedly eccentric view of modern science, it quickly comes to seem more like a pre-emptive strike against hard-core detractors, an apology in advance that while Fuller's jargon may be scientific, his method is not. For while his engineering and visual thinking was highly creative, his science was suspect.

It would be tedious to trawl Fuller's epic *oeuvre* for all its scientific solecisms. A few examples will suffice. These may be categorized into confusions to do with physical nature, with geometry, and with number.

Despite the fact that he claims to have taken a chemistry option in his freshman year at Harvard,[25] Fuller speaks a lot of chemical rubbish.[26] He employs the terminology of quantum mechanics as a kind of mystical jargon, presumably to persuade credulous non-scientists of dubious arguments, although in this at least he is far from unique. Fuller makes many more specific misconnections in his writings. He implies there is some kind of equivalence of single chemical bonds to the gaseous state, of double bonds to liquids, and so on.[27] He describes the process of metals forming alloys when "symmetrically stable arrangements of atoms – invisible to naked human eyes – happen to come into critical proximity under the right heat conditions to produce their common liquidity."[28] An account of the particle nature of light is littered with physical non-sequiturs in a bizarre attempt to evaluate its price per pound: "Photons retail at approximately $1^1/_2$ billion dollars per lb." He goes on to suggest that electromagnetic radiation (outside the range of visible light) may be "visible": "The width of heat waves may be as great as $^1/_{100}$ of an inch and are [sic] perceptible to some eyes...."[29]

On geometry, Fuller refers to the principle of chemistry that "tetrahedra represent the way that atoms cohere",[30] a statement that is quite simply meaningless in most chemical circumstances. He asserts that: "The closest packing of spheres characterizes all crystalline assemblages of atoms".[31] It does not; many crystals adopt other packing schemes. He finds cause for delight in the fact that it is possible to close-pack fourteen spheres, in five layers, 1, 3, 6, 3, 1, to form the "first closest-packed, omnitriangulated, ergo structurally stable, but non-nuclear, equiradius-sphered, cubical agglomeration" and suggests – speciously – that this is the arrangement of protons and neutrons in the Carbon-14 nucleus.[32] In one place, "all organic chemistry is tetrahedrally configured".[33] In another, he asserts that: "Organic chemistry begins with the cube: *carbon.*"[34]

Most of what has been written about Fuller is notable for its forgiveness of these transgressions. But one book at least admits the suspicion that he has gone off the rails here. It features a "dialogue with a skeptic" who takes Fuller to task for his latitude with chemical imagery:

> Look at *Utopia or oblivion* [a 1969 book by Fuller]. On page 101, we find pictures of linked tetrahedra, labeled 'Chemical bonding.' We are told that point-to-point tetrahedra – single bonded – resemble a gas. Hinged edge to edge, they are like water. Face to face, they are rigid, like crystals. Interpenetrating, they have the hardness of diamonds. Alas, Linus Pauling, who is mentioned on that very page, tells us (*The nature of the chemical bond*, page 559) that 'The presence of shared edges and especially of shared faces in a coordinate structure decreases its stability,' since the positive ions the tetrahedra enclose dislike coming so close together. Pauling's picture of tetrahedra touching only at corners depict crystals, not gases. Bucky's is a pretty intuitive model, but it is not chemistry.[35]

Fuller desperately wishes to find some precedent in the natural world for the numbers that recur in his geometry. He abjures astrology, but finds in numerology "tantalizingly logical theories".[36] He was much taken with the discovery that certain "magic numbers" recur in tables of the relative abundance of certain isotopes, but he seems to confuse atomic number (the number of protons in the nucleus) with atomic mass (the number of protons and neutrons in the nucleus).[37]

Fuller reserves most of his attention for one number: twenty. He concludes an exposition that ranges over the amino acids, the icosahedron, the fact that twenty spheres can stack as a tetrahedron (start with the ten balls in a pool table triangle and add six, then three, then one to complete the pyramid), and the knowledge that the Mayans used base 20 in their calculations with this aphorism:

> This may be why twenty appears so abundantly
> In the different chemical element isotopes.[38]

This is an optimistic distortion of the facts. The most abundant isotope of calcium has nuclei with twenty protons and twenty neutrons, the most abundant isotope of potassium has twenty neutrons, and the most abundant isotope of neon has a total of twenty protons and neutrons combined. But isotopes of other elements with twenty neutrons (sulphur, chlorine, argon) are not the most abundant and there are no elements other than neon with a total of twenty protons and neutrons combined.

Several authors have taken the trouble to paint a general portrait of the pseudo-scientist. By their measure, it is hard to except Fuller from this label.

John Casti, in his book, *Paradigms lost*, lists as the "Hallmarks of pseudoscience" the following qualities: anachronistic thinking (for example, making use of theories discarded by mainstream science); the seeking out of mysteries; the appeal to myths (for example, taking as legitimate observation what are in fact mythic accounts); a casual approach to evidence; the use of irrefutable hypotheses; the pointing out of spurious similarities, not least with the findings of mainstream science; explanation by scenario without the support of scientific law or theory; the taking of scientific literature as interpretable to one's own cause regardless of the facts it expresses; the refusal to revise. And a final trait: "Write only vacuous material replete with tautologies; make sure that your statements are so vague that criticism can never get a foothold; simply refuse to acknowledge whatever criticism you do receive."[39]

Martin Gardner's pseudo-scientist:

> stands entirely outside the closely integrated channels through which new ideas are introduced and evaluated. He works in isolation. He does not send his findings to the recognized journals, or if he does, they are rejected for reasons which in the vast majority of cases are excellent.... the reputable scientist does not even know of the crank's existence unless his work is given widespread publicity through non-academic channels.... The eccentric is forced, therefore, to tread a lonely way. He speaks before organizations he himself has founded, contributes to journals he himself may edit, and – until recently – publishes books only when he or his followers can raise sufficient funds to have them printed privately.
>
> A second characteristic of the pseudo-scientist, which greatly strengthens his isolation, is a tendency towards paranoia....

> There are five ways in which the sincere pseudo-scientist's paranoid tendencies are likely to be exhibited.
>
> (1) He considers himself a genius.
>
> (2) He regards his colleagues, without exception, as ignorant blockheads....
>
> (3) He believes himself unjustly persecuted and discriminated against.... He likens himself to Bruno, Galileo, Copernicus, Pasteur, and other great men who were unjustly persecuted for their heresies. If he has had no formal training in the field in which he works, he will attribute this persecution to a scientific masonry, unwilling to admit into its inner sanctums anyone who has not gone through the proper initiation rituals....
>
> (4) He has strong compulsions to focus his attacks on the greatest scientists and best-established theories....
>
> (5) He often has a tendency to write in a complex jargon, in many cases making use of terms and phrases he himself has coined....[40]

Fuller does not meet all these requirements. But the damage is done. Although he called himself a scientist and sought the intellectual company of scientists, it was as a pseudo-scientist that he was regarded by many of them. Some scientists mocked Fuller. Others patiently heard him out. A few were more favourably disposed. Among those who grew enthusiastic about Fuller's ideas were the biochemists Aaron Klug and Jonas Salk, the materials scientist and historian Cyril Stanley Smith, and the crystallographer Arthur Loeb, who became an active protagonist for Fuller's views and a contributor to the *Synergetics* volumes. Through these people Fuller met other notable scientists, including Albert Einstein, Leo Szilard, who turned from the physics of the Manhattan Project to biology, and Harold Urey, the chemist who had shown that simple amino acids could be created in the conditions thought by many to have prevailed in the earth's primeval atmosphere.

Fuller made much of these associations, to judge from a memorandum on the subject by Ed Applewhite.[41] The note reveals how Fuller received and recorded encouragement from scientists, while deliberately maintaining his distance from the scientific establishment. The account seizes upon "real" scientists' favourable or even neutral comments as validation of Fuller's research, even putting the onus on them to come round to Fuller's views. The tone is maintained in more recent commentary, in *Trimtab*, the bulletin of the Buckminster Fuller Institute, on the discovery and potential applications of buckminsterfullerene.[42]

It was in seeing the big picture, not the details, that Buckminster Fuller made his real contribution to science. He was a declared enemy of specialization. In *Operating manual for spaceship earth*, one of his most famous books, published in 1969, Fuller devotes an entire chapter to the "origins of specialization", a trait which he saw as a kind of cancer of the intellect:

> Of course, our [social and economic] failures are a consequence of many factors, but possibly one of the most important is the fact that society operates on the theory that specialization is the key to success, not realizing that specialization precludes comprehensive thinking.[43]

He later notes with pleasure the apparent convergence of biological, chemical and physical science brought about by collaborative research programmes during the Second World War. Fuller died in 1983, two years before the discovery of the molecule that would bear his name. It would have pleased him that the discovery was facilitated by cross-over between the disciplines of chemistry, physics, and astronomy, and that it has continued to draw the attention of "specialists" in subjects from mathematics to biology.

Making the usual allowance for Fullerian hyperbole, this postscript to the poem "No more secondhand god", dated 1961, the year before Fuller was elected Charles Eliot Norton Professor of Poetry at Harvard, could serve as chorus for Kroto and Smalley's discovery:

> Because scientists today are specialists
> rather than generalists,
> they have tended to avoid consideration
> of the total significance of their interrelated work.
> The recent decade's evolution of science
> finds its specialists continually surprised
> to discover their respective specializations
> bringing them into unexpected proximity
> to other specializations theretofore considered
> almost infinitely remote.
> We have come to the era of the hyphenated sciences –
> bio-chemistry, mathematical-physics, physio-chemistry [sic].
> Suddenly the chemist and physicist discover
> that they are both dealing essentially with the atom.
> So they now agree to divide their common field
> and the physicist deals with the internal affairs
> and the chemist with the external affairs
> of the atom....
>
> In these coming decades
> the scientists will express their amazed discovery
> not only of the total orderliness
> discovered throughout universe

> but also of the total interrelatedness
> and nonsimultaneous interaccommodation regularities
> of universal evolution.[44]

Fuller's principal agent in realizing connections between specialisms was the use of visual models. These can transcend barriers put up by difference of scale or habit of education. He urged students of architecture to learn chemistry which "is the basic structure, *ergo* architecture", adding of chemistry, truthfully and instructively this time:

> Nature has made certain things which we call natural, and everything else is 'man-made', ergo artificial. But what one learns in chemistry is that Nature wrote all the rules of structuring; man does not invent chemical structuring rules; he only discovers the rules. All the chemist can do is find out what Nature permits, and any substances that are thus developed or discovered are inherently natural.[45]

Fuller's fondness for models led him to attack science that he saw as too abstract. He criticized the astronomers James Jeans and Arthur Eddington for "making their living from elegant mystifications".

> Science, he says, lost touch with nonscientists, and engendered the famous 'two cultures', when it gave up the use of models, thus letting us suppose it was talking about nothing real.[46]

Applewhite concurred:

> Ever since the discovery of electromagnetics around 1850 the mathematicians in effect 'went on instruments'. It is Fuller's goal to reintroduce the importance of models.[47]

Fuller's models, such as his tetrahedral picture for the interaction between atoms, may not be chemically accurate. Some are fanciful, obscurantist, or mystical which does not always help his cause. But they have an intuitive appeal. It was this aspect of Fuller's thinking that attracted distinguished scientists such as Aaron Klug.

This instinct of Fuller's makes it more than likely that he would have recognized the specialness of this molecular form of carbon (even if it did not bear his name) rather more rapidly than many scientists did.[48] According to Applewhite:

> RBF would have regarded the advent of the fullerenes as an exhilarating vindication of his long struggle to present a new view of nature's coordination, as well as a validation of his specific icosahedral structural strategies.[49]

The reaction of Fullerians to Kroto and Smalley's discovery has indeed been predictably enthusiastic; and it would probably have been so even if the molecule had not been named for their hero.

For some, there is a sense of *déjà vu* about this. In the 1960s, when Caspar and Klug and others discovered the icosahedral symmetry of the protein shells called capsids of some viruses, their similarity to Fuller's then highly fashionable domes was much commented upon. Fuller felt that the opportunity had been missed by the world at large to make a more holistic, "synergetic", connection between these microbiological discoveries and his work on a larger scale.[50] There is no reason to think that a similar opportunity will not be missed again.

Perhaps Fuller's real vindication lies not in the aftermath but in the genesis of the discovery, in the unconscious homage that Jim and Carmen Heath, and Rick Smalley and Harold Kroto, paid to Fuller's belief in the power of models as they toiled through that Monday night in September 1985 in their desperate effort to give shape to the shapeless and explain the inexplicable.

6

THE CHEMICAL SENSES

Most of us lack the wherewithal to confirm any but the most basic scientific "fact" that is vouchsafed to us. We take what scientists tell us quite literally "on good authority". Put more crudely, we believe. We believe, for example, that the moon is not made of green cheese (contrary to the proverbial characterization of credulousness). But how would we set about proving it? Quite simply, we could not, although we could perhaps begin to make a little progress.

If we were of an analytical frame of mind, we might divide the question into two parts to test separately the proposition that the moon is green (at least to the same extent that cheeses are green) and the proposition that it is made of cheese. We could confidently report that the moon is not green, because we can see that it is not. As for its cheesiness, our first inclination would surely be to take a good sniff – something rather ruled out by the quarter of a million miles of vacuum that lie between the moon and our noses. We would have to launch a rocket and collect a sample of lunar material in order that we might smell it, rub it between our fingers, and taste it. Only then could we come to a preliminary conclusion regarding its constitution.

However we were to attempt this assay, we would have to use our senses – they are all we have. Between them, these five senses, of sight and hearing, of smell, taste, and touch, are what allow us to make our way in the world and to experience its pleasures and pains.

Spectroscopy is no more than the extension of our mostly highly developed sense – our sight. The award of several Nobel prizes for the development of its multifarious techniques testifies to the central position that spectroscopy has come to occupy in modern chemistry. Spectroscopy is the theory of the interaction of electromagnetic radiation and matter and its practical interpretation in order to deduce the nature of the matter. It has its root in the first attempts by science to analyse how we see. Today, it has blossomed into an array of techniques as versatile as our own senses. Like the human senses, these chemical senses possess an elegant complementarity. Just as we cannot readily distinguish a Rembrandt from a Vermeer by feel (we would like to take a look), or a burgundy from a claret by sight alone (a sip would be nice), so it is that one form of spectroscopy is valuable in determining the contents of a forensic sample, say, while another is ideal for probing the mysteries of space.

Like our human senses, the various techniques of spectroscopy work by indirect means that have come to seem direct through their familiarity and utility. Our senses are indirect in that we require relays of chemicals in order to interpret a smell or an electromagnetic go-between in order to identify a colour. All true forms of spectroscopy rely similarly on the passage of electromagnetic radiation between the object under analysis and a detector.

Matter and radiation interact in two basic ways. Matter either absorbs energy or it releases it. It does both in precise measures determined by the laws of quantum mechanics, and it is this precision that makes radiation at particular energies characteristic of particular substances, and hence spectroscopy useful as an analytical technique. Both absorption spectra and emission spectra are important in chemistry and in fields where chemical analysis is required such as metallurgy and astrophysics. In most cases, the device used to record a spectrum contains a dispersing element such as a prism or a diffraction grating that spreads the transmitted radiation out into a spectrum. In general, the instrument also incorporates its own source of radiation and a space in which to place the sample for analysis, although in some cases either or both of these items may be remote from the detector. An emission spectrum will reveal bright lines or bands at the wavelengths at which light is being emitted; an absorption spectrum will show dark lines or bands interrupting an otherwise bright continuum. Historically, these lines have been recorded on photographic film. Today, electronic equipment monitors the intensity of this radiation and plots it graphically against the frequency of the radiation to produce a trace with peaks in place of lines. For short, these traces are also known as spectra. Often, lines group together in which case they are called bands; peaks overlap to become broader humps. We are familiar with this terminology already from the discussion of such features as the diffuse interstellar bands and the "camel hump" ultraviolet spectrum in Chapter One.

Each atom or molecule has its own distinctive spectral signature. Taken across the range of the electromagnetic continuum, no chemical species produces the same signature as another. In practice, the various techniques of spectroscopy home in on small stretches of the continuum, each generally associated with a particular type of atomic or molecular action. For any given species, some techniques will convey useful information. Others will not. Many chemical species have signatures within which there are very similar parts. It might require very careful examination of a single spectrum – of what is effectively only one part of a whole signature – to determine exactly what is responsible for it. Often, it will prove impossible to make a reliable assignment without the confirmation of a complementary spectroscopic technique. (The same is usually our experience with the human senses.)

Most laboratory work is absorption spectroscopy. The sample for analysis is interposed between a radiation source and a suitable detector. Radiation that is not absorbed by the sample reaches the detector and records the absorption spectrum. In the case of absorption spectra recorded of matter in space, it is necessary that there be a source of radiation directly behind the matter still further out in space.

Emission spectroscopy is more direct but less frequently encountered. In order to produce an emission spectrum, the emitting species must have excess energy; it must be in an excited state. One form of emission spectroscopy is familiar from the school laboratory. When small samples of salts are placed in the flame of a bunsen burner, the flame becomes coloured. The new colours are the emission spectra of metal atoms from the salts. (Of course, the existing colours of the flame are also emission spectra – of excited atoms and molecules formed from the burning fuel.)

Although in principle quite simple – everything has its characteristic spectrum – spectroscopy is in practice a sophisticated art. Even lone atoms have many lines in their spectra. Only very simple molecules have anything like as few lines – and by their nature these are usually not molecules of anything other than specialized interest. The day-to-day task of the spectroscopist is to identify complex molecules with very many more features in their spectra.

To make matters worse, spectra are distorted by the nature of the medium in which the species are present. Most generally, the peaks or lines of a spectrum are broader for molecules that are in closer proximity to each other and to other molecules. Spectra of molecules in the gas phase thus have the sharpest resolution. Those for molecules in solution or in the liquid phase are broader – and correspondingly more ambiguous. The spectral signals of solids are muddied still further. Spectra are distorted, too, by the media they must traverse between sample and detector. These must be chosen where possible to eliminate the superimposition of their own spectral events on the desired signal. It is for this reason that a sample is sometimes placed in a cell made of quartz rather than ordinary glass, or that a spectroscope is sent aloft in a satellite in order to bypass the earth's atmosphere.

Spectroscopy is a powerful tool which can provide incontrovertible evidence for the presence of this, that, or the other chemical species. On the other hand, it can also seem very fickle, like a temperamental old radio. Sometimes you can get a station and sometimes you cannot. You have to have a hunch that you ought to be able to record a particular spectrum, and then the persistence to fiddle with the controls, arrange and rearrange the sample, and adjust gas pressures or solution concentrations or substrate thicknesses, to coax it out.

In practice, then, spectroscopy is seldom a simple matter of identifying a signature as a cashier does in a bank. It is a challenging intellectual exercise, more like solving a cryptic crossword puzzle. This quality of the spectroscopist's work is further betokened by the contribution made by the theoreticians who frequently meet the experimental chemists halfway with the results of calculations that predict the position of spectral lines. These predictions allow the experimentalists to focus their search. Such a dialogue, it will be remembered, was conspicuously successful when Harry Kroto and his colleagues set about identifying interstellar chain molecules.

The theoreticians are able to make their predictions on the strength of the laws of both classical and quantum mechanics that govern the atomic and molecular actions that give rise to spectra. Quantum theory is so often made to seem mysterious outside the world of science. Even in works that ostensibly seek to proselytize on behalf of science, it seems that there is an undercurrent of mystification, of delight in the paradoxes, oddities, and uncertainties of the theory, that is designed not to bring about understanding, but rather to conjure up quasi-religious awe or perhaps to replenish our wonderment at nature (the wonderment that it is constantly supposed that science is out to destroy).

Not that quantum theory is simple or that it produces expected results. Certainly, it does not. Nevertheless, quantum theory is entirely familiar so far as science is concerned; for spectroscopists, it is nothing more than a tool of their trade.

Buckminster Fuller was one of many who have complained that science has become too abstract and mathematical (as if it had a choice). With the discoveries of subatomic structure, and the subsequent articulation of the laws of quantum mechanics, much of science was widely perceived as having grown remote from everyday experience. Events could be sensed only by means of instruments; they could never be visualized; some of them could not even be imagined. The complaint is a not unfamiliar one made of science in the twentieth century. However, J.D. Bernal knew better:

> To the scientist the picture of the disappearing reality of phenomena is so much nonsense; he knows that he can handle materials the better for the

> quantum theory, or biological preparations for a knowledge of bio-chemistry and genetics; he also knows that he cannot get this across because publicity for these views is so much harder to get than for the opposite ones. The result is that the public is deluded into believing that idealism rules in science at a time when materialism is winning all along the line....[1]

Spectroscopy is living proof that the scientists can indeed do their job the better for quantum theory. Quantum theory is rendered suddenly legible when bathed in the bright neon (or sodium or mercury or halogen) light of atomic and molecular spectroscopy.

Many kinds of spectroscopy were to prove important in the characterization of buckminsterfullerene – and in proving that it had the soccer ball structure posited by Harry Kroto and Rick Smalley. During the wilderness years, Smalley had written of the want of a conclusive method of confirming the structure. "To most, such a proof would be spectral, involving high-resolution IR or UV spectroscopy, or (ideally) NMR spectroscopy...."[2]

Infrared, ultraviolet, and nuclear magnetic resonance spectroscopy form the core repertoire of the analytical chemist. They are described in brief below together with two further techniques – in all making a portfolio of five chemical senses in loose analogy with the five human senses. The additions are microwave spectroscopy and mass spectrometry. This last technique, with which we are already familiar from the experiments on carbon clusters, is not truly a form of spectroscopy because it does not involve a fundamental interaction between radiation and matter. It is included here because it is, however, an important physical analytical technique that provides information of a kind that forms a useful complement to that provided by the true forms of spectroscopy.[3]

Although spectra rely for their detailed interpretation upon an understanding of quantum mechanics, the means by which they arise do not lie beyond the grasp of intuition. It *is* possible to visualize the processes at work. Vibrations and rotations are graphic enough. The transition of electrons between energy levels within atoms and molecules, either when they are given energy by absorbing radiation and become able to step up the ladder or when they release energy by stepping down, is equally easy to picture. Nuclear magnetic resonance spectroscopy, in many ways the most powerful technique, is a little more challenging to the imagination. On the other hand, the reward for making the effort to understand its workings is correspondingly greater, especially in the case of our special molecule.

MASS SPECTROMETRY

Mass spectrometry detects the ions of molecules and fragments of molecules and sorts them according to their mass. Knowledge of the mass of the parent ion confirms the molecular weight of the sample under examination. Knowledge of the mass of the fragment ions tells us how that molecular ion breaks up. This in turn allows us to deduce information about the structure of the parent molecule.

The mass spectrometer first ionizes the sample for analysis in a suitable way (in AP2, this was the job done by the second laser). Electric fields then focus and accelerate the ions in a beam away from the ion source and towards a mass analyser. Acceleration imparts the same kinetic energy to each ion, as a consequence of which fragments of different mass acquire different velocities. A magnetic field (sometimes supplemented by another electric field) deflects these ionized fragments to an extent determined by the ratio of their charge to their mass. Heavier ions are deflected less than lighter ones. In principle, the scattered ions could be detected as they fan out across a range of angles away from their original line of flight. In practice, a varying magnetic field is used to guide ions of correspondingly varying mass in turn through a narrow slit placed at one fixed angle of deflection. These ions are then counted at a detector.

A variant of the apparatus, the time-of-flight mass spectrometer, was used in most of the experiments with the fullerenes. This machine dispenses with magnetic deflection and incorporates a long "drift tube" sufficient to separate the ions of different mass as they travel at their different velocities. The lightest ions acquire the highest velocities and arrive first at the detector.

In both cases, the mass spectrum sets out the mass-to-charge ratio of arriving particles in sequence. The height of each peak provides a measure of the abundance of each ion. Since the bulk of the ions carry a single positive charge, the recorded spectrum is in effect a chart of the mass of the fragments alone. In the analysis of molecules, the ions produced are those corresponding to the most stable constituent chemical groups of the molecule in question. The position (mass) and size (abundance) of the peaks enables quite detailed information to be extracted about the structure of many molecules.

Chapter One took ethyl alcohol as an archetypal molecule in its discussion of the complexities of molecular science. We shall co-opt it again here in order to reveal the power of spectroscopy in making sense of these complexities. The mass spectrum for ethyl alcohol shows among other peaks a peak corresponding to a mass of 29 atomic units suggestive of the $C_2H_5^+$ ion with two carbon atoms with an atomic mass of twelve apiece and five hydrogen atoms with a total atomic mass of five units. The information helps to distinguish between ethyl alcohol and, say, formic acid, a molecule with a very similar but not identical molecular weight, or between ethyl alcohol and methyl ether, a molecule with the same molecular weight and the same empirical formula (C_2H_6O) as ethyl alcohol, but a different structural formula (methyl ether is CH_3OCH_3; ethyl alcohol is CH_3CH_2OH). The value of this technique resides in the importance of knowing such structural formulae, because it is these that allow chemists to plan the chemical reactions that form steps in the chemical synthesis of new molecules. The peak at a mass of 29, though not a large one, is sufficient indication of the presence of an ion that is an obvious fragment of ethyl alcohol, but is quite impossible as a product of the direct fragmentation of methyl ether. The technique is not always as conclusive as one would wish, however: the most prominent mass fragment in the spectrum for ethyl alcohol has a mass of 31 corresponding to the CH_2OH^+ ion, which is indistinguishable on this evidence from the CH_3O^+ ion fragment of methyl ether.

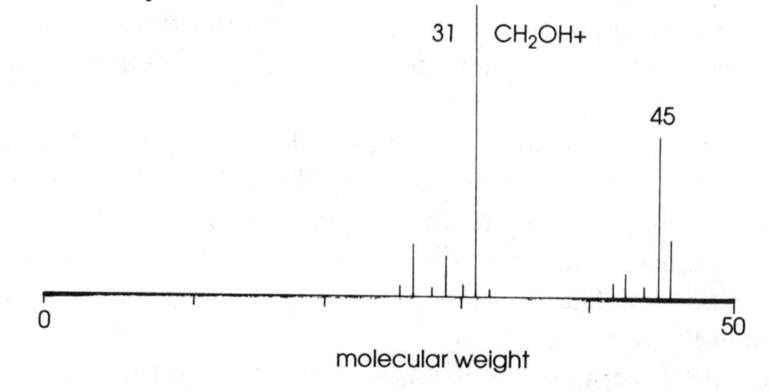

The mass spectrum of ethyl alcohol. The molecular weight of the unfragmented molecule is 46

OPTICAL SPECTROSCOPY

The history of the true forms of spectroscopy begins, unsurprisingly, in the visible region of the electromagnetic spectrum. Almost everything we know of the universe beyond the earth (and, since landing on it in 1969, the moon) comes from the study of the radiation that reaches us from more distant objects.

The standard texts on spectroscopy used by students and researchers give details of the physical principles in operation and some description of instrumentation and laboratory procedure.[4] What they lack, however, is any historical perspective of the evolution of this versatile tool, the *sine qua non* of today's practising chemist. One title that begins to compensate for this omission and conveys some of the poetry inherent in the subject is *The analysis of starlight.*[5] I have consulted this and one or two other histories of spectroscopy in order to reveal the intimate link between the development of spectroscopic techniques and the progress made not only in chemistry but also in astronomy.

Until the sixteenth century, man had nothing but unaided observation to help satisfy his curiosity about the universe around him. The eye is a fine instrument. At a pinch, it can detect single photons and can describe colours with great, if subjective, accuracy. But it cannot separate light of many colours or provide any further analysis of what it sees from the quality of the light alone. By eye, therefore, it was quite impossible to prove that the moon was not made of green cheese, although it was assumed even at this time that its composition as such would be absurd (the saying is at least as old as the sixteenth century). The advent of the telescope revealed the existence of satellites around the planets in our own solar system, of nebulae and galaxies, and of the motion and variations in brightness of the stars. But the composition of these bodies remained as mysterious as ever. Were they even made of the same stuff as the earth? It was not until the nineteenth century that rudimentary spectroscopy began to provide some answers. Eventually:

> The astronomical observatory became an extension of the physics laboratory, and the precise measurements of quantities such as wavelength and brightness made it possible to apply sophisticated new theories such as quantum mechanics to explain how stars worked; the subject of astrophysics was born.[6]

Today, instruments such as the Hubble Space Telescope and the Cosmic Background Explorer, COBE, are essentially glorified spectroscopes. The task of the former is to measure ultraviolet absorption spectra; of the latter to map the entire pattern of infrared and microwave emission across the sky. One or another kind of spectroscopy now lies behind half of all astronomical observations.[7]

Spectroscopy has its conception, if not its birth, in the famous discovery by Isaac Newton in 1666 that sunlight may be split into its constituent colours by means of a prism. However, it was not until nearly 150 years later that the implications of Newton's experiment were investigated. Even then, progress was sporadic. The first absorption spectra of sunlight were observed not by Newton but in 1802 by William Hyde Wollaston and 1814 by Joseph Fraunhofer.

> Newton failed to notice that the light from the sun is not perfectly homogeneous, and it was Wollaston who first discovered this by observing the rays of sunlight admitted through a narrow slit in a window blind. Wollaston then noticed a number of black lines, which crossed the spectrum in a direction parallel to the slit. For some reason he did not investigate these lines further, and it was reserved for Fraunhofer, the celebrated optician of Munich, to study them thoroughly, and point out their immense importance – an investigation which laid the foundation-stone of the modern science of spectroscopy.[8]

Fraunhofer observed "almost countless strong and weak vertical lines, which however are darker than the remaining part of the colour image; some seem to be nearly black."[9] The strongest of these lines he labelled alphabetically A to H. Fraunhofer knew what he had detected were no experimental artefacts. Nor were they some kind of natural boundaries between the colours of the rainbow. However, he forbore to speculate on the cause of the lines. While others set in motion a tradition that lasts to this day of overzealous "interpretation" of astronomical spectra, Fraunhofer went on to look at the spectra from the planet Venus and from various stars. He found that light from Venus had enough of the same lines to be considered sunlight-like, but that light from the stars was both different from sunlight and different from star to star.

The significance of Fraunhofer's black lines was only explained more than thirty years later. From space, we must return to earth for the background to the discovery. It had long been known that certain metal salts produce a distinctive colour when placed in a flame; indeed, this knowledge formed a cornerstone of chemical analysis. In 1822, John Herschel arranged for light from such coloured flames to pass through a prism whereupon he discovered sharp bright lines of emission rather than a rainbow continuum of light. Common salt produces a particularly bright single line (in fact, a closely spaced pair of lines) that is yellow-orange in colour. Fraunhofer, of course, knew about this, and furthermore noted the coincidence of the salt flame colour with the D line of his solar absorption spectrum, but again did not speculate. In 1848, Jean-Bernard-Léon Foucault showed there was some kind of connection between the emission line and the absorption line by in effect combining both previous experiments and shining the white light from an electric arc through a sodium flame and recording the spectrum. Now, a black line appeared in the D line position showing that a given medium can give rise to both emission and absorption at particular frequencies.

It was, however, more than ten years until Gustav Kirchhoff established that both signals have a common source, namely, in the case of the D line, the element sodium. With this new comparative tool at their disposal, Kirchhoff and his collaborator, Robert Bunsen, were able to establish that many elements known on earth are also present in the sun. Their efforts to undertake the chemical analysis of the solar spectrum excited scientists and the public alike. "The concept of a chemical analysis of a distant heavenly body such as the sun aroused the admiration and wonder of many."[10]

The goal quickly became to determine whether matter in space generally was fundamentally the same as that on earth. It seems almost obvious today that this should be the case. But in the context of the nineteenth century, it was by no means clear that the earthly elements were the same as the celestial ones. As William Huggins, the founder of stellar astrophysics, wrote retrospectively in 1909:

> One important object of this original spectroscopic investigation of the light of the stars and other celestial bodies, namely to discover whether the same chemical elements as those of our earth are present throughout the universe, was most satisfactorily settled in the affirmative; a common chemistry, it was shown, exists throughout the universe.[11]

Elsewhere, Huggins did not fail to point out the relevance of some of the elements seen in the stars to the processes of life, just as some contemporary scientists have made much of the detection of molecules in space that are of biological significance on earth.

The early spectroscopists did more than suggest the universality of the chemical elements. They also added to their known number. In 1860, during the course of their experiments in solar spectroscopy, Bunsen and Kirchhoff discovered the elements caesium and rubidium. William Crookes discovered thallium in the following year. The names of all three derive from the colour of their flame spectra[12] – from the Latin *caesius*, a bluish grey, and *rubidus*, red; and, most poetically, from the Greek, θάλος, *thallos*, a green shoot. Not long after Bunsen and Kirchhoff had identified two new alkali metals, spectroscopy claimed an even more notable discovery. In 1868, Norman Lockyer and Edward Frankland detected a characteristic spectrum in the atmosphere of the sun. This they attributed to a new element for which they chose name helium, from the Greek god of the sun, Helios, ηλιος, although proof that it was indeed an element had to await its discovery on earth in 1895 when William Ramsay found that the gas was released by radioactive minerals.

During the course of the nineteenth century it was also realized that the electromagnetic spectrum stretches in both directions beyond the visible region. The first invisible spectrum was detected by William Herschel in 1800. Experimenting with the ability of light of different colours to produce a heating effect, he found that the warmest area of the rainbow formed by his prism lay beyond the red band of light. In 1840, William's son, John Herschel, proved that both the spectrum and the Fraunhofer black lines continued into this "infrared" region by painting the paper on which the spectrum impinged with gum and lampblack to absorb heat and then dipping it in alcohol before positioning it near the prism. The alcohol evaporated from the warmer regions, leaving a moist violet end of the spectrum and a dry red end with narrow spectral lines of unevaporated liquid. Scientists made similar inroads into the ultraviolet region of the spectrum using photographic paper to make a record of this invisible radiation. In 1842, Antoine-César Becquerel showed there were Fraunhofer lines here too.

Despite these pioneering discoveries, spectroscopy did not blossom. One obstacle was the difficulty of ascertaining the origin of a given spectrum. It was quickly realized that anything along the line of sight of a spectroscope could account for its signal. John Herschel, for example, had pondered whether Fraunhofer's lines might not be the result of absorptions by atoms and molecules in the earth's atmosphere rather than in the sun.

A tougher challenge was to replicate astrophysical observations in laboratory experiments using known materials in order to establish their cause. At a time when the nature of ions – the source of many signals – was still unknown, it was not surprising so little progress was made. A famous episode in this history illustrates the difficulty. Moving on from stars to collect spectra of various nebulae, Huggins noticed a prominent pair of green lines in some of his results. Despite intense efforts, it proved impossible to reproduce the lines in the laboratory. Reluctantly, for there were no suitable vacancies in the periodic table, he concluded that it might be a new element. The name nebulium was coined. Only much later, in the 1920s, was it found that the source of the lines was a doubly ionized oxygen atom, a species impossible to study in Huggins's day.[13]

The absence of adequate spectroscopes had not dampened scientists' enthusiasm for trying to explain such spectra as they were able to detect, and the advent of more sensitive instruments encouraged chemists and astronomers to make new observations. The scientists in turn pressed instrument-makers to make further improvements. "Makers seemed poised to respond to the new challenge and chemists and astronomers were keen to collaborate with them in extending the new techniques and applying them as widely as possible."[14] This synergy continues today, and the wish to probe the fullerenes in due course provided its own impetus for the construction of more sensitive spectroscopes.

Scientists had used their primitive instruments to amass a welter of spectra, of both astral and terrestrial origin. According to histories of the period:

> Astrophysics was undoubtedly the major spectroscopic field in the second half of the nineteenth century, and most spectroscopists – including Kirchhoff, Ångstrom, Cornu, and Rowland, all of whom produced maps of the solar spectrum – had some active interest in the subject. Its literature was enormous....[15]

But:

> ... much of stellar spectroscopy was no more than the collection and classification of data. It could hardly be otherwise, for how could interpretation play a significant role, considering that J.J. Thomson and the electron, Albert Einstein and quantum theory, Niels Bohr and the electronic theory of atomic structure and M.N. Saha's theory of ionisation were all in store for the future?[16]

Nevertheless, this copious record was to prove of use:

> It was because of this great interest in astrophysics that abundant spectroscopic materials were readily available for a Balmer or a Rydberg, whose tools were pencil and paper rather than spectroscope and photographic plate.[17]

Progress in understanding the spectra accumulated during the nineteenth century depended upon the articulation of a theory. Several fundamental discoveries were key – of the electron by Thomson in 1897, of the particle nature of electromagnetic radiation and the fact that it is absorbed and emitted not in a continuous way but in quanta in 1905 by Max Planck and Einstein, and of Bohr's theory of the atom with electrons in "orbits" of different energies in 1913.

The Balmer series was one of a number of series of spectral lines for atomic hydrogen that were prominent among the solar spectra recorded in the nineteenth century. In the 1880s, Johannes Rydberg had found an equation whose solutions gave the position of the lines in these series. His empirical work lacked a theoretical base, but was proved spectacularly prescient when it was found that the equation in effect described the transition of electrons between orbits in the Bohr atom.

A classic text from this exciting period makes it clear that spectroscopy was now on the threshold. Various editions of Baly's *Spectroscopy* were printed between 1905 and 1924, right at that moment when spectroscopy was gaining a sound theoretical foundation, and as its

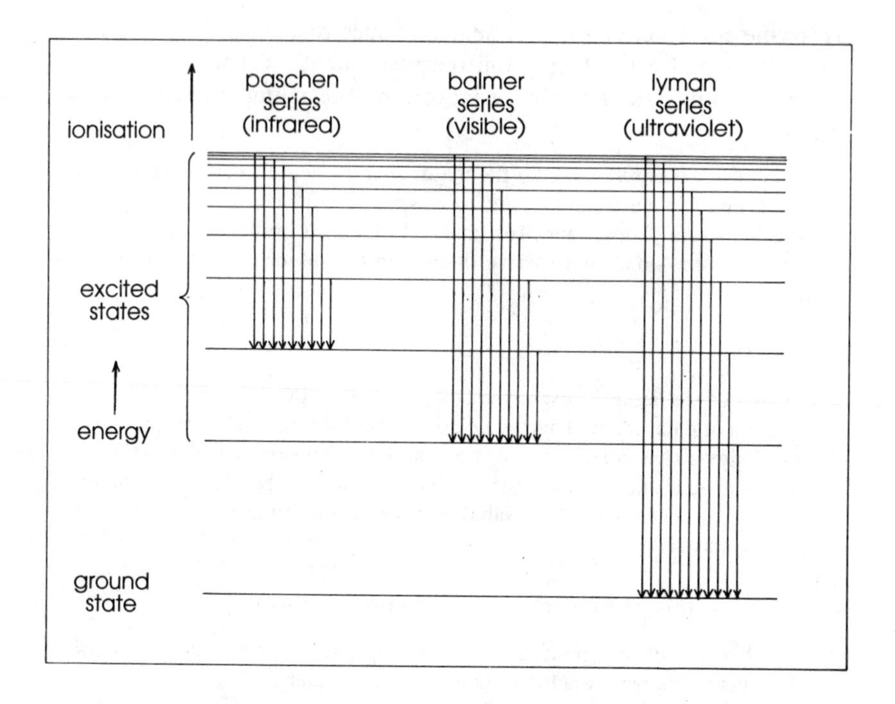

Three series of lines from electronic transitions in the hydrogen atom. The visible Balmer series is in the centre

techniques were extending their range still further. Numerous atomic spectra could now be explained, but equally, it was clear that there was a growing number of unexplained spectral lines lying across an ever greater span of frequencies. Musing on what spectroscopy in the far infrared and far ultraviolet might reveal, Baly wrote:

> At the present time the fascinating work based on the Bohr theory of the atom and the modern developments of X-ray spectroscopy have tended to focus the attention of scientific workers on to the most refrangible end of the spectrum, but I firmly believe that the time will soon come when the long-wave region will equal the short-range region in interest.... Up to the present no direct connection has been found between these very small frequencies and the very large frequencies dealt with in the Bohr theory. This connection will doubtless soon be discovered and I am brave enough to prophesy that the key to the problem of the absorption and radiation of energy by elementary atoms will be found in the infra-red.[18]

Baly was right in principle in that spectroscopy of the long-wave region would indeed come into its own. He was wrong, however, in his prophecy. Atoms absorb and radiate light mainly in the ultraviolet and visible regions of the spectrum. Something quite different lay behind the infrared spectra.

With some system at last shown to govern the spectra of atoms, attention turned to the probing of molecules where matters were clearly far more complicated. Friedrich Hund and Robert Mulliken made a start by imagining the atomic orbitals of two hydrogen atoms stretched out along the length of a molecular bond to form the molecular orbitals of a hydrogen molecule. This model was gradually developed, by Erich Hückel among others, to the extent that it could cope with moderately complex polyatomic molecules. The transition of the electrons between molecular orbitals was essentially just like that between atomic orbitals. Both sets of transitions take place largely in the visible and ultraviolet regions, although in the case of molecules sharp spectral lines are replaced by broad bands.

There are other ways in which molecules can interact with electromagnetic radiation, ways that are not open to single atoms. The promotion of an electron from a low-energy molecular orbital into a higher one requires a large amount of energy. However, molecules also absorb and emit radiation with lower energy. This is the lower-frequency, longer-wavelength radiation about which Baly had speculated, and it corresponds to molecules' characteristic frequencies of vibration and rates of rotation.

The development of the various forms of molecular spectroscopy is again bound up with astrophysical discovery. A Jesuit priest, Angelo Secchi, identified the first carbon stars from their atomic spectra in 1868. During the middle decades of this century, Alec Douglas, Gerhard Herzberg and others showed that other strong spectral lines associated with these stars were attributable to diatomic molecules such as CH, CO, CN, C_2, and their ions. Herzberg was later awarded the Nobel prize for chemistry for his contribution to the development of spectroscopy. Such results confirmed not only that the elements but also that the processes of stellar chemistry – and a little later, interstellar chemistry – were fundamentally the same as those on earth. Exploration of the chemistry of stellar and interstellar regions has continued to gather pace to this day.

The various optical methods of spectroscopy are named for convenience by the wavelength of the radiation they employ. The three important methods discussed here lie in the ultraviolet, infrared, and microwave regions of the spectrum. They are known also by the kind of molecular (or also, in the case of ultraviolet spectroscopy, atomic) change with which the radiation is typically associated. Thus, ultraviolet (and visible) spectroscopy deals mostly with electronic transitions between energy levels within atoms and molecules, and is customarily abbreviated as electronic spectroscopy. Ultraviolet photoelectron spectroscopy, encountered in Chapter Three, is an extreme form of this spectroscopy which measures not the transition of an electron from one energy level to another but its outright expulsion from the atomic or molecular orbitals in question. Infrared spectroscopy is associated with molecular vibrations and is often termed vibrational spectroscopy. Microwave (and some far infrared) radiation has at frequencies associated with molecular rotation; thus, microwave spectroscopy is to all intents and purposes the same thing as rotational spectroscopy.

THE ULTRAVIOLET REGION

An electronic spectrum provides a record of the transition of electrons between different energy levels. In the case of atoms, the spectral lines are distinct. In molecules, broad humps made up of a myriad of overlapping lines often take the place of sharp peaks because electronic transitions between molecular orbital energy levels are complicated by molecular vibration and rotation. Electronic spectra may be recorded in absorption or emission, reflecting respectively the promotion of an electron from a low to a high energy level or its relapse from high to low.

These electronic transitions lie behind a number of familiar phenomena. Sodium street lights shine light at the wavelength of Fraunhofer's D lines. Emissions from excited oxygen and other atoms colour the aurora and airglow occasionally seen in the upper atmosphere. Absorption cannot be observed with such ease, but we are all acutely aware of the protective value of the ozone layer whose absorption of ultraviolet light is made possible by the promotion of electrons in the orbitals of the ozone molecule.

As a practical tool in the laboratory, ultraviolet spectroscopy is of considerable utility for metallurgists who use it to look at metals in their atomic state rather than in molecular combination. For the chemist, it is less powerful. Most molecules that one wishes to study have many atoms and many molecular orbitals. Many have a propensity to fall apart under the impact of ultraviolet light. For those that do not, ultraviolet spectroscopy provides information about the energies of the orbitals. Whether or not a transition is observed depends upon the symmetry of the molecular orbitals before and after the transition, and upon other factors. Some transitions are "forbidden" by symmetry considerations arising out of quantum theory. Others are allowed. These so-called selection rules also aid in describing a molecule under analysis.

It is electronic transitions which give organic dyes (and most other things) their colour. The fact that transitions are relatively unperturbed by variations in the surrounding chemical structure helps to explain why molecules of the same family, anilines, for example, serve as dyes of different colours; the influence of the different chemical substituents on the core molecule is sufficient to alter the colour that is observed but not to push it outside the visible region of the spectrum altogether.

THE INFRARED REGION

Most molecules have $3n - 6$ characteristic modes of vibration where n is the number of atoms in the molecule. This seems a surprising, even arbitrary mathematical rule, but it is easily comprehended. Begin by considering a lone atom. This atom has three degrees of freedom; that is to say, it is free to move along any of the orthogonal axes of three-dimensional space. Now consider the various ways atoms in a molecule might try to move. If they all move in the same direction, the molecule as a whole slides off in that direction. Three degrees of freedom are thus given over to translation of the molecule in the three dimensions. In other circumstances, atoms at one end of a molecule might move "up" and at the other end "down". The result this time is a rotation of the molecule. There are three axes around which the molecule may rotate. Deducting these translations and rotations leaves $3n - 6$ degrees of freedom remaining in which atoms are pulling against each other with no net motion of the molecule as a whole. On these occasions, vibration is the result. The water molecule has three atoms and hence possesses three modes of vibration – a symmetric bending which changes the angle between the molecule's two bonds, and a symmetric stretching and an asymmetric stretching, both of which change the lengths of the bonds. For larger molecules, the number of vibrations multiplies in proportion to the number of atoms in the molecule. Twisting, rocking, wagging, and other motions join the simple bending and stretching. But soon it becomes impossible to describe the vibrations accurately in words. This is so even in a fairly simple molecule such as ethyl alcohol, with its nine atoms and twenty-one modes of vibration, although it is the case that certain vibrations are clearly based on simpler actions such as the stretching of the carbon-carbon or carbon-oxygen bonds.

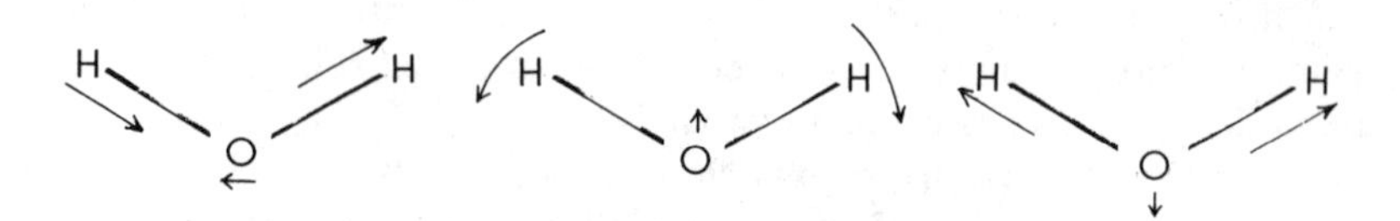

The water molecule may vibrate in three ways. Larger molecules have many, more complex, vibrations

It is possible in some cases to calculate the frequencies of molecular vibrations. In the simplest case of a diatomic molecule, the bonded atoms may be compared with two masses connected by a spring. Hooke's law then states that the period of oscillation of the bond increases as the square root of the mass of the bonded atoms. This principle is employed not so much to predict the frequency of molecular vibrations outright, but to check that a vibrational spectrum of a molecule is correctly assigned, which may be done by substituting an isotope of one of its constituent atoms and then noting whether the spectral frequency is shifted by the predicted amount.

The remit of infrared spectroscopy runs from wavelengths around 800 nanometres to one millimetre. For standard laboratory analysis, a narrow range between about two and sixteen micrometres, the so-called fingerprint region, encompasses the most frequently encountered vibrations. Spectra in this region are thoroughly documented. The spectrum for our pet molecule of ethyl alcohol, for example, has peaks characteristic of the stretching of the oxygen-hydrogen bond at one end of the molecule and of the opening and closing "umbrella" vibration of the methyl group at the other end. This time, the former feature is sufficient to distinguish this molecule from its isomer, methyl ether. In larger molecules, the elucidation of structure is a greater challenge. Analysis is hampered by the multiplicity of similar vibrations as well as by the fact that the strength of a vibration signal bears no direct relation to the strength of the vibration itself, and the practical difficulty that sensitive signals are frequently swamped by presence of water and solvents. It is rarely possible to deduce the structure of a molecule of any complexity from infrared spectroscopy alone.

It is also possible to probe molecular vibrations by a technique known as Raman spectroscopy, which provides a useful counterpoint to conventional infrared spectroscopy. In 1928, C.V. Raman observed that a small proportion of the radiation scattered by a sample emerges with a small increase or decrease of frequency as a result of "inelastic" collisions with the sample molecules. Most of the light passes straight through the

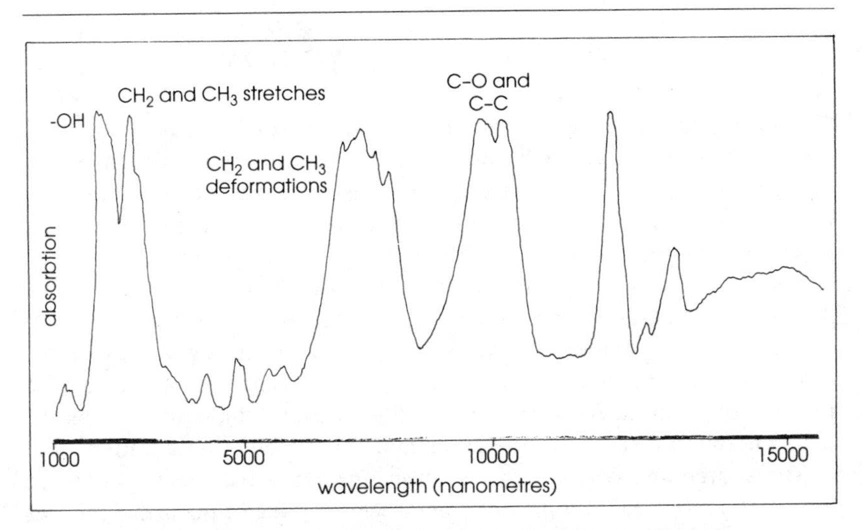

The infrared spectrum of ethyl alcohol exhibits peaks for a number of distinctive vibrations within the molecule

sample, and most of the rest is scattered "elastically" as photons bounce off molecules without experiencing a change in frequency – a process called Rayleigh scattering. The "inelasticity" of the remaining few collisions arises because some energy is exchanged between the photons and the molecules. It so happens that these energy quanta are those of the molecular vibrational (and also rotational) transitions.

Raman spectroscopy would be nothing more than a perverse way of collecting the same information as is available from conventional infrared and microwave spectroscopy but for one thing. The selection rules that apply to electronic transitions also apply to vibrational transitions, and furthermore they apply differently for vibrational transitions detected directly by absorption or emission and those detected indirectly by Raman scattering. A transition is only permitted if the symmetry of the energy states both before and after is in accordance with the rules. Specifically, symmetric vibrations in certain symmetric molecules do not give rise to an infrared spectrum, but they do produce a Raman spectrum. For a highly symmetrical molecule like buckminsterfullerene, Raman spectroscopy will thus prove a vital probe.

In practice, Raman spectroscopy is highly demanding. The effect takes place among a tiny minority of molecules at any moment. It requires the detection of a tiny fraction of scattered light which is in turn a tiny proportion of the light entering the sample cell. However, the advent of lasers has greatly advanced this technique by increasing the intensity of light available to experimenters.

THE MICROWAVE REGION

Transitions between the quantum rotational levels of molecules occur at lower frequencies than vibrations, in the far infrared through to the microwave region of the spectrum. Rotational changes are often observed as a side-effect in electronic or vibrational spectroscopy when they appear as fine lines in an otherwise smooth spectral feature.

Just as Hooke's law yields the frequency of vibration of some simple molecules, so classical mechanics can help to predict rates of molecular rotation since the angular momentum of any molecule is given by simple arithmetic from the mass of the constituent atoms and their distance from the axes of rotation. As before, this enables a useful dialogue to be established between theoretical prediction and experimental observation.

The source and detector of radiation in microwave spectroscopy are essentially like a radio transmitter and receiver. One of the principal uses of the technique, as we have seen, is in radioastronomy for the study of molecules in outer space. Elsewhere, rotational spectroscopy is of limited utility. The rotational behaviour of most molecules is hideously complex. Molecules in liquids and solids are not free to rotate. Only a few simple gas molecules with relatively high symmetry are susceptible to the full treatment. These are the sorts of molecules that can be easily imagined as rotating, as is indicated by the labels that microwave spectroscopists have given them – linear molecules are called rotors, molecules such as ammonia are tops. Some of these small molecules are so well cheracterized that they are of only trivial interest in the laboratory; others are, for all their simplicity, hard or impossible to make. Fortunately, they are exactly the type of small molecules that are found in space (although their being found there is of course consequent upon the capability of the technique used to detect them!). The most symmetric molecules, however, do not yield rotational spectra just as they do not yield vibrational spectra. It is thus impossible to study such important molecules as hydrogen, oxygen molecule, methane, and benzene in space or anywhere else by this means. As before, Raman spectroscopy can be used to summon up rotational data for these symmetrical species, but this is not an option available to astronomical observers.

NUCLEAR MAGNETIC RESONANCE SPECTROSCOPY

At still lower radio frequencies, nuclear magnetic resonance (NMR) spectroscopy is a probe not of events involving a number of atoms such as the stretching and bending of bonds or the spinning and tumbling of entire molecules, but of an intrinsic property of selected constituent atoms of a molecule. Because this property of the atom in question is susceptible to the influence of neighbouring atoms, this is a form of spectroscopy that provides specific information not only on what atoms are present in a given molecule but also on where they lie in relation to one another.

NMR spectroscopy is today the single most powerful probe of the molecular structure of many compounds. "The subject has grown so fast that it has almost become a scientific discipline in its own right", according to an expanded section on the subject in a recent edition of a standard text on spectroscopy in organic chemistry (whose authors add that the technique can fortunately still be employed without the help of experts).[19] Indeed, NMR spectroscopy is uniquely at home in the bourgeois world of the molecule and the technician. Unlike some other forms of spectroscopy, it cannot be used to identify transient species in harsh environments or at great distances. It is a homely procedure, involving generous samples in ordinary test tubes which are then placed inside an instrument looking a bit like a giant washing machine.

NMR spectroscopy relies on processes less readily visualized than those involved in other forms of spectroscopy. On the other hand, its diagnostic powers are so great that it repays the effort taken in understanding its principles. This technique will, in due course, provide the most compelling proof of the structure of buckminsterfullerene.

Some atomic nuclei possess a "spin" which can have a number of energy states determined by a nuclear spin quantum number. NMR spectroscopy registers transitions between these states of a sample held in a magnetic field. The hydrogen nucleus or proton has a spin quantum number of one half and energy states of plus one half and minus one half representing spin of the nucleus in the direction of the magnetic field and against it. Carbon, the other element of most interest to organic chemists, unfortunately has a nuclear spin quantum number of zero and consequently exhibits no energy transitions. Its heavy isotope, carbon-13, however, has, like hydrogen, a spin quantum number of one half, a fact that becomes useful when proton NMR spectroscopy is inappropriate.

An NMR spectrometer contains a sample chamber located within a powerful adjustable magnetic field. Coils are arranged so as to expose the sample to radio-frequency radiation and to detect any absorption of that radiation. The sample is subjected to the magnetic field and simultaneously irradiated. Resonance is detected when the energy of the radiation is the same as the difference in energy of the spin states created by the magnetic field. In other forms of spectroscopy, the difference in energy

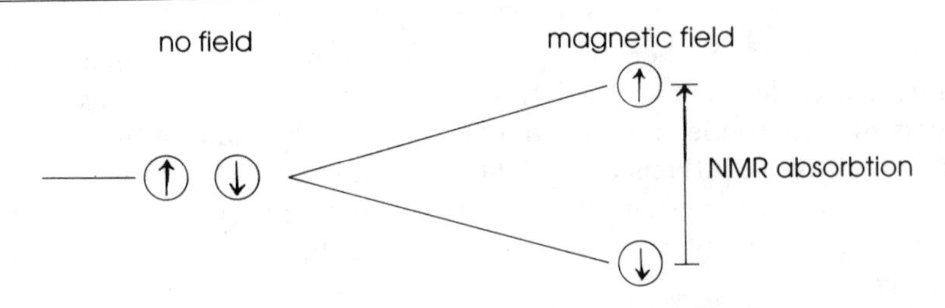

NMR signals arise when radiation energy matches the energy difference between spin states in a magnetic field

levels is a fundamental quality of the atom or molecule. Here, the difference is determined partly by a constant of the atomic nucleus – a quantity called its magnetogyric ratio – but also by the strength of the applied magnetic field. In practice, then, it is possible to obtain a nuclear magnetic resonance spectrum by holding the frequency of radiation constant while varying the strength of the magnetic field, or by holding the field constant and scanning through a range of radio frequencies. In fact, either procedure is often wasteful of time, and a variety of refinements have been developed in order to pick through the dross and home in more quickly on resonant signals of interest.

This is all very well, but the only information we can glean so far from this interaction between magnetic field, electromagnetic radiation, and sample matter is some knowledge of what atoms are present. Admittedly, the strength of the signal is proportional to the number of resonating nuclei, so this information is quantitative. But on the other hand, many nuclei do not show up at all because they have no spin.

This form of spectroscopy relies on local colour for its real interest. The first local effect is the distortion of the external magnetic field experienced by the nucleus in question by the magnetic influence of electrons in nearby bonds. Of course, all nuclei bound into molecules experience the magnetic field in slightly different ways from the ideal isolated nucleus. The spectral peak for each nucleus appears accordingly in a position removed to some degree from a suitable reference peak – an amount described as the chemical shift. In proton NMR spectroscopy, protons "shielded" from the external magnetic field by regions of high electron density exhibit small chemical shifts. Methyl groups, for example, appear near the reference peak. Hydrogen atoms attached to chemical groups that tend to draw electrons towards them, such as oxygen atoms or carbonyl groups, are described as "deshielded" and appear at the opposite end of the spectrum. Carbon-carbon double bonds and conjugated and aromatic carbon systems have a moderately deshielding effect; thus, the proton signals for molecules such as ethylene and benzene appear in the middle part of the spectrum. In the case of benzene and similar molecules, the deshielding of the hydrogen atoms bonded to the carbon ring is accentuated by the "ring current" induced in the molecule by the magnetic field as the clouds of delocalized electrons in the pi bonds above and below the plane of the carbon hexagon are caused to circulate.

The other local effect is the coupling that takes place between the spin of one nucleus and that of its neighbours mediated by the electrons in the bonds between them. This chain of charged particles transmits the effects of the magnetic field between nuclei with the result that a spectral peak for protons in one environment may be split into several smaller, closely spaced peaks by the presence of protons nearby. Consider, for one last time, our molecule of ethyl alcohol. The NMR spectrum reveals that there are two protons with a chemical shift characteristic of a CH_2 group and three protons with the characteristic chemical shift of the methyl group, CH_3. Well away from these features, there is a further peak for the hydrogen atom in the hydroxyl group with its very different chemical shift. Now look more closely: The two former protons feel the distant influence of the spins of the three protons in the methyl group. These three spins (each with a value of plus or minus one half) can be summed in four ways: $^3/_2$, $^1/_2$, $-^1/_2$, $-^3/_2$. They thus split the CH_2 peak into a quartet. Conversely, the protons in the methyl group feel the influence of the two protons and are split into a triplet reflecting the influence of net spins of 1, 0, −1). (The methyl ether isomer by marked contrast, produces one strong peak for the six equivalent protons in its two identical methyl groups.) For present purposes, it is not necessary to understand exactly how all this happens, but merely to appreciate how very powerful these nuances are in allowing spectroscopists to assemble a complete portrait of a molecule under analysis.

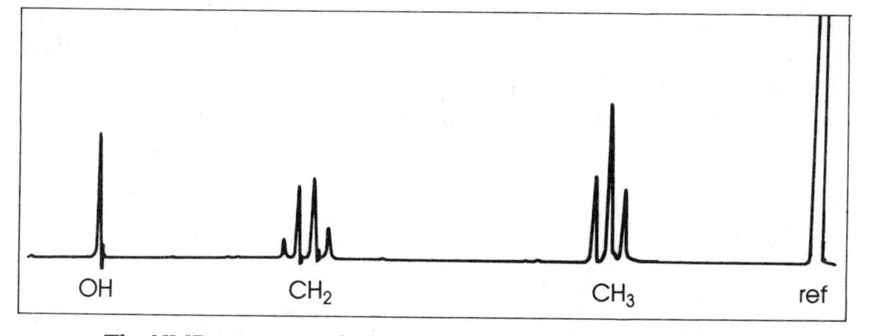

The NMR spectrum of ethyl alcohol shows the three different proton environments present in the molecule

Protons have a large magnetogyric ratio and give correspondingly strong NMR signals. Fortunately, many organic compounds have sufficient hydrogen atoms distributed in a sufficient number of distinct environments that it is often possible to deduce the structure of a molecule given no more than its empirical formula and its NMR spectrum. This is certainly true for ethyl alcohol.

However, some other organic compounds do not. They do, of course, have carbon atoms. Unfortunately, the natural abundance of the carbon-13 isotope is about one per cent of all carbon. Furthermore, the magnetogyric ratio of the carbon-13 nucleus is a quarter that of the proton. Carbon-13 resonance signals are in all six thousand times weaker than proton signals.[20] Not surprisingly, the development of carbon NMR spectroscopy has lagged behind proton NMR spectroscopy. Only recently has it come into its own thanks to sophisticated ways of accumulating spectra, and it is now almost as easy to perform as proton NMR spectroscopy. Carbon-13 NMR spectroscopy relies on information that may be inferred from chemical shift data. Like protons, carbon nuclei experience chemical shifts according to the nature of their surroundings. It is useful in the analysis of larger compounds, where there are carbon atoms in a number of distinct environments, and where there are often too many hydrogen atoms to be able to deduce the structure from this information alone, or alternatively too few hydrogen atoms to tell much at all. In the case of buckminsterfullerene, of course, it is the only option.

There are many more specialized forms of spectroscopy in addition to the main ones described here. There exist also a number of powerful analytical techniques, most notable among them X-ray and electron diffraction, which provide information about the chemical composition of matter, by other means. My purpose is not to sing the praises of one technique or family of techniques above others, but to highlight their complementarity. The key to success in analytical chemistry lies in using these techniques in concert, and in knowing which ones to use, and in which order, the better to make known what was unknown.

7

THE CHEMIST-STYLITES

Cosí avviene, dunque, che ogni elemento dica qualcosa a qualcuno (a ciascuno una cosa diversa), come le valli o le spiagge visitate in giovinezza: si deve forse fare un'eccezione per il carbonio, perché dice tutto a tutti, e cioè non è specifico, allo stesso modo che Adamo non è specifico come antenato; a meno che non si ritrovi oggi (perché no?) il chimico-stilita che ha dedicato la sua vita alla grafite o al diamante.[1]

It happens, then, that each element says something to someone (to each person something different), like the valleys or beaches visited in youth: one must perhaps make an exception for carbon, because it says everything to everyone, and it is not specific, in the same way that Adam is not specific as an ancestor – lest one discover today (why not?) the chemist-stylite who has dedicated his life to graphite or to diamond.

Why not indeed. For as the chemist and author, Primo Levi, surely knew when he wrote his autobiographical sketch, *The Periodic Table*, there is such a man. The stylites were medieval religious ascetics who withdrew from their communities by living atop pillars. The chemist-stylites who have dedicated their lives to the study of carbon are not the organic chemists – their work has been at the cutting edge of chemistry for more than a hundred years – but those who study elemental carbon and the combustion of hydrocarbon fuels. In the shorthand of the profession, these people make up the "soot community".[2]

The soot community wears its history on its grimy sleeve. There are hundreds of combustion science laboratories around the world, but the centre of this study in the United States is located amid the vast coalfields of Appalachia. This is a field of science implicitly or explicitly supported by the energy industries. It is applied science made manifest. Its aim is not the study of one of the ninety-two naturally occurring elements *per se*, but the study of the one element that happens to bulk large in our fuels.

Not many elements of the periodic table (Mendeleev's now, not Levi's) warrant their own journal. But along with *Sulphur*, *Fluorine*, and one or two metals, this one does. *Carbon* is published, fittingly between black and grey covers, by the American Carbon Society and Pergamon Press. (*Sulphur* is resplendently bound in yellow.) A regular series of volumes, *Chemistry and physics of carbon*, is published by the same society under the same editors. Both journal and monographs emanate from the Department of Fuel Science at Pennsylvania State University. (It was in these volumes that Harry Kroto, Jim Heath, and Sean O'Brien found the micrographs of spheroidal graphite that encouraged them to put forward their spiral model for the growth of soot particles.)

This series did not acknowledge the existence of buckminsterfullerene until 1991, when, in the first sentence of the introduction to the twenty-third volume, the editor, Peter Thrower, reminisced on the enduring fascination of carbon science over the quarter of a century of this series, and then added: "The recent emergence of Buckminsterfullerene as a new form of carbon is yet another indication of the unmatched diversity of forms this element can assume." He then went on to list work that could only seem mundane (to non-stylites, anyway), like the mayor in a council meeting pausing briefly to acknowledge that a UFO has landed in the town square before returning to "any other business".

Carbon in the more or less pure form of carbon black has been used at least since the third century AD in inks and dyes. In more recent centuries, it became a key ingredient in printing and in vulcanizing rubber. During the present century, its fortunes have continued to rise. "The chemistry and physics of this material formerly thought to be 'inert' is far from being a dull subject."[3] Carbon research enjoyed a boost during the Second World War with the use of artificial graphite in early atomic reactors. Graphite of various kinds is the basis of many carbon materials. Others, such as carbon fibres, are not graphite-like but are derived from hydrocarbon polymer fibres or from precursors spun from resins and pitch (originally, by Thomas Edison, in order to create filaments for his electric lamps, from cotton or linen). The turn came for carbon fibre technology in the 1950s with the advent of widespread jet propulsion and a need for more extensive aircraft materials research.[4]

Like so many other sciences, the study of carbon has grown to become a specialism, a combination of physics and chemistry, fluid dynamics and chemical engineering, with dashes of astrophysics and geology to spice the mix. Any self-respecting specialism has its own journals. These exist to publish new knowledge in the field. But if the new knowledge is especially noteworthy, then this role will be usurped; the work is seen to be of more general interest and is published somewhere that it will get a more general readership. The first paper on buckminsterfullerene appeared in *Nature*, and subsequent ones in well read chemical journals. Nobody thought for a minute to publish it in *Carbon* despite the fact that what was at issue was a novel compound made entirely of this element. (It is arguable whether *Carbon* would have published the paper even if it had been submitted, for reasons that will be made clear.)

One scientist's specialism is another's backwater. If some revolution in science happens to take place within a field that is widely perceived as a backwater, then something apparently quite unfair happens. The novel discovery makes a big splash in the larger scientific pond, the backwater remains unruffled. The backwater is condemned to remain a backwater. Thus, when it is finally shown how the fullerenes assemble and how that process of assembly relates to the various ways in which soot forms, one can be sure that this work will not first be published in *Carbon* or in *Combustion and Flame* or any similar title. The backwater is self-perpetuating. No really significant discovery concerning the science of carbon will ever be published in the journal of that name.

It is not my intention in sketching this portrait to single out combustion scientists as especially dull or recondite. Perceptions are relative. To others, cluster science is a backwater. It too has its own journals, and here too the most significant work is published elsewhere, in journals of chemical physics or general science. Now, the fullerenes have their own journal too, so even this rushing stream of new science has its very own backwater forming as surely as an ox-bow lake. On the shores of this lake, there will grow up a community that will probably never make a revolutionary discovery, but will spend its days doing what Thomas Kuhn calls "normal science". This is the natural process of consolidation of scientific discovery.

And yet, carbon science is no ordinary backwater. The roll call of those who have studied the element reads like a scientific hall of fame. The formation of soot has been studied since before Michael Faraday published *The chemical history of a candle*. Humphry Davy originally ascribed the incandescence of a luminous yellow flame to particles of elemental carbon. Nearly two centuries later, however, the details are still sketchy. Even the determination of such very basic properties of carbon as its melting point and boiling point are current hot (!) topics. In this century, the Braggs, Bernal, Rosalind Franklin, and George Porter have been among the distinguished figures to investigate the element in Britain alone. Carbon science may be a backwater, but it is one that attracts the best class of tourist!

Hearing this roll-call, it is hard to believe that the fullerenes were not discovered much sooner than they were. Diamond and graphite have been known for millennia, and have been known to be forms of pure carbon since the nineteenth century. Regular geometries have been observed in carbon micrographs continuously since the 1950s. The crude technology that was soon to be employed to make quantities of buckminsterfullerene had been available for fifty years. Mass spectrometry for large mass fragments is of similar vintage, developed as it was during the atomic bomb programme of the Second World War. It would have been possible to perform supplementary spectroscopic analysis sufficient to reveal the molecule's structure perhaps as much as thirty years ago.

The chemist-stylites of carbon are accustomed to imperfection. Not for them the elegant pathways of organic synthesis or the passable accuracy of Hückel calculations. They know that while diamonds may sparkle, there is no such thing as the perfect honeycomb graphite that is illustrated in textbooks. Even the best that is commercially available – a type called "highly oriented pyrolytic graphite" – is disordered. The run-of-the-mill graphite used in electrodes or for laser vaporization experiments is even messier.

When it is not present as diamond or as large expanses of graphite, carbon forms small particles. The existence of these particles has been known for a long time. With the advent of electron microscopy, it was revealed in the 1930s that these particles, whose diameter ranges from about ten to more than a hundred nanometres, are often roughly spherical. They generally coalesce to form aggregates, in the form of large clumps or strings like beaded necklaces, except in particular types of carbon black where they remain discrete. These particles are not pure like our idealized diamond and graphite. The particles formed in flames generally contain at least one per cent of hydrogen by mass, corresponding approximately to an empirical formula of C_8H, though carbon black can contain as little as one part in a thousand of hydrogen.

Within these particles, it is believed that the carbon is arranged in crystallites, each typically comprising three or four graphite-like layers, stacked, not neatly as in drawings of graphite, but with sheets randomly thrown down one on top of another. Each crystallite might contain perhaps one hundred carbon atoms in these "turbostratic" (randomly stacked) layers, spaced on average rather more generously than in graphite. Within the crystallites, the layers are presumably hacked off at the edges, with their dangling bonds satisfied with the hydrogen or oxygen atoms abundant in solvent or flame conditions. X-ray diffraction studies show that each particle might contain ten thousand or so crystallites. Thus, an average particle might contain a million carbon atoms.

Why is it that these crystallites arrange themselves in such a way as to construct particles that are spherical? If these particles were composed of concentric fullerene-like shells, a typical particle of carbon black might be equivalent to a buckminsterfullerene nucleus with one or two dozen

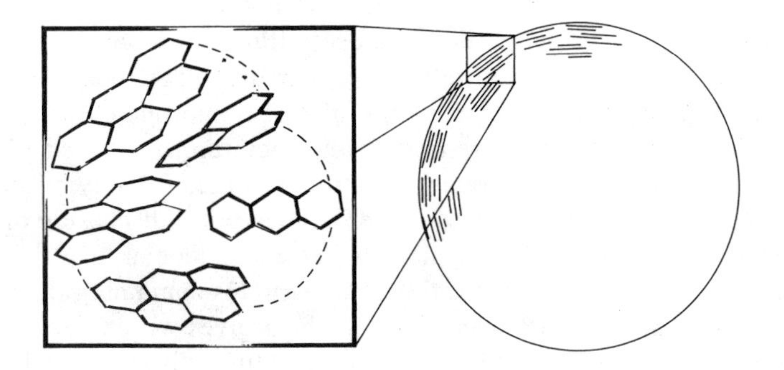

Standard models of soot formation show untidily layered crystallites of carbon agglomerated into particles

concentric carbon shells wrapped around it at radial increments equal to the layer spacing in flat graphite. But apparently, they are not. Some proposed mechanisms of formation invoke liquid droplets as precursors to the spheroidal carbon particles. In such cases, the sphericity might stem not from the geometry of chemical bonding but from the physical laws of surface tension. Perhaps because they are the creatures of physicists and materials scientists, these mechanisms do not dwell on how part of the carbon network might be terminated at the rim of the graphite planes.

Spherical carbon particles a hundred times this size, where physical effects are that much more powerful, do take shape in the liquid phase – a process known as spherulitic crystallization. This form of carbon is the key ingredient in a form of steel called "ductile iron" (a material, incidentally, which was employed by the architect Renzo Piano in the roof trusses of the De Menil Collection, an art gallery barely a mile from the campus of Rice University in Houston).

Precise details of the structure of any of these spherical particles and their constituent crystallites are sketchy. In the crystallites, the graphite-like layers may be relatively neatly stacked like a club sandwich, or inconsistently oriented and unevenly spaced, or more thoroughly amorphous. Despite these variations, the standard texts offer no speculation on the molecular structure. Electron microscopy and X-ray diffraction of the crystallites confirm the crude graphite-like stacking of carbon sheets and provide some indication that they are connected to each other within the particles by means of single planes of carbon. In order for the crystallites to line up across the surface of a spherical particle, these connecting planes must bend, but this need to bend is not acknowledged. Indeed, schematic diagrams showing how the crystallites align within spherical particles omit these layers entirely rather than draw them bent. The Flatland paradigm persists in discussions of larger morphologies such as fibres and sheets. There are reports in the literature of twisted ribbons and hollow filaments, even onion-skin structures made simply by stirring pitch. But in none of these cases is there any explicit discussion of how the curvature that is hinted at is mediated at the molecular level.[5]

Then there is the mystery of how these particles containing thousands or millions of carbon atoms can assemble themselves within a fraction of a second in a flame. Since the nineteenth century, different models for soot formation have vied for attention. Since the Second World War, it has become possible to study the dynamics of fast reactions between highly reactive transient species, and the attention of the soot community has shifted somewhat from the dissection of large particles to the investigation of the assembly of small particles. Single atoms of carbon, two-carbon units, three-carbon units, acetylene, hydrocarbons, and carbon monoxide are all proposed as agents in the process. In the 1950s, it was discovered that C_2 and C_3 were important constituents of carbon vapour of the kind found in flames and around certain stars. (It had hitherto been assumed that atomic carbon was the key player.) By trapping these species in low-temperature matrices, freezing them out of the reaction arena in order that spectroscopy might be undertaken, it was subsequently found that C_3 was in turn more prevalent than C_2 under many conditions.

With this new knowledge of the building blocks of the sooting process, the next priority was to consider how these small, highly reactive molecules join up. Studies of carbon clusters by means of mass spectrometry had shown clusters of up to fifteen atoms in the 1940s and more than thirty by the 1960s. In a key paper published in 1959, Kenneth Pitzer and Enrico Clementi at the University of California at Berkeley calculated the heat of formation for "all" C_n molecules (in fact, their graph extends only as far as C_{14} – how things can change). The graph is a zigzag, oscillating sharply between the more stable odd-numbered molecules and the less stable even-numbered molecules. As the molecules become larger, the oscillation decreases, implying that the difference between odd and even rapidly becomes unimportant at least so far as the heat of formation is concerned. The authors predicted that larger clusters would be important at high temperatures.[6]

Such studies were of only marginal use in understanding the full cycle of growth of soot particles. It was known or assumed that clusters with single figures of carbon atoms are linear in form, that those with between ten and, say, twenty, atoms are either linear or rings, and that the hexagonal graphite structure begins to emerge in rather larger clusters. Beyond this lay more than one kind of blackness. In the fourth volume of *Chemistry and physics of carbon*, Howard Palmer and Mordecai Shelef, two scions of Pennsylvania State University, could only say where the work needed to be done, ending their article with a plea for more mass spectrometry experiments and more detailed analysis of the resulting spectra (two challenges of which it might now be said that the Exxon group has met the first and the Rice group the second). They concluded in naked frustration: "The uncertainty that exists is quite intolerable and must be cleared up."[7]

The priorities of the soot community, like those of any community, are determined by context. Working as they do in close professional – and, as we have seen, even geographic – proximity to the oil and coal industries, the scientists of the soot community are accustomed to dealing with materials that, although basically of a type (rich in carbon), are heterogeneous, impure and, above all, vastly abundant. From this perspective, there were some good reasons – and some less good – for these scientists to regard Kroto and Smalley's discovery with indifference.

The most obvious reason is the question of scale. While Smalley and his team remained incapable of producing their new molecule in tangible form, in wagon-loads, the soot community could continue to regard it as an esoteric object far removed from their everyday concerns. Then there was the fact that the molecule had been produced in conditions where carbon is the only reacting element. Although the soot community studies carbon in all its elemental forms, these forms are all imperfect, either doused with hydrocarbons or infiltrated by hydrogen and oxygen contaminants. Paradoxically, the idea of a pure, molecular form of carbon was almost alien to them.

Among the less good reasons for ignoring the discovery were the traditional barriers of the scientific disciplines. If buckminsterfullerene was not a fragment of some bulk material but a discrete molecule, especially one that looked as if it might possess aromatic properties, then surely it fell within the remit of organic chemistry. Or if it could only be made in a molecular beam apparatus, then perhaps it was for the cluster scientists to consider. If there was no way of making quantities of it and no demonstrable use for it, then wasn't it up to the fundamental scientists to study it? Overlaying everything was the pique that the soot community had not discovered what might be described as the first perfect molecule of soot.

Small wonder, then, that, in the introduction to one of the first monographs on the fullerenes, it was noted that "few were inclined to 'count' the species as carbon allotropes. The new species, like small molecules such as C_2 and C_3, were thought of as transients in dilute vapor phases, hardly what one classifies as 'materials' based on elemental carbon."[8] Eric Rohlfing, a former colleague of Cox and Kaldor's at Exxon, subsequently at the Combustion Research Facility of Sandia National Laboratories in Livermore, California, wrote more diplomatically that "to many these molecular beam experiments represent curious results obtained in an esoteric context". By the time that Smalley's group had performed their impressive series of experiments on the reactivity of the fullerenes and Klaus Homann had demonstrated their presence in hydrocarbon flames, Rohlfing felt that the forging of a connection between the physical chemists and the soot community was overdue. It was not only the observation of fullerenes in flames that advertised the need for such a connection, but also the fact that laser vaporization experiments had been used not only to produce fullerenes, but also, on other occasions, hydrocarbons, polymers, and soot particles. Alluding to the position the Exxon scientists had taken, Rohlfing pointed out that: "Regardless of one's subjective view, the objective fact is that large carbon clusters are intimately connected with carbon particles, whether in supersonic nozzle expansions or in flames."[9]

It was not until 1992, two years after it had been discovered how to make large quantities of fullerenes, that the journal *Carbon* embraced the new developments in its field. Its editors invited Harry Kroto to serve as the guest editor of a special issue on the topic. In his introduction, Kroto wrote carefully but pointedly of the experience that he and his colleagues at Rice University had undergone as they had sought to prove their result: "During the period 1985–90 *unequivocal evidence* for the exceptional stability of C_{60} was amassed, which provided *convincing evidence* for the cage structure proposal" (italics added). The choice of words was a reminder that during much of this time Kroto and Smalley had felt that they had unassailable proof of the stability of C_{60}, but also an admission that the proof that it was the shape of a soccer ball was still circumstantial. Kroto could not resist biting the hand that was feeding him at this moment. Mischievously, he added: "There is food for thought in the fact that this fascinating, new round world of synthetic chemistry and materials science was discovered as a consequence of the desire to understand the role of carbon in space and stars, rather than the probing of carbon's possible material applications."[10]

What really rankled with the soot community was not the report of the fabrication and detection of buckminsterfullerene, nor the proposal that it has soccer ball symmetry, but the role proposed for it as a new paradigm in the growth of soot particles. While the physical chemists stuck within their parish, studying small, sparse, transient clusters, their work could be ignored. In all probability, the fullerenes barely changed the accepted order of things. In their view, the state of knowledge was as follows: two to nine carbon atoms – linear chains (as before); ten to twenty carbon atoms – rings or chains (as before); twenty-something atoms – unknown, but probably polycyclic fragments (as before); thirty-two atoms and up – fullerenes (a sideshow, probably irrelevant to the soot formation process); hundreds to millions of atoms – soot particles (as amply documented in the existing literature). But when they offered up their speculation that the fullerenes might be some kind of missing link in soot formation the cluster scientists were seen as committing a professional trespass. The soot community objected to the pitch that fullerenes could explain everything they had spent their careers puzzling over. It was not so much a scientific dialogue as a territorial dispute.

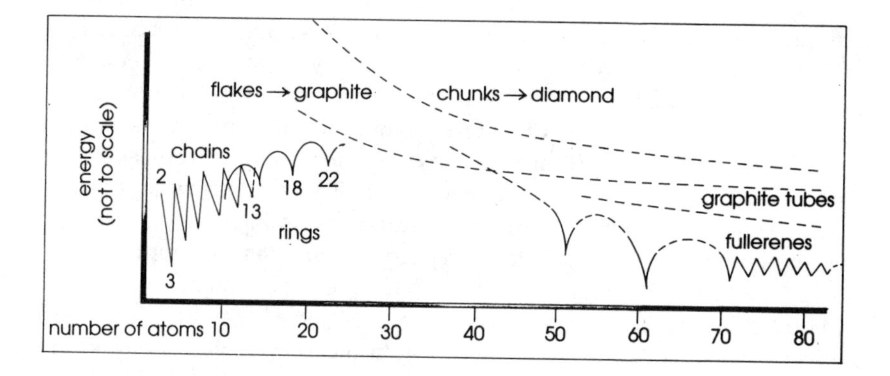

The energy stability of different carbon morphologies is strongly dictated by the number of atoms present

On the face of it, the spiralling carbon shell mechanism postulated, with various modifications, by Smalley's group and by Kroto seems an attractive model for soot formation. Soot forms incredibly quickly, building up large, ordered particles – usually roughly spherical and of roughly constant diameter – out of a regime where the tendency would seem to be toward chaos, where the regular structure of the fuel being burned is fragmented into tiny ions and radicals of one or two or three carbon atoms. The great pace with which these radicals knit together suggests that they are very reactive and that the growing particles continue being very reactive until they reach their ultimate size.

To the chemist, this speed suggests the presence of dangling bonds. The beauty of the spiralling shell is that it cannot rid itself of a leading edge of dangling bonds. Indeed, the edge grows in length as the particle increases in size, automatically maintaining the growth impulse. Furthermore, carbon-carbon bonds are stronger than carbon-hydrogen bonds which favours the building of a carbon network rather than small hydrocarbon molecules.

Though made in all good faith, this model was rejected outright by many in the soot community. The Rice group and Kroto had not consulted with them, had not referenced their papers, and had no solid evidence for what they proposed. Even after these first skirmishes, the fullerene camp would continue to incur the wrath of soot community for its selective use of evidence from the combustion literature and for its cocksure ability to rediscover what others had long known.

On this occasion, it was another scientist from Exxon, Lawrence Ebert, and Michael Frenklach, a collaborator based at Pennsylvania State University, who gave voice to the community's feelings with a detailed critique of Kroto and Smalley's ill-informed theorizing.[11] While Smalley and his colleagues were searching for the yellow vial, Frenklach proposed a "detailed chemical reaction mechanism" whereby soot might form from polycyclic aromatic hydrocarbon molecules present in flames (just one example of work not cited by the Rice group). Frenklach and Ebert maintained that this mechanism could also account for the formation of fullerenes. Their simulation of this process indicated, however, that the growth of shells would be far slower than that of graphite-like planes. In flame-like conditions, it predicted clusters of no more than ten carbon atoms whether planar or otherwise. It followed that this molecular growth mechanism alone could not explain the very rapid formation of soot particles in flames. Frenklach argued that a process of physical coagulation takes over from chemical accretion and that this keeps the engine of growth stoked. Such coagulation is a key part of traditional models of soot formation. Hence, he concluded, "... there is no compelling need to invoke spherical clusters to account for the structure of soot on a molecular, crystallite, or particle level."[12]

Kroto countered that his icospiral shell explained more aspects of soot formation than Frenklach's model based on the condensation of polycyclic aromatic hydrocarbon molecules.[13] The strong chemical bonds in the shell explained why soot particles were able to bear extremely high temperatures, for example, whereas the weak van der Waals bonding in Frenklach's coagulations of very small clusters implied that they should be comparatively volatile. The fact was that just as conventional mechanisms for soot formation can explain aspects of the fullerenes, so the fullerene growth mechanisms put forward by the Rice group, by Kroto, and by Heath can explain some aspects of the formation of soot. The problem lay in gathering evidence that would favour one model over another. However, it is fair to say that the real relationship between the fullerenes and their growth and soot particles and their growth is clearly more complex than the level of debate at this time was prepared to admit.

What are the main bones of contention?

The *speed* of particle growth is perhaps the most important issue. Here, the urge to form strong chemical bonds in a regular growing spiral shell seems to have the edge over ill-defined physical coagulation, although neither scheme nor a combination of the two offers a complete explanation of events.

Related to this is the question of *scale*. At fullerene sizes, it is to be expected that chemical forces will predominate. For larger carbon particles, these will gradually lose out to weaker, but longer range, physical effects such as van der Waals forces and surface tension. The soot community is divided over the relative importance of these two influences at the scale of typical soot particles. The advent of the fullerenes tips the balance in favour of chemistry.

Whatever the forces at work, they are such that soot particles generally grow to within a narrow range of sizes. This *uniformity* suggests that the growth mechanism is self-limiting. Something must stop the particles growing as they reach a particular size. The chemical growth of the spiral shell requires high temperatures to form the carbon-carbon bonds. As the growing particle cools, for example as it is carried upward in a flame, it gradually loses the ability to add further carbon atoms. However, there is still sufficient energy to allow carbon-hydrogen bonds to form, and these may then terminate the leading edge of the particle shell. It is less easy to envision a similarly convincing physical check on particle size in these conditions.

The most contentious matter is the need for *curvature*. We have seen how the soot community has fudged attempts to reconcile the sphericity of soot particles with the planarity of graphite in traditional models. The "folk memory" of these models undoubtedly provided a barrier to the acceptance not only of the spiral growth model but also of fullerenes themselves. However, we have also seen how electron micrographs and other data strongly suggest that graphite sheets may be bent and curved, and how this information has been inadequately recognized.

Finally, there is the fly in the ointment in the form of *hydrogen*. One reason advanced for the fullerenes' being unlikely as a major constituent of soot is the omnipresence of this element in flames. Why form a pure carbon ball when it seems so much easier to form any number of amoeba-shaped hydrocarbons? The answer may lie in the fact that at these temperatures hydrogen atoms are courted not only by the dangling bonds of the growing soot particle, but also by other hydrogen atoms in order to form hydrogen molecules. Thus, the presence of the element would not hinder the growth of a pure carbon structure.

New light was shed on these matters with the discovery that buckminsterfullerene and other fullerenes could be made in ordinary flames. The first such announcement came in 1987 when Klaus Homann and coworkers at the Technische Hochschule in Darmstadt found strong signals for the buckminsterfullerene ion and other "magic number" fullerene ions in flames of benzene and oxygen and acetylene and oxygen. Homann's group concluded that fullerene ions are formed not by the addition of carbon to curling shells but from the spontaneous wrapping-up of polycyclic precursors of the same or greater mass than the final fullerene. The result gave Kroto and Smalley hope that the fullerenes might ultimately prove easy to make and be widespread in nature, as they had maintained (it also supported the idea that fullerenes might be present in the carbon-rich atmosphere surrounding red giant stars and elsewhere in space). However, it stymied their spiral growth model. Homann held to the conventional wisdom of the soot community that the nature of large carbon particles is consistent with their formation from polycyclic precursors. He believed that the fullerenes evaporate in effect ready-made from the surface of nascent soot particles. In other words, the fullerenes are formed as a consequence of soot formation or, at best, in parallel with it, not in advance of the soot particles or as a precursor to them.[14]

So far as the soot community was concerned, Homann's result, like that of Kroto and Smalley before it, was little more than a diversion. As in Smalley's cluster beam apparatus, the fullerenes were detected only in the form of tiny quantities of ions. The moment when fullerene science finally came in from the cold so far as this group was concerned was almost four years later. In early 1991, Jack Howard at the Massachusetts Institute of Technology performed a similar experiment to Homann's but on a much larger scale. One of the senior figures in combustion science, Howard has spent a career studying flames and their contents. He and his colleagues found they could produce three grams of fullerenes per kilogram of carbon burned when benzene was burned in oxygen flames. They could even vary the flame conditions to produce relatively more C_{70} or C_{60}. Furthermore, they found that it was not the sootiest flames that produced the greatest volume of fullerenes. This indicated to Howard that the preferred mechanism for the formation of fullerenes was different from that for ordinary soot. Like Homann, Howard stood by the standard model for soot formation, ruling out any role for spiral carbon shells: "Soot formation does not involve the development of the curved or bowl-shaped structures required for fullerenes," he wrote, "nor would these curved structures permit the rapid stacking required in the reactive coagulation of the relatively flat sheet-like soot precursors."[15] But unlike Homann, he saw no problem with there being some shared features behind the separate but parallel formation of both fullerenes and soot particles, namely the assembly of small carbon radicals and subsequently of polycyclic components. Fullerenes were not causative agents of soot formation as Kroto and Smalley had been taken to imply. They were merely points of termination in some of the many processes of soot formation. Howard's work brokered something of a truce between the soot community and the cluster scientists and brought the fullerenes into the fold of carbon studies.

The larger truth about what goes on in a flame remains as mysterious as ever. The difficulty, now as in Faraday's time, is that what one observes are the products of combustion, not the reactive intermediates that are its key.[16] Science still awaits its answer. Considering the progress made in the field so far, it could be a long time in coming.

8

SEPTEMBER 1990

The long hiatus during which the yellow vial remained out of reach came to an end during the summer of 1990. At last, the chemist-stylites could contemplate stepping down from their pillars. Exactly five years after Smalley and Kroto made their discovery, it was physicists rather than chemists who showed how buckminsterfullerene could easily be made in quantity.

The physicists' announcement kicked the soccer ball molecule back into play with a vengeance. With the knowledge of how to make buckminsterfullerene, many groups now worked in parallel, racing to catalogue its vital statistics. The molecule became perhaps the fastest characterized new substance in history. Anyone with any technique was now able to play a part in the chemical discovery of the decade, perhaps of the century. All that was needed was access to one of the myriad analytical instruments that could reveal some hitherto unknown quality and the people to work it. The result was a flood of papers, many with crowds of authors, a phenomenon common in particle physics or molecular biology, but practically unheard of in chemistry. The stampede revealed the great intrinsic interest in the new form of carbon. But also, in the view of at least one respected chemist,[1] it betrayed the sense of stagnation that had pervaded chemistry in general that so many people felt able to drop what they had been doing to turn their attention to buckminsterfullerene.

The large commercial laboratories were especially well placed to join battle. They sensed the industrial potential of a novel class of molecules. They were also well equipped, and accustomed to assembling teams of scientists from diverse disciplines.

The speed with which events now unfolded, however, was not down to such commercial pressures, but because there was glory to be won. Science is a race. Scientists, though they may deny it, are in competition. Artists may rest secure in the certainty of the uniqueness of their personal art. One composer's symphony cannot pre-empt another's, but one scientist's discovery of some natural phenomenon certainly can and often does pre-empt another scientist's making the same discovery. In the race of science, second is nowhere. We are about to be spectators at such a race.

Let the words of that portraitist of an earlier form of carbon, J.D. Bernal, sound the off:

> Science is one of the most absorbing and satisfying pastimes, and as such it appeals in different ways to different types of personality. To some it is a game against the unknown where one wins and no one loses, to others, more humanly minded, it is a race between different investigators as to who should first wrest the prize from nature. It has all the qualities which make millions of people addicts of the crossword puzzle or the detective story, the only difference being that the problem has been set by nature or chance and not by man, that the answers cannot be got with certainty, and when they are found often raise far more questions than the original problem.[2]

It is not so much that there are only so many discoveries out there to be made, but more that there is a limited number which our mechanical and mental equipment have prepared us to make. The discovery of buckminsterfullerene illustrates both points perfectly. The fact that this cluster of a particular number of atoms of a particular kind should prove so alluring is no evidence of one glaring anomaly, but an open invitation to shed our ignorance of all those other atomic combinations and permutations we have not yet explored. Yet the discovery of buckminsterfullerene required not only the existence of apparatus that could both make it and detect it but also, at the right moments, the suspension of disbelief.

The negative aspect of such a frenzy is that much work is duplicated. In the months that followed the announcement of a way to make buckminsterfullerene, many scientists would record measurements of the same fundamental properties of the molecule. A certain amount of duplication is useful in confirming a result, but thereafter the effort serves no purpose (other than to add to an author's list of papers published). This may seem to represent poor use of scientific resources, but the alternative to this free market is some form of centrally directed research administration which most scientists would find irksome.

The community of science runs on an odd mix of jealousy and generosity.[3] Teams of researchers may be rivals but they are beholden to – and dependent on – the unwritten rule that scientific knowledge should be made public without delay in order that others may most economically advance understanding from the point of greatest understanding yet reached. To this end, protocol dictates that the literature describe an experiment sufficiently to enable someone else to repeat it without further ado. If this means detailing the apparatus to the level of brand names, then so be it. In reality, matters are seldom so cut and dried. It is of course unethical to withhold scientific information, but many factors conspire to compromise the information that is presented. This might be expected when new data might have commercial value, but it is a more general condition. Many journals place constraints on the length of papers, for example, and authors, understandably reluctant to cut the presentation or discussion of results, elect to economize on experimental details.

The speed of publication is also a subject of contention. Some journals offer rapid publication (taking perhaps four to eight weeks to publish rather than as many months) and are favoured for the communication of important results in fast-moving fields. Very occasionally, slow publication may be desirable. (Like Kekulé's dream, this can be a device to demonstrate in retrospect that you had a controversial result without the embarrassment of public association with it while it still seems dubious.)

All reputable journals send submitted papers to referees. A scientist called upon to referee a rival's paper gains privileged access to its contents and perhaps a useful impetus to his or her own efforts. Two dates are thus important in the life of a paper. The date of its publication, which is when most of the scientific community will see it, is one. The other, which many journals publish in order to establish priority and to compensate for delays occasioned by slow referees or for other reasons, is the "received by" date. It is never easy to verify to the satisfaction of all interested parties the date of a discovery. The "received by" date places the responsibility for establishing priority in the hands of the independent journal editor. In a fast-moving field, it is this date which is often the more significant.

All these factors played a role in the rush of research that surrounded the publication of the key paper on how to make buckminsterfullerene. However, like the seismographic trace of an earthquake, the formal manoeuvres evident from the literature provide only a hint of far greater activity. When new science is breaking as fast as it was here, the journals become almost irrelevant for those most closely involved. Drafts and preprints of papers circulate around the globe by means of overworked fax machines. Rivals read each other's papers, perhaps as referees, but also for the mutual benefit of being able to compare notes. When the literature cannot keep up, it is occasionally the case that a fortuitously timed conference is hijacked to become the forum at which the major players come together, each either with hot news to communicate or hungry for intelligence of another group's latest results. This is what happened in the September of 1990.

This chapter is laid out in a way that will, I hope, convey a sense of the excitement that drives each group, but also illustrate how ideas and news of discoveries are passed (or, occasionally, not passed) between groups. Five groups are principally involved: the physicists from

Heidelberg and Tucson who discovered how to make macroscopic quantities of buckminsterfullerene; Kroto and the chemists at Sussex; an interdisciplinary group of spectroscopists and physical chemists at IBM's research centre in San Jose; a duo of physical and organic chemists at the University of California at Los Angeles; and Smalley's cluster group in Houston. They work for the most part independently towards similar goals, but with critical moments of mutual interdependence. We track their progress in turn.

HEIDELBERG, GERMANY AND TUCSON, ARIZONA

PRELUDE

At the University of Arizona in Tucson, Donald Huffman had not learned of Smalley and Kroto's discovery by the expected route.[4] In the fall of 1985, he had taken on a doctoral student, Lowell Lamb. A physics graduate grown tired of working as a computer engineer on Wall Street, Lamb had come to the Arizona desert to reassess his career priorities. His doctoral degree programme with Huffman was to look at the way light scatters from the surfaces of small particles. The project was of astrophysical interest to Huffman and Lamb, but also had the not inconsiderable benefit that it was easy to make a more earthly case for its potential "relevance".

One habit Lamb had brought with him from Wall Street was taking the *New York Times*, and it was here that he read of the buckminsterfullerene discovery. "Here's a small particle," he had quipped to Huffman that day a few weeks after he had embarked upon his studies. The remark had set Huffman thinking. Perhaps this was what he and Wolfgang Krätschmer had been making at Heidelberg when they had recorded that two-humped ultraviolet absorption spectrum at 220 nanometres – the camel spectrum – in 1982.

Don Huffman pursued his long-running – and always unfunded – obsession with carbon only intermittently, when he got together with Krätschmer, or when he could sidetrack one or other of his students. Lamb had showed the requisite interest but was still only able to work on carbon when he was not doing other things – attending classes, teaching, doing funded research, mending equipment. It was 1988 by the time Lamb could reproduce the camel spectrum on demand. But it was no subject for a thesis. He could not graduate with nothing more to his name than one lumpy spectrum of rather dubious provenance. So for his doctoral project he abandoned the bootlegged carbon experiment and began work on a more realistic project for which he had succeeded in rounding up a consortium of semiconductor companies to fund him.

Huffman and Krätschmer next bumped into one another at a conference on interstellar dust during the summer of 1988. During the course of delivering his paper to the conference, Huffman confessed his hunch that the camel spectrum at 220 nanometres might be caused by buckminsterfullerene. The scientists in the audience were surprised, while even Krätschmer, who was already familiar with Huffmans's idea, was not wholly won over. It was the first he had heard of this. He saw the logic of Huffman's argument, based on the correspondence of the observed absorption frequency with the frequencies predicted by the theoreticians and on the apparent absence of other suitable candidates, but he remained deeply sceptical.

In Heidelberg over the same period, Krätschmer worked with his student, Bernd Wagner, independently homing in on the experimental conditions that would produce the best ultraviolet camel spectrum. To his surprise, it was now possible to produce this feature on a regular basis. But it might be just some persistent artefact.

Krätschmer had reason for his scepticism. In 1987 Smalley's group had reported that buckminsterfullerene absorbed light in the near ultraviolet region of the spectrum most strongly at 386 nanometres. Huffman and Krätschmer's carbon soot showed no sign of this absorption. Indeed, its absence was one reason why they only worked on carbon in a desultory fashion even after Huffman had dropped his bombshell. Accepting the conventional wisdom at the time that Smalley had succeeded in producing only the minutest quantities of the new carbon molecule – and that under the very special conditions made possible by his fancy apparatus – they thought it most unlikely that it could be made in a crummy old bell-jar. Nevertheless, they tried. This casual attitude would shortly prove to be Krätschmer's salvation. Unencumbered by the knowledge that the Rice group had amassed of the chemistry of the fullerenes and by that group's fixation that such molecules could only be produced in cluster beam apparatus, Krätschmer set to work in an ambivalent mood, half hoping that Huffman's crazy prediction might turn out to be right, but just as happy to believe that the spectrum was just "junk", some artefact of the experimental set-up, that could later be explained and erased.

Another reason for the general lack of urgency was that Krätschmer had to fit this work in between contract work on an infrared satellite observatory to be launched by the European Space Agency. This was a commercial project with a real deadline; understandably, it took precedence. One merit of the project, however, was that Krätschmer's institute acquired a sophisticated new infrared spectrometer.

There were by now a number of spectral predictions for buckminsterfullerene in addition to the troublesome Rice observation. Theoreticians at Los Alamos National Laboratory and the University of Arkansas had performed calculations, greatly simplified by the symmetry of the molecule, predicting the frequencies at which the molecule would absorb or emit light in the infrared region of the spectrum – bands arising from energy consumed or released in the transition between different vibrational states, their number whittled down according to the rules of quantum mechanics and symmetry to just four major features from the many conceivable modes of vibration of a molecule with sixty individual oscillating atoms.[5] Another team, from universities in Göteborg in Sweden, predicted the frequencies of ultraviolet absorptions.[6]

With the new spectrometer and Wagner's optimized soot conditions, Krätschmer and Huffman now began to see distinct peaks emerging from the undergrowth of their infrared absorption spectra. The four strongest of them lay close to the American predictions. The camel spectrum was right where the Swedish group predicted strong ultraviolet absorptions for buckminsterfullerene. They were not absolutely sure of what they had, but they were sure enough to use the information as a basis of a paper that would provide a good excuse to go to a conference – a pleasant summer jamboree on the island of Capri. The proceedings of the 1989 conference, on "Dusty objects in the universe", would not be published for some while, leaving time in which to get better results that could be published elsewhere. As for now, it didn't really matter. It was only astrophysics, after all.

Krätschmer and Huffman announced their result under the guarded title: "Search for the UV and IR spectra of C_{60} in laboratory-produced carbon dust".[7] It was a footprint in the sand. If later work proved them right, it would be there for them to point back to and say: "We were there first." If wrong, it would soon be washed away and quietly forgotten.

Among the audience in Capri was Mike Jura, an astronomer from the University of California at Los Angeles who was an occasional collaborator of Harry Kroto's. He picked up a copy of the paper and scribbled across the top of it: "Do you believe this?" and sent it off to him.

Krätschmer and Huffman were excited by the appearance right on cue of four bands in their infrared absorption spectrum. But it could still be a misleading coincidence. Because of the proximity of one molecule to another, spectra of solid compounds are more complex than those in the ideal gas state favoured by theoreticians. The oil used in the vacuum pump could easily contaminate the soot and hence the spectrum. There were many potential complications. They needed to do an experiment that would prove that the bands belonged to buckminsterfullerene.

They followed a standard practice of spectroscopists when they want to confirm the identity of a spectral band. If natural carbon, largely made up of the isotope carbon-12, produces a spectrum with bands at certain frequencies, the same compound made of carbon-13 in place of carbon-12 will produce a spectrum with bands at certain other frequencies, shifted by a predictable amount from the initial bands. The shift is easily calculated for vibrational spectra. A heavy mass oscillates on the end of a spring with a period proportional to the square root of its mass – the heavier it is, the slower it oscillates. It is the same with atoms in a molecule. Carbon-13 atoms vibrate about four per cent slower than carbon-12 atoms and give rise to spectral bands four per cent lower in frequency.

Krätschmer's difficulty was that he only had access to carbon-13 powder, not solid graphite enriched with the heavier isotope. With some difficulty, his graduate student, Konstantinos Fostiropoulos, succeeded in making up some rods from the powder and they redid the experiment. The resulting spectrum that they then recorded was shifted by the full four per cent indicating that they did indeed have a molecule entirely composed of carbon atoms. At the same time, the chemists at Sussex communicated their independent confirmation of the infrared spectrum obtained from soot made with their own carbon vaporizer. Krätschmer was able to take heart from this knowledge. There was no more argument about "junk". (Admittedly, the absence of a peak at 386 nanometres was still giving them grief, but they took and retook spectra, until they believed more strongly in their own result than in Smalley's.) This result they could submit for publication in a hard science journal, and with an unequivocal subtitle: "evidence for the presence of the C_{60} molecule".[8] They even included in the paper a nonchalant hint of things to come: "... it may be possible to extract bulk quantities of this molecule from smoke samples."

Rick Smalley received the paper for review on 1 May 1990 in his capacity as an advisory editor of *Chemical Physics Letters*. He accepted it immediately, called and congratulated the authors, but did not act on its implications. Others did, but not until the paper appeared in print two months later.

MAY–JUNE 1990

The famous university city of Heidelberg is reached through the industrial conurbation of Ludwigshafen and Mannheim, thick with the chimneys of chemical works. After dark, the district glitters with white gantry lights. Effluent plumes hang overhead unreal as the clouds in a painting by Henri Rousseau. On a clear night, the sight is almost inspiring. By day, it is less edifying. Heidelberg itself benefits from the creature comforts that flow from such ugliness but screens itself off behind the hills that line the River Neckar that runs through the city.

The Max Planck Institute for Nuclear Physics caps one of these hills. Two miles of road wind up through pines, then beeches and oaks, and finally orchards before disclosing a number of pavilion-like modern brick and glass buildings grouped informally amid the woodland. In his laboratory in one of these buildings, Wolfgang Krätschmer had constructed an apparatus for making carbon vapour. It consisted of a large glass bell-jar to which a vacuum pump had been attached. The bell-jar was surrounded by a wire mesh like a fencing mask to protect workers against implosion. Inside it, two clamps held two carbon rods which were connected to an electricity supply. Around the apparatus, nearby work surfaces were filthy and the floor was slippery with graphite.

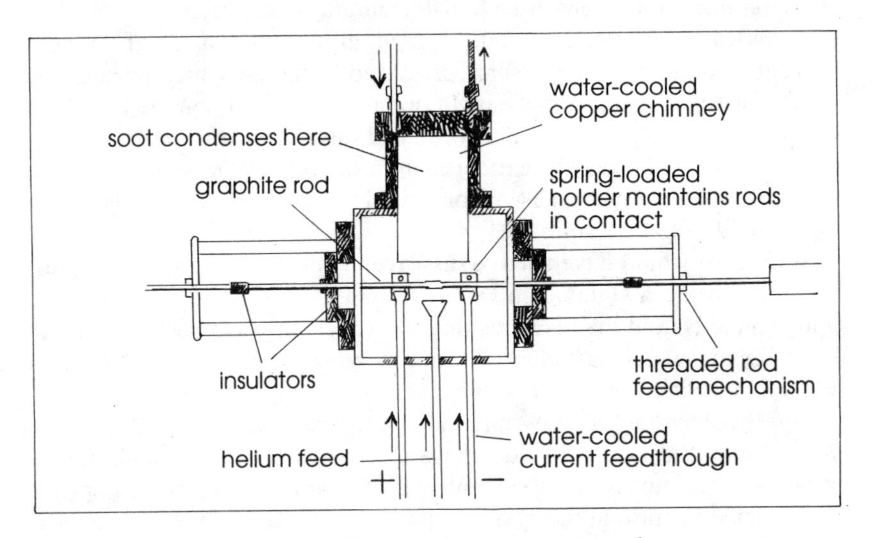

The Krätschmer–Huffman apparatus creates carbon vapour when graphite rods are heated by an electric current

Krätschmer has thinning sandy hair beneath which a twinkling eye and cherubic face give him an air of being faintly amused at everything in general. Donald Huffman has Mount Rushmore features made stronger by a desert tan and a grizzled beard. It has long been one of their amiable conceits that, despite their positions at reputable institutions, both men like to paint themselves as outsiders. They almost revel in the fact that their studies of the physics of carbon particles were unfunded and had to be squeezed in between other projects, making use of borrowed time and bootlegged equipment. This shared sense of battling against the odds fuelled their friendship and was the foundation of their intermittent collaboration.

As their paper on the infrared spectrum hinted, Krätschmer and Huffman were indeed on the brink of being able to produce macroscopic quantities of buckminsterfullerene for the first time. In their Heidelberg laboratory, Fostiropoulos and Krätschmer adjusted the pressure of the gas – helium or argon – in their bell-jar until they could produce the most soot. About one eighth of atmospheric pressure was found to be best. They varied the diameter of the carbon rods, the shape of the tips of the rods, the magnitude of the current passing through them, and the closeness of their contact, either producing a bright arc or – which they thought better although it was harder to set up – merely making the current leap the gap in a way that produced heat but no light, vaporizing the carbon to fill the jar with a bluish cigarette smoke haze.

The smoke particles eventually settled, all over the inside of the bell-jar, but also on transparent surfaces placed in the jar which became the plates for spectroscopic analysis. In order to try and obtain the buckminsterfullerene that the spectroscopy had hinted was present, it was necessary to find a way to extract it from the rest of the soot. If indeed, that was where it lay, because no one at this time knew whether the molecule would even be a solid.

A chemist would have tried to dissolve the buckminsterfullerene preferentially using a suitable solvent, drawing the molecules into solution and leaving behind the black residue. However, Krätschmer's knowledge of chemistry was insufficient for him to know what solvent to choose, even though there was one solvent that cried out to be tried, one that would be obvious to any undergraduate chemist. Being physicists, Krätschmer and Fostiropoulos chose a physical method – sublimation. Whereas the addition of a solvent is quite natural to a chemist, it is a messy complication in the eyes of a physicist. If you are trying to extract one substance from another, why slop in a third? Sublimation was cleaner, a physicist's process. They placed a sample of soot in a tube and heated it in a stream of argon. Sure enough, a little way downstream, in a cooler part of the tube, a new film was deposited. Upon analysis, it showed the four infrared peaks more clearly than ever. Buckminsterfullerene molecules behaved very differently from the rest of the soot and could be separated with almost incredible ease.

Later, they obtained the solvent they needed – benzene – and found that this too worked. Some of the soot dissolved to form a burgundy solution which could be drained off and evaporated to leave crystals of fairly pure buckminsterfullerene, crystals not like those of diamond, with strongly bonded atoms in a lattice, but with whole individual molecules lining up to form a regular array in three dimensions and held in place by weak van der Waals bonds. It was the first time that anybody had actually seen the new, third form of carbon. Krätschmer quickly called Huffman and told him of his success.

Within an hour, Huffman and Lamb repeated the Heidelberg procedure and had their own crystals. It was now very clear to both pairs of scientists that they were sitting on the major discovery of their careers. They had the means to produce visible, usable quantities of the substance that every chemist most wanted to lay hands on. And they could do it every day.

With the exhilaration of the find came the realization that much work lay ahead, refining the experimental procedure, improving the purity of the crystals, and characterizing them so as to establish beyond anybody's doubt that they were indeed the new form of carbon. As luck would have it, Huffman was committed to leave Tucson for a brief tour of duty at an observatory in Paris. What had looked a most appealing trip – Paris in the spring – suddenly became an irritating chore. At short notice, Huffman could not escape his obligation, so he called once more upon Lowell Lamb, who was by now within a few months of finishing his own research project, to carry on the necessary work of characterizing the new solid. Little did Lamb know then, as he set about checking off the items on the list that Huffman had left him, that he would not be able to return to write up his thesis for a full year. Huffman, meanwhile, made the most of his European trip, stopping by in Heidelberg to sit down with Krätschmer and start drafting the most important paper they would ever write.

JULY–AUGUST 1990

The report of the spectral evidence for buckminsterfullerene appeared in *Chemical Physics Letters* at the beginning of July, by which time the proceedings of the "Dusty objects" conference had still not been published. So it was this paper that was most people's reintroduction to Kroto and Smalley's molecular curiosity of 1985. Both reports described the method of making the carbon vapour in brief, almost casual terms. Anyone who read them with a professional eye might think that the spectra were almost incidental. They seemed to be saying that it was *easy* to make buckminsterfullerene.

This paper was a signal to many to initiate their own attempts to make buckminsterfullerene and explore its properties. To those already on their way, not only at Sussex, where they had taken heed of Krätschmer and Huffman's progress from their earlier, more tentative paper, the one annotated and forwarded by Mike Jura from the Capri conference, but also at IBM's Almaden Research Center in San Jose, California, where scientists had initiated an independent programme to make fullerenes, it served to stimulate redoubled effort. Having in effect given away their secret in these early announcements, Krätschmer and Huffman knew these two groups were quite possibly hot on their heels. They had reason to believe that Smalley might be too.

Despite this pressure, the paper that Huffman and Krätschmer had sat down to write in May was not ready for final submission until August. Why? It was not enough for Krätschmer and Huffman to say that they had crystals of buckminsterfullerene. They would have to prove that the crystals were what they said they were. First of all, however, they needed to perfect the method of preparation. The first crystals were impure. Micrographs showed blobs of liquid lying between them. Spectra showed a telltale carbon-hydrogen bond vibration that confirmed that there was contamination by hydrocarbons. Once the impure compound had been washed in ether and then resublimed, it recrystallized in elegant hexagonal plates, easy to identify under a microscope and an ideal shape for analytical purposes.

Krätschmer and Huffman and their students made the basic observations. The crystals are reddish-brown, not the buttery yellow near liquid that Smalley had predicted (although thin films of the material are yellowish). Because of the empty space incorporated in each spherical molecule, it is considerably less dense than either diamond or graphite. Like graphite, it is fairly compressible, as external forces push the balls up against one another.

Where Kroto and Smalley claim discovery of the buckminsterfullerene molecule, Krätschmer and Huffman could claim to have discovered its form as a molecular crystal. They decided to submit their findings to *Nature* where Kroto and Smalley had published.[9] The paper was refereed first by Harry Kroto and then by Bob Curl at Rice.

Carbon comes in many forms – most of them somewhat indeterminate, like glassy carbon and amorphous carbon, but hitherto only two true crystal forms. In order to consolidate their claim to the third crystal form, Huffman suggested the name "fullerite" for the bulk solid form of buckminsterfullerene.

Having given the basic physical description of "fullerite", the Heidelberg–Tucson team faced a bewildering choice. They were in possession of the only macroscopic samples of a unique material. Every conceivable experiment was new. Huffman's instinct was to call round to colleagues in different fields who had access to various analytical techniques and invite them to join him in characterizing the new substance. Some failed to rise to the bait; others responded with alacrity, and it was not long before they had more descriptions of solid C_{60} than they knew what to do with.

They recorded ultraviolet and infrared absorption spectra. Where the fullerene soot had shown the camel humps in the ultraviolet and four little peaks poking up shyly in the infrared, the pure material produced bold, clean bands. What else? A mass spectrum, perhaps. In their original submission to *Nature*, the authors had omitted the mass spectrum. *Nature* places strict limits on the number of figures that may be published with a paper. Krätschmer and Huffman used their allocation to display new kinds of identifying data, mentioning only in the text that they had recorded a mass spectrum that corresponded with those seen before for buckminsterfullerene. Refereeing the paper, Curl requested that one be added, however. This would permit everyone to compare the bulk substance with former buckminsterfullerene data by the only measure available to them both.

Krätschmer had disappeared for one of those European vacations whose length causes American jaws to drop. So it was left to Huffman to wrestle with the referees' demands. They already had a mass spectrum that they had recorded in Heidelberg, but it was not the best. In Tucson, Huffman sought out a suitable higher-resolution mass spectrometer in another department. He showed the technician the Heidelberg spectrum and asked if he could better it. The job was duly booked in and bootlegged to be billed under some grant or other. The technician recorded the spectrum but was so pleased with what he saw that he announced he would only release it for publication if Huffman put his name on the paper. Huffman balked at this. The Heidelberg spectrum was the one that went into print.

Spectra were all very well. The infrared and ultraviolet spectra established the link with Krätschmer and Huffman's previous papers. The mass spectrum showed that they were dealing with the same material as Rice University and Exxon. But they wanted something more persuasive, pictorial evidence of these most symmetrical of molecules stacked in crystalline array. The traditional proof of a crystalline structure is the pattern created by its diffraction of X-rays or electrons. The Braggs and J.D. Bernal had resolved the crystal structure of the diamond and graphite forms of carbon in the pioneering days of this technique. The third crystalline form should produce a pattern no less distinctive.

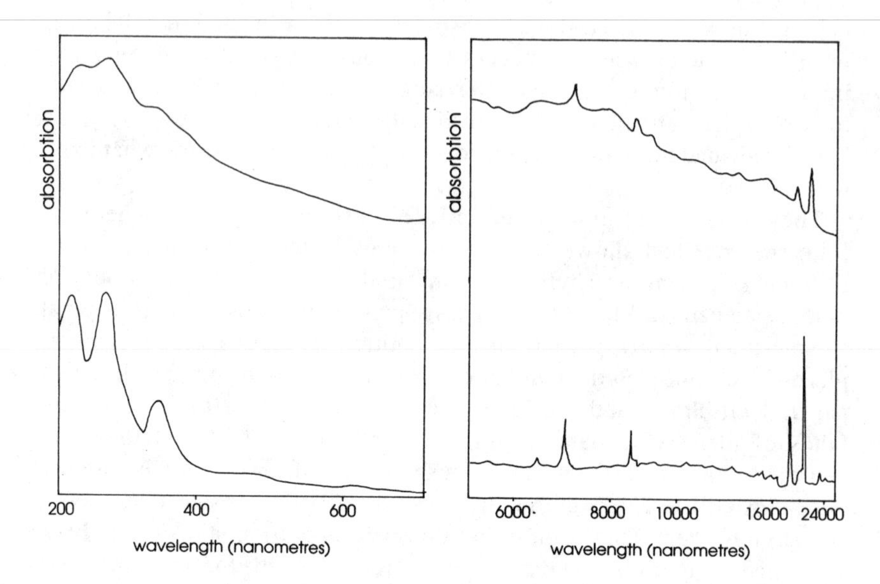

Early infrared and ultraviolet (camel) soot spectra (top). Later spectra (bottom) agreed with theory for C_{60}

Once again, Huffman and Lamb strolled across the campus, this time to the geology department, handed over their sample, paid their money, and waited while the technician recorded the diffraction pattern. As the technician commented, it was indeed quite unlike any seen before. As a crystal of identical repeating spherical units, it might be expected that "fullerite" would exhibit a rather simple diffraction pattern like that of a pure metal. Gold, for example, crystallizes in a face-centred cubic array of atoms, one of the two most tightly packed arrangements for spheres in space. The diffraction pattern for buckminsterfullerene might be just like that for gold but rescaled in proportion to the size of the carbon molecule. The obvious novelty in the pattern that was recorded was indeed the unusual spacing between diffraction signals generated by the larger crystal units. But some reflections seemed to be missing.

It took the best part of a month to make sense of the anomaly. Huffman calculated that it might be explained by a stacking disorder, a change of the layering sequence at intervals through the crystal, but they remained uneasy about it.[10] Fortunately, it did not require an unequivocal resolution of these questions in order to calculate the spacing between the fullerene molecules. This turned out to be just over ten ångströms, confirming the fullerene diameter as about seven ångströms which was what had been calculated from a knowledge of typical carbon bond lengths and the geometry of the truncated icosahedron structure of the molecule. This left just over three ångströms to account for the spacing between the spheres, nicely close to the spacing due to the van der Waals force between flat layers of graphite.

Buckminsterfullerene balls stack to form a molecular crystal which Krätschmer and Huffman named "fullerite"

The X-ray and electron diffraction patterns were pictures of a kind, but still not direct images of the stacked fullerene footballs. These solid-state physicists still wanted to see something solid. Huffman contacted colleagues at the University of Missouri who had a scanning tunnelling microscope. This advanced version of an electron microscope records a sort of topographic map of atomic surfaces by jiggling a scanning probe up and down across the surface while maintaining a constant current between the surface and the probe tip. No one had been able to record an image of an individual fullerene molecule. With the crystal array holding the molecules in place, Huffman and his colleagues hoped that they might be able to do this and even discern the hexagons and pentagons on its surface. What they saw in the end was no more than its general roundness. This was a psychological boost, but it had little scientific merit. The X-ray data gave more accurate crystal stacking dimensions. Despite being the first direct view of "fullerite", the scanning tunnelling microscope pictures fell foul of *Nature*'s guillotine.

This was all icing on the cake. The real content of this paper was the recipe for making fullerenes. After five years of frustration, the answer lay not in the synthesis of an organic chemist, nor in the high technology of lasers and molecular beams. The chemical aesthetes could argue whether it was more elegant to extract this most beautiful molecule in one step from the black chaos of a sooty bell-jar than to make it by tedious assembly. For everybody else, the way forward was now clear.

Or almost so. For Krätschmer and his colleagues had merely written that pure graphitic carbon soot "is produced by evaporating graphite electrodes in an atmosphere of ~100 torr of helium". This was the extent of their experimental description of the soot production. They offered no details of the current passed through the electrodes or of their exact shape and orientation. Rick Smalley was frustrated by the lack of detail. Others were not. Within weeks, half a dozen laboratories had repeated the experiment. Indeed, it was a source of considerable satisfaction to Huffman and Krätschmer that people generally found it easy to reproduce their results. They felt from their own experiments that the details they appeared to have omitted were not critical. Other experimenters used different combinations of conditions and still produced C_{60}. The most elusive of molecules was suddenly available to any who wished to make it.

In order to appreciate the significance of this breakthrough, it is important to recall the widespread scepticism that had grown up as year after year had gone by since 1985 with no way of making macroscopic quantities of buckminsterfullerene. The introduction to one of the first monographs on fullerenes stated the general position:[11]

> ... few were inclined to "count" the species as carbon allotropes. The new species, like small molecules such as C_2 and C_3, were thought of as transients in dilute vapor phases, hardly what one classifies as "materials" based on elemental carbon.

The Krätschmer–Huffman method changed this. However, the ease with which buckminsterfullerene could now be made was not entirely an unmixed blessing. There were ominous parallels for the new carbon chemistry during these years. In 1969, the world of physical chemistry had been excited by reports from Soviet scientists of an "anomalous", supposedly polymerized, form of water that came to be known as polywater. Like buckminsterfullerene, polywater represented an unfamiliar form of a very familiar substance. Polywater too was said to exhibit physical properties very different from ordinary water, and it too had only been obtained in minute quantities. Polywater was eventually exposed as a myth, but not before a few scientists claimed to have confirmed the anomalies. One of them was Brian Pethica of the Unilever laboratories in Port Sunlight who nevertheless issued this caveat in an article in *Nature*:

> The existence of "anomalous" water as a stable and distinct molecular entity will only be established when it becomes available in amounts sufficient for an unequivocal characterization.[12]

Scientific American put it more bluntly: "If polywater is so stable, why has it never been found in nature?"[13] Before the summer of 1990, this was equally the paradox of buckminsterfullerene.

There is a still more notorious parallel that demands attention. Another episode of pseudoscience unfolded precisely contemporaneously with Krätschmer and Huffman's work. Consciously or unconsciously, it conditioned their actions and the actions of many scientists at this time.[14] The demon was the alleged phenomenon of cold fusion. The story broke in March 1989 when Stanley Pons of the University of Utah and Martin Fleischmann of the University of Southampton held a press conference in Salt Lake City to announce their "discovery" that they could induce nuclear fusion to take place at room temperature. The fusion of atomic nuclei can be brought about only in the most extreme conditions such as in the sun or, artificially, in a hydrogen bomb. Control of nuclear fusion promises an answer to many of the energy needs of humankind. Fleischmann and Pons claimed to have produced fusion not only at room temperature but in a tabletop apparatus of preposterous simplicity built around an electrolytic cell containing nothing more esoteric than heavy water.

Krätschmer and Huffman would have been preparing their contribution to the "Dusty objects" conference at the time of the announcement. Controversy raged through the summer of 1989 as scientists in laboratories around the world tried and failed to reproduce Fleischmann and Pons's results. The charged atmosphere that the debate produced in the physical sciences community was perhaps one factor that led Mike Jura to write his "Do you believe this?" on the paper that he sent to Kroto at this time. By June 1990, cold fusion had collapsed in scientific and financial scandal leaving a bitter aftertaste for many chemists and physicists and besmirching the reputation of pure science in the eyes of the public.[15]

Acutely conscious that their simple bell-jar fullerene factory could look quite as naive as the cold fusion cell, Krätschmer and Huffman determined to do everything by the book. Their experiment was quite definitely reproducible. Krätschmer had done it in Heidelberg; Huffman had done it in Tucson. What's more, it could be done under broadly similar rather than strictly identical conditions. Nevertheless, in the poisoned atmosphere of the cold fusion episode, they had good reason to support their results with as much evidence as possible. Certainly, they made sure they upheld scientific protocol by waiting until their paper appeared in *Nature* before the discovery was announced to the press.

The relative ease of fabrication of buckminsterfullerene "fullerite" and its appeal to the media and the general public have obvious parallels with recent episodes of pseudoscience, not only with polywater and cold fusion, but also with the notorious "high dilution" report in 1988 by Jacques Benveniste and colleagues of the INSERM unit for immunopharmacology and allergy at the Université Paris-Sud to the effect that an aqueous solution diluted to an such an extreme extent that it became to all intents and purposes indistinguishable from pure water could still somehow "remember" the properties of the more concentrated solution. The "discovery", made in the conventional way within the priesthood of hard science, was seized upon by the partisans of homeopathic medicine as grist to their mill. In all cases, people were presented with a sensational story of an everyday substance reconstituted in an unusual form and exhibiting highly unusual properties. Buckminsterfullerene appears to fall into the same category. The only difference was that the ubiquitous ingredient was now carbon rather than water.

It is instructive to explore a little further the superficial similarities and differences between these phenomena. John Huizenga, the co-chairman of the Energy Research Advisory Board Cold Fusion Panel set up by the US Department of Energy to investigate the fiasco, made a number of comments in his detailed report which bear on moments in the history of the emergence of the fullerenes.

Huizenga believed that the interdisciplinary nature of the work leading to the reporting of both polywater and cold fusion added to the controversy associated with them. Buckminsterfullerene is also the outcome of the efforts of diverse disciplines. Though controversial for a time, as we have seen, it became rapidly less so (though certainly no less fascinating) following the announcement of Krätschmer and Huffman's results.

Huizenga went on to highlight the irregularity of the cold fusion experiments by describing the circumstances that typically prevail at the brink of a major discovery.

> One of the characteristics of real discoveries is that they are often made by a number of independent groups fairly near in time. Similar lines of forefront research are usually being conducted by more than one group, such that confirmation of a discovery is soon established by other related groups.[16]

Kroto and Smalley had the Exxon and AT&T Bell Laboratories' mass spectra of carbon clusters as their independent near parallels. Confirmation of Krätschmer and Huffman's experiment was both rapid and multiple. In this respect, the fullerene history is unlike that of cold fusion or polywater but not unlike the discovery of, say, high-temperature superconductivity.

On the other hand, the discovery of buckminsterfullerene and of "fullerite" are to some extent discoveries by "outsiders", something that Huizenga regarded as suspicious in the case of cold fusion. He contended that it is those intimately involved in a discipline who usually make its discoveries and that the cold fusion episode illustrated how science may be set back when outsiders apply the models of other disciplines to a new field.[17] It may be that those who are intimately involved *usually* make the discoveries, but it was quite clearly not the case here where the chemist-stylites of the discipline of combustion science have had little to say. One may conclude rather that the fullerene story illustrates precisely the reverse, that significant advances may be made by just such interlopers.

While these comparisons are enlightening, it is important not to forget for a moment the most important difference between these phenomena, which is that while polywater, cold fusion, and the high dilution effect are discredited, fullerenes are very much a genuine phenomenon.

There were also more personal reasons why Krätschmer and Huffman were able to feel pleased with themselves. Both men, as I have said, paint themselves as renegades, poorly funded, working in fringe institutions (an image that is readily contradicted by the facts, but no matter). They were David to Smalley's Goliath. Where Smalley had struggled to find a way to make buckminsterfullerene and the other fullerenes with the supposed advantage of sophisticated apparatus, they succeeded with a low-tech lash-up that it would not be beyond a well equipped high school to emulate. They celebrated this fact in their public lectures. This was not merely triumphalism. They believed there was a moral in the story that discoveries are not made in proportion to the money thrown at them. This is a matter of some bitterness to many workaday physicists who see funding for their subjects bled off by the particle physicists ever hungry to construct more powerful and expensive accelerators.

There was also a sneaking satisfaction for Krätschmer and Huffman in being physicists and knocking on the head this chemists' conundrum. There was perhaps even an element of settling a score. In April 1989, W.H. Breckenridge, Professor of Chemistry at the University of Utah, stated in a letter to the *Wall Street Journal* that the adversarial relationship between chemists and physicists over cold fusion stemmed from "the fact that two chemists have made an unusual discovery with profound implications to an area of physics in which most of the active physicists never talk to chemists."[18] Now, two physicists had made an unusual discovery with profound implications to an area of chemistry. And a real one at that.

SUSSEX, ENGLAND

MAY–JUNE 1990

It was the summer of 1989 when Jonathan Hare, just completing an undergraduate degree in physics, wrote to the astronomy department at the University of Sussex enquiring into the possibility of doing a doctorate that combined astronomy with experimentation. For obvious reasons, experimental astronomy is a discipline of rather limited opportunities, and Hare was referred to Harry Kroto. Kroto did most of the talking in the interview, telling Hare all about buckminsterfullerene. (As a physicist, he had never heard of it.) Hare was won over by Kroto's enthusiasm and the obvious fascination of the subject and Kroto agreed to take him on. Hare began at Sussex in October, ostensibly to study the way acetylenes form soot, using the chemistry school's newly commissioned cluster beam apparatus.[19]

But Hare never got to work on this. By the time he arrived, Kroto had received Krätschmer's paper from the "Dusty objects" conference with Mike Jura's comment on its doubtful buckminsterfullerene spectra. After several years of struggling[20] to produce fullerenes by various means following the deterioration of the collaboration with Smalley, here was the impetus Kroto needed to give it one more go.

They salvaged the clapped-out carbon arc apparatus that Kroto's former graduate student, Ken McKay, had used without success a couple of years earlier. With Amit Sarkar, a final-year undergraduate chemical physicist, Hare laboured to reproduce Krätschmer's result. After a few days tinkering with the apparatus, they had an infrared spectrum but could not reproduce it with any consistency. Others in the chemistry school doubted the validity of the "Dusty objects" paper. Hare hoped it was true but, as an experimentalist, was more absorbed with the physical manipulation of the apparatus. After a week of varying the parameters for soot production, the tired old generator burned out. It took another week to rebuild it.[21] A couple of months later, the results coming off the machine were good enough for Sarkar to retire from the scene with a couple of spectra for his undergraduate project.

Hare continued to adjust the many variables, spending much time setting up each run, filing down and sharpening the carbon rods after each run for re-use. He wanted to be able to present his results at an astrophysical conference coming up at the end of March. Finally, with just one month to go, he was compelled to write to Krätschmer for help, with the excuse that he could ask for the correct citation data for Krätschmer's "Dusty objects" paper (which was still unpublished) to include in his own conference presentation. He told him of his limited success in confirming Krätschmer's infrared spectrum and asked for further hints as to the conditions under which he had carried out his experiments. In doing so, he set Krätschmer a problem for they were now in competition. Krätschmer wrote back in his ingenuous way with encouragement and advice, putting the pressure back on Sussex by revealing just how easy it was to make C_{60}, so easy, in fact, that other laboratories were almost bound to be in the race as well:

> The "C60" production is no black magic; [it] is in fact very easy. After a few trials you will find out for yourself what is important. Here is how we do it:[22]

Krätschmer went on to describe his experimental conditions in more detail than he had given at the conference, and indicated that the soot from anywhere inside the bell-jar was good, not just that deposited on a special collection surface near the electrodes. He signed off by wishing Hare luck. In the meantime, Hare himself had discovered that he could get better results by collecting the soot deposited in the cooler regions further from the carbon arc. Hare spent the next few months firing up the carbon vaporizer at every available opportunity in between seeing to other tasks.[23] Over and over, he set up the graphite electrodes, switched on the current, and watched the smoke fill the chamber before settling. Then he carefully scraped out the soot and added it to his growing stock. By the end of May there was at last enough soot to contemplate some spectroscopy.

JULY–AUGUST 1990

Attempts to record new data – good infrared spectra and nuclear magnetic resonance spectra – were dogged by equipment failure. Kroto's group fell back on what they knew they could do. One day in mid July, Hare handed a sample of the soot over to a postdoctoral fellow, Ala'a Abdul-Sada, for mass spectrometry and went off for a week's vacation. Upon Hare's return, Abdul-Sada announced that the spectrum had peaks in all the right places but that it could not be repeated. This instrument too had let them down. Hare came back also to find Krätschmer's superior infrared and ultraviolet spectra in *Chemical Physics Letters*, and was thrilled to see himself cited for the first time for the infrared data that he had put in with his letter to Krätschmer.

The Sussex group had no reason to think that the Heidelberg scientists were on the verge of being able to mass-produce fullerenes, however. Had they done so, they might have redoubled their efforts. As it was, Hare thought it might be worth trying to extract some C_{60} from the soot, and he was enough of a chemist that he had no trouble in selecting the appropriate solvent.

Hare put half the soot he had made into a vial of benzene and left it over the weekend. When he arrived at the laboratory on Monday morning, he was stunned to see that the benzene had turned red. He filtered the solution and concentrated it by evaporation. It was now a deeper red, and he hawked it around the laboratory, shaking the vial in people's faces and boasting ironically that he had C_{60}. But scepticism remained strong. Red was the worst colour for the solution. It was quite likely a colloidal suspension of very fine soot particles rather than a genuine solution. (Colloids are frequently red because the suspended particles scatter red light less efficiently than other colours. Like the sunset, this is a manifestation of Mie scattering.) Kroto was still of a mind that fullerenes only account for one part in a million or so of the soot. Nothing he had heard from Rick Smalley or Eric Rohlfing or others led him to believe otherwise. It would be almost impossible to assay this concentration in a solution or to extract it from that solution. Hare was in any case unable to check the content of the solution because of the faulty spectrometer. At the end of the week, the fax of Krätschmer's paper came through from *Nature* for Kroto to referee. Kroto quickly read the paper and faxed back that it should be accepted, nominating Bob Curl as the second referee. The game was up.

In his personal account of this research Jonathan Hare wrote:

> The following days were not what one would call easy, one can imagine what it must be like to spend years working on an idea and then be proved correct by someone ELSE.
>
> Personally, my feelings were mixed, on the one hand all this work I had been doing was not in vain; C60 really did exist in the soots, which was rather satisfying. On the other hand it looks like we had been so close! My view is that we could have proved the structure of C60 at Sussex but I would have done it cautiously in my own time, probably in the remaining two years of my degree. However this was not to be the case because the fullerene story had entered a new phase.[24]

The Sussex group recovered quickly from their despondency at seeing Krätschmer and Huffman's submission to *Nature*. The physicists had won the race to make solid C_{60}, but they had not done it the way a chemist would have done it. Roger Taylor thought he could extract the pure buckminsterfullerene more efficiently from the soot Hare had been collecting. Like many organic chemists, he was disappointed, even offended, that this paragon of their craft, had been manufactured by such an incredibly crude process. "Unsporting" was how Roald Hoffmann had described it.[25] Taylor took a chemist's umbrage that Krätschmer and Huffman had separated their product by sublimation rather than by solvent extraction and chromatography.

Kroto surrendered a portion of his meagre supply of soot. Taylor begged a little more and then more still from him. When he felt he had enough, he shook it up in some chloroform to see if it would dissolve and then put it into a Soxhlet extractor, an apparatus which draws out the soluble portion of a substance by continuously cycling boiling solvent through it. Soon a rich red liquid began to run from the apparatus. Taylor collected it, evaporated it, and dissolved the residue in another solvent, hexane, more suited to the next stage of the process – column chromatography. As the new solution filtered through the white alumina column that would separate out the ingredients of the solution, a pink band appeared. Taylor was perplexed. He did not expect buckminsterfullerene to be this colour. The conventional wisdom still favoured Smalley's yellow vial. Surely the band should be yellow. Conscious all the time that the Rice group might be hot on their heels because Curl had received the same tip-off, Taylor hurried on with the chromatography until a number of differently coloured bands grew distinct. There was now a red band. Was that C_{60}? And a weaker magenta band. Could that be the first real hint of a second fullerene, C_{70}? Or perhaps it was the other way around.

SAN JOSE, CALIFORNIA

MAY–JUNE 1990

The decision of IBM Corporation's Almaden Research Center to start a fullerene research programme came as the result of one of those rare fusions of basic research and creative management.[26] By the end of 1989, Donald Bethune, a physicist at the San Jose centre, had spent two years investigating aspects of surface science, a well ploughed furrow at IBM and elsewhere. Seeking to move him out from under the shadow of a good co-worker, his line manager proposed he take a look at something new. Bethune had attended a Fermi summer school in Italy on clusters in

1988 and announced his wish to turn his attention to clusters. He had heard Smalley speak about C_{60} at a Gordon Conference (a series of residential conferences held every summer in schools across the state of New Hampshire at which scientists can discuss their latest work in mutual confidence), and had gathered together some key references on the subject. It was sufficient to launch him in a new direction.

Bethune's proposal dwelt on the paradox that buckminsterfullerene, supposedly so stable, had so far proven so hard to make. There were reams of literature *predicting* this or that property, but only a few results of real experiments, notably the series of photofragmentations done by the Rice group and their spectroscopic results, the ultraviolet photoelectron spectrum and the ultraviolet absorption at 386 nanometres. No one had made enough of the stuff to establish any more. At first, Bethune suggested constructing an apparatus that would be a sophisticated version of Smalley's cluster beam machine. It would be a slow and expensive way to get off the ground, but a reliable, evolutionary approach building on the known facts of fullerene fabrication. But when Bethune's colleague Heinrich Hunziker looked the proposal over, he noted that even the simple property of the ionization potential, the energy required to expel the first electron from the molecule, was not known. That value, he thought, could be found quite simply by using an existing laser desorption apparatus, a piece of equipment quite similar to Smalley's apparatus, except that where Smalley used a high-energy laser to blast part of a surface to vapour, this used a low-energy laser pulse gently to loosen a cluster or molecule that had settled on a surface and lift it off without damaging either the surface or the cluster. In collaboration with scientists at the nearby NASA Ames Research Center, the IBM physicists had used the apparatus to look at carbon-rich meteorites, "desorbing" from their surfaces polycyclic aromatic hydrocarbon molecules larger than any detected there before.

Bethune, Hunziker and their colleagues, Mattanjah de Vries and Gerard Meijer, reasoned that if the fullerenes were as stable as they were cracked up to be, they should sit happily on a suitable surface once deposited there. Then, using the lighter touch of the desorption laser, they could be lifted off at leisure for more controlled analysis. The difference should be rather like a geologist studying mineral samples by gently taking them from a specimen cabinet rather than trying to examine them as they hurtle through the air after a dynamite blast in a quarry. With this goal in mind, they set about adapting the apparatus.

By May 1990 they were ready. Their first problem was how to form a carbon deposit that might contain the fullerenes for the laser to desorb. A Smalley-type cluster beam seemed the obvious way, but a spontaneous phone call to Eric Rohlfing, who had been the first named author on the 1984 Exxon report of large carbon clusters, produced a different recommendation. A combustion scientist, he advised Bethune to avoid the high-tech route and simply to deposit some ordinary soot which would be easier and, he thought, just as likely to contain fullerenes.

> We thanked him, thought "Why not?" and started to look around the lab for a way to make soot. We found an empty peanut can, filled it with alcohol, stuffed a rag through a hole cut in the plastic top as a wick, placed the can in a laboratory fume hood and lit the rag. The alcohol gave an almost invisible, smokeless, sootless flame. But when the plastic lid of the can inadvertently caught fire, thick black soot was produced. When I saw the soot I grabbed a copper sample bar and stuck it into the smoke. In seconds it was covered with black soot. We then put it into the mass spectrometer. The mass spectrum we measured went out to very large masses with signal at every mass. But the spectrum showed strong modulation with a period of 12 mass units. This meant that the species in the soot were nearly pure carbon, with small numbers of heteroatoms filling in the spectrum. It also had a monstrous peak at mass 23. Sodium! from the huge dose of salt provided with the nuts....[27]

Bethune and his colleagues quit fooling around and began to think a little more seriously about the problem in hand. A cleaner approach was called for. Turning back to laser technology, they set up a simple apparatus in which a laser would vaporize carbon from a graphite target in an inert argon atmosphere. The resulting soot would collect on a sample bar like the one that Bethune had thrust into the plastic lid smoke, from which they could desorb any buckminsterfullerene molecules for analysis. On its maiden run on 18 June, the mass spectrum of the film of soot deposited on the bar showed peaks for sixty and seventy carbon atoms and for every larger even number out to two hundred atoms and more. And the films were stable; they could be carried across the laboratory in the open air. Here was C_{60}, still in tiny quantities, in a film so thin that it barely clouded a transparent surface, but many orders of magnitude more than had been captured before. Could it really be this simple to make fullerenes?

With this preliminary result, Bethune had a reason to attend a conference on clusters coming up in the fall that he had had his eye on. He duly sent off an abstract of the paper he might give. There was ample time between now and then to sort out the minor details that would allow him to give an astonishing presentation with a polished set of results. Meanwhile, Bethune gave an internal presentation. He entitled his talk: "World Cup madness or a new approach to the study of carbon clusters?" In honour of his colleague, the Dutchman and soccer fan

Gerard Meijer, Bethune wore the Dutch soccer strip, clogs and a wig complete with dreadlocks in echo of the star of the Dutch team, Ruud Gullit. He brought his talk to a triumphant conclusion with one more prop – a soccer ball.

JULY–AUGUST 1990

As they began to write up their paper, Bethune and Meijer suddenly realized that they had not in fact proved that they had buckminsterfullerene molecules on their deposition surface at all. It was possible that they were only being created at the last moment by the desorption laser from some precursor species deposited on the bar. Before they could submit their results, they had to disprove this possibility.

They devised an elegant experiment using samples of two isotopes, carbon-12 and carbon-13. They recorded a mass spectrum for the desorption of clusters made from the carbon-13 isotope for comparison with the ordinary carbon-12 spectrum which they already had. This had a large peak for the carbon-13 buckminsterfullerene at 780 atomic mass units rather than at the usual 720 units. Then they vaporized carbon from a powder made up of a half-and-half mixture of carbon-12 and carbon-13. If the fullerenes were formed in the initial vapour and deposited ready-made on the sample bar the spectrum would be a statistical average of the two individual isotope spectra. The peaks at 720 and 780 atomic mass units would vanish, reflecting the low probability that buckminsterfullerene molecules purely composed of one or other of the carbon isotopes could form from a vapour of mixed isotope particles. A new large peak would appear with a broad maximum around 750 for fullerenes with some carbon-12 and some carbon-13 atoms. They did the experiment and duly found that this was the case.

This was the groundwork. Now came the key part of the experiment. They deposited on the sample bar successive layers of soot of each isotope, a carbon-13 layer on top of a carbon-12 layer and so on like coats of different coloured paint. If it was the vaporization laser that was causing the fullerenes to form before settling on the sample bar with the desorption laser merely supplying the energy to lift these molecules free of the sample bar, then each fullerene layer should be isotopically pure and the mass spectrum would have the 720 and 780 peaks. If, on the other hand, it was the desorption laser that was actually causing the clustering, then the layers of isotopes would mix in the new vapour and it would be the blended fullerenes that would form with a modal mass of 750. Again, they did the experiment. The results supported the former scenario; they had final confirmation that buckminsterfullerene formed as a stable material in the solid state and could finally submit their paper to the *Journal of Chemical Physics.*[28]

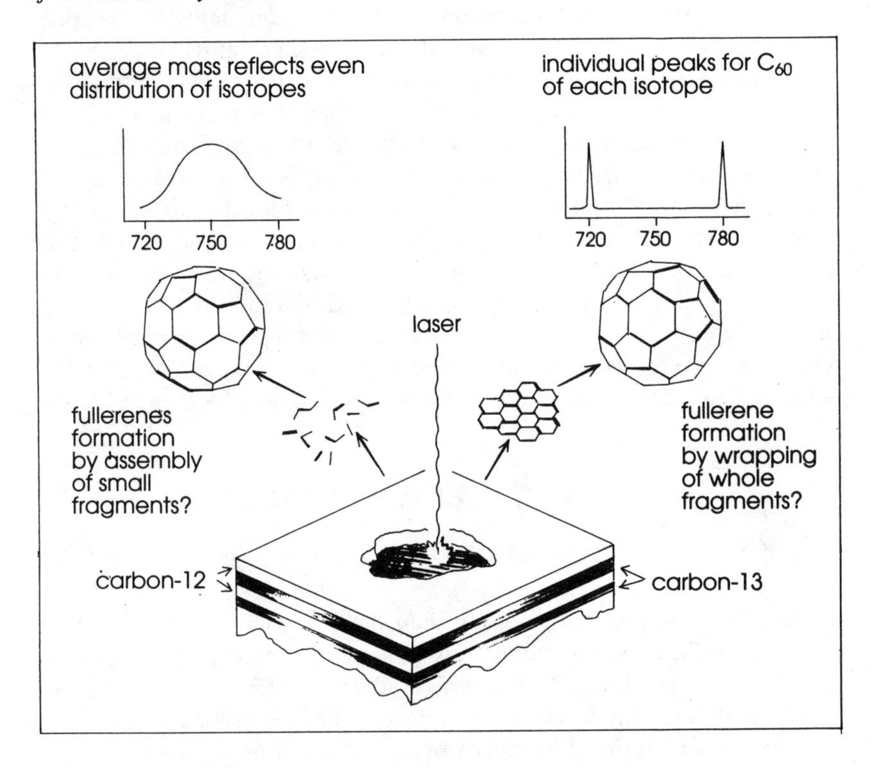

The IBM group's experiment showed that C_{60} formed from atoms in carbon vapour, not from graphite sheets

Unfortunately, they still only had some few hundreds of nanograms of the material, rather less than was convenient for other forms of spectroscopy. Despite having given the matter much thought, it was not until the day in August when their attention was drawn to Krätschmer and Huffman's report of the infrared and ultraviolet spectra that had appeared some six weeks before in *Chemical Physics Letters* that Bethune and Meijer realized that it was totally unnecessary to vaporize the carbon by means of lasers. They raced around assembling the equipment needed to emulate the experiment. Within two days they had done so. They called Don Huffman. But for Huffman in Arizona and Krätschmer in Heidelberg it was old news. Not only were they now entirely confident that their data showed the presence of buckminsterfullerene, but they had discovered how to make it in bulk. Huffman mentioned the forthcoming paper that had been received by *Nature* by chance on the same day that Meijer and Bethune's isotope experiment paper had arrived at the *Journal of Chemical Physics*. He went on to mention the referees' comments, their demand for a mass spectrum, and his row with the Arizona technician who wanted his name on the paper just for recording this spectrum, and the likelihood that as a result there might be no mass spectrum published. As their excitement that they might have been the first to make solid C_{60} ebbed away, Meijer and Bethune were left with the prospect, encouraged by Huffman, at least of publishing their mass spectrum. They submitted the data to Rick Smalley, the appropriate advisory editor of *Chemical Physics Letters*. Smalley had only one question: how did the IBM researchers know that the buckminsterfullerene molecules were genuinely deposited on the sample surface and were not created by the desorption laser? Bethune explained the isotope experiment that was awaiting publication and Smalley promptly accepted the paper.[29]

LOS ANGELES, CALIFORNIA

JULY–AUGUST 1990

Robert Whetten had more reason than most to join in the scramble to make the fullerenes. The young physical chemist had spent a year doing postdoctoral research at Exxon with Andrew Kaldor and Donald Cox at the time of their carbon cluster experiments before moving on to take up an appointment at the University of California at Los Angeles.[30]

Whetten's principal contribution to the story so far had been a 1986 paper which had attempted to pour oil on the troubled water of the dispute between the Rice and Exxon groups, highlighting the merits of each group's case, and concluding that there was a need for experiments which avoided the use of high-energy sources to generate the fullerenes, and which by their subtlety might clear up for good and all the confusion over exactly what was being made, where, how, and why it was so stable.[31] In his conversations with his former colleagues at Exxon and with Smalley's group, Whetten realized however that he was out of his league and in no position to compete experimentally to prove his point.

On his arrival in Los Angeles in 1985, Whetten had met François Diederich, a Luxembourger who was returning from Europe to the university having done his postdoctoral research there under Orville Chapman, synthesizer *extraordinaire* of molecular oddities. As the sorcerer's apprentice, Diederich believed the physical chemists were barking up the wrong tree. He had C_{60}, C_{80} and other species in his sights as targets to be made by conventional organic synthesis. Where Whetten, a physical chemist, was calling buckminsterfullerene and its family clusters, he was already calling them all-carbon molecules.

Constructing a synthetic route for such a molecule is no small feat. First, they had to locate suitable precursors which themselves often had to be synthesized. Over the next few years, in between pursuing funded research topics, Diederich and Whetten struggled to make suitable building blocks. One of these was a ring molecule containing eighteen carbon atoms. Their success was limited. By 1989, they had evidence that they had made the ring, but no more than that, and certainly not enough of the stuff to proceed with the next stage of the synthesis in which it was hoped that a number of such rings would snap together under the influence of heat or light to produce the target fullerene. This result did, however, have an important consequence of procuring substantial new funding. By April 1990 Diederich and Whetten had something no other group had – a National Science Foundation grant specifically to look for all-carbon molecules.

The publication in July of Krätschmer and Huffman's spectral evidence for buckminsterfullerene made simply by heating rods of graphite threw them into a quandary. Should they persist with organic synthesis or should they follow this new lead? During a group meeting they decided to compromise and assigned a student to reproduce the Krätschmer–Huffman result.

HOUSTON, TEXAS

JULY–AUGUST 1990

Word did not reach Rick Smalley of Krätschmer and Huffman's tentative claim made at the "Dusty objects" conference in 1989 that their carbon soot spectra contained buckminsterfullerene. Neither did he act when the confirmed results were submitted and published in *Chemical Physics Letters*. His cue for action came when Bob Curl received Krätschmer and Huffman's *Nature* paper to referee.

Smalley's first response was one of delight. Here, at last, was the vindication he had sought for five years. The news immediately confirmed the importance of his and Kroto's discovery. If a company had made this announcement, its share price would have shot through the roof. (And if buckminsterfullerene itself had been quoted on the commodities market, its price would have dropped through the floor.) His second thought was more ruminative. Starting to think, in the light of this new and very different approach, why he had not made the breakthrough himself, he was forced to conclude that it came down to his own bone-headedness.

Smalley was more than a little vexed that he had not been able to produce solid buckminsterfullerene with his lasers. He himself had long argued that the fullerenes form by a cooking process, during a period of annealing in moderate heat rather than under the full force of the laser blast, but now it appeared he had never really fully pursued the implication of this idea which was that lasers were not necessary to the process. Nature had triumphed in its usual messy, complex way where Cartesian man had failed.

Smalley swallowed his pride, put his lasers aside, and set about reconstructing Krätschmer and Huffman's humble bell-jar.

KONSTANZ, GERMANY

10–14 SEPTEMBER 1990

The nexus for these researchers was to be the Fifth International Symposium on Small Particles and Inorganic Clusters. This conference became the stage upon which the players could exchange news and views at a pace which matched the unfolding of the story. Rick Smalley would be there, so would Rob Whetten, both scheduled to speak on topics other than buckminsterfullerene. Don Bethune would be there from IBM. This was the meeting for which he had submitted a poster, and for which, as their results had got progressively more compelling, he had tried in vain to persuade the conference chairman to find time in the schedule for him to give an oral presentation. Tony Stace, the Sussex cluster-beam specialist, would be in the audience. So would Don Cox from Exxon.

Conspicuous by his absence was Wolfgang Krätschmer. He was in his laboratory barely a hundred miles to the north. Unbeknown to almost all the delegates, his paper on the production of solid C_{60} had been accepted for publication by *Nature* just the week before. It would not appear until the end of September.

As they converged on the pretty lakeside town of Konstanz on the German-Swiss border, the chemists and physicists put out their antennae. Bethune found that a speaker had cancelled and that he could after all give an oral paper. Whetten worked the conference floor comparing notes with Bethune and Smalley and others. Hints were dropped and brains were picked over the weekend before the conference began.

Monday was given over to metal clusters. On Tuesday morning, it fell to Don Bethune to be the first to discuss carbon. He gave his speech, describing the unexpected stability of the small quantities of buckminsterfullerene that he and Meijer had made, and giving details of how they had separated, purified, and begun to characterize the substance. The audience lapped it up. Without knowing it, they had savoured as an *entrée* what was in fact a filling appetizer. Smalley thought they should taste real meat of the discovery. Aware of what would shortly appear in *Nature*, he had contacted Krätschmer the week before and suggested that he should come down to Konstanz before any more of his thunder was stolen. He offered to donate a few minutes of his allotted time so that Krätschmer could announce his breakthrough. Krätschmer and Konstantinos Fostiropoulos arrived that evening.

The next morning Krätschmer gave what was of necessity a brief and matter-of-fact description of how he and Fostiropoulos had made C_{60}. He rattled through the experimental technique, he showed the spectral evidence, one type after another, the infrared, the mass spectrum, the X-ray diffraction pattern. It was in essence the *Nature* paper spoken aloud. There were no questions. There was no time for anything but a round of table-rapping, the restrained convention by which scientists show their admiration at German meetings.

Smalley then stood to deliver his own talk in what time was left to him. Departing from his prepared text, he delivered a spontaneous paean to the skills of the team that had made the 1985 discovery, and especially to Jim Heath and Bob Curl.

Then it was Don Cox's turn. He had the hardest task of all, publicly to recant his earlier scepticism regarding the spheroidal structure of the sixty-atom carbon molecule. He did so with grace.

At the end of the week, everybody dispersed full of renewed zeal with new leads and instructions for their laboratories.

SUSSEX, ENGLAND

SEPTEMBER–OCTOBER 1990

Towards the end of August, Roger Taylor and his colleagues finally obtained the spectra that revealed the separation products of the chromatography. The two bands of colour, the magenta and the red that had emerged from the single band of pink, were indeed buckminsterfullerene and its new cousin, C_{70}, which, in a happy homophony with the generic, fullerene, they christened "falmerene" after the Sussex village site of the university. They set to writing up this first confirmed report of a new fullerene and what they believed to be the first chromatographic separation of different forms of the same element. To celebrate the event, Jonathan Hare did some back-of-an-envelope arithmetic to come up with Hare's Universal Constant: the earth is in proportion to a soccer ball as a soccer ball is to buckminsterfullerene and as buckminsterfullerene is to an electron![32] Meanwhile, Harry Kroto, *en route* to Yugoslavia for a conference, took the opportunity to stop off at Krätschmer's laboratory to bring him up to date with the Sussex work.

Taylor's was to be another paper that was economical with details. "Chromatographic separation" was as far as he wished to go in describing what he had done, in order to buy time in which to learn more about the two fullerenes before any rival group could catch up. A referee demanded that he identify the hexane solvent and the absorbent alumina. The revised paper was received by the Royal Society of Chemistry journal, *Chemical Communications*, on 10 September, the Monday of the Konstanz conference.[33]

SAN JOSE, CALIFORNIA

SEPTEMBER–OCTOBER 1990

The following day, many time zones to the west, *Chemical Physics Letters* received the first in a fusillade of papers from the IBM group seeking to provide a full spectroscopic album for buckminsterfullerene.[34] In order to speed the race to publication, Don Bethune and his team decided to knock the results out one by one, submitting each as it became available. They were still working with the tiny quantities of soot available by means of laser vaporization, but felt that IBM's top-of-the-range equipment gave them a good shot at being first in with several key spectroscopic observations.

This first result was the Raman spectrum which depicts the symmetrical vibrations of a molecule. The IBM researchers tentatively assigned the spectral peaks to the most symmetrical distortions – an even "breathing" of the spheroid, a "squashing" action, and a more complex vibration whereby the pentagonal rings of carbon atoms shrink as the hexagonal rings expand and *vice versa*. Further papers followed at regular intervals over the next few months, some reporting results achieved with minute amounts of fullerenes, others, following the news from the Konstanz conference, benefiting from the fabrication of the more substantial quantities made possible by a visit to the local hardware store to buy an arc welder.

There were a number of prizes to be sought. Perhaps the most glittering of them for the physicists was still to confirm the shape of the molecule literally by seeing it. The Almaden laboratory had the tool for the job in their own scanning tunnelling microscope. They duly recorded images of the stacked fullerene balls deposited on a gold surface and submitted them to *Nature* where they were received just three days ahead of Krätschmer and Huffman's images which they had finally got from the University of Missouri. The two papers were published back to back in the same issue of the journal.[35] Like Krätschmer and Huffman, the IBM scientists had hoped the pictures would reveal the pattern of hexagons and pentagons on the surface of each molecule, but again they only showed each fullerene as a smooth sphere. They did, however, confirm the spacing of the array of around eleven ångströms, in agreement with Huffman's X-ray diffraction measurements. Both groups found the occasional larger sphere interrupting the neat arrays of buckminsterfullerene molecules and speculated that these were larger fullerenes.

LOS ANGELES, CALIFORNIA

SEPTEMBER–OCTOBER 1990

As the revelations tumbled forth at the Konstanz meeting, Robert Whetten and François Diederich learned that their colleague, Yves Rubin, was having no luck back in Los Angeles with his assignment to reproduce Krätschmer and Huffman's fabrication of buckminsterfullerene. A look at Krätschmer's lab would be invaluable, Whetten decided. After the session broke, he asked to come and visit him at Heidelberg. To his surprise, Krätschmer agreed.

Diederich joined him and they set off northward through the Black Forest. Their aim was merely to examine the apparatus that Krätschmer had used to make his "fullerite". As a bonus, they got a sneak preview of the recent progress at Sussex from the draft that Kroto had left with Krätschmer as a courtesy. Diederich even had the temerity to beg a party favour, a sample of "fullerite" crystals.

Armed with the precious "fullerite" and an appetite for competition sharpened by the sight of the Sussex results, they returned to California. With first-hand knowledge of Krätschmer's set-up they could save far more time than if they had worked from the meagre details published in *Nature*. Their efforts were not without difficulty, however. The crystals were imperfect and did not produce the X-ray diffraction pattern they had hoped for. Efforts to purify their own buckminsterfullerene were hampered by the incomplete description of the chromatography process that Roger Taylor had used at Sussex. Their frustration was intensified by the impending embarrassment of having to explain to the National Science Foundation how it was that the only research group in possession of funding specifically to investigate all-carbon molecules was running last in the race to make and characterize the most conspicuous member of the class. While other members of the group experimented with different chromatography materials, Diederich fired off terse salvoes into the electronic mail pressing for more details from Kroto's group. In the end, they were better placed than they dared hope. Their paper characterizing C_{60} and C_{70} appeared in the American *Journal of Physical Chemistry* soon enough that American readers saw it ahead of the Sussex publication.[36] In it, they took a subtle vengeance over both Krätschmer and Taylor by describing in some detail not only the way they had made their own "fullerite" but also the exact materials and methods they had employed in the chromatographic separation of the purified fullerenes.

HOUSTON, TEXAS

SEPTEMBER–OCTOBER 1990

The Los Angeles group were not the only ones to experience difficulties. Rick Smalley, too, felt that Krätschmer and Huffman had been less than forthcoming in describing their method of making buckminsterfullerene. What diameter were the graphite electrodes, he wanted to know? What held them in position as they sooted away? And what prevented them from bursting into a luminous electric arc rather than calmly evaporating away to form the fullerene-rich soot? Smalley felt that answers to such questions were vital. He had tried using quarter-inch rods with no success. Comparing notes with Robert Whetten, he learned that eighth-inch rods had worked for him. The size of the rods seemed an important factor in determining how much of the current gushing through them went into heating them to the point of vaporization whereupon fullerenes would be produced and how much was wasted as radiant energy if an electric arc began to bridge the gap.

In the absence of precise information, Smalley's group built a device that could control the gap between the electrodes. What they found was that the precise gap thickness was not critical after all, and that arcing conditions were no hindrance to producing fullerene-rich soot. This gave Smalley's team the information they needed to set up a fullerene generator using large rods. One graphite rod was sharpened with a pencil-sharpener to provide a focus for the electric current in order to generate the heat that would start the vaporization process; thereafter the rod was gently driven onto an opposite flat graphite electrode in such a way that the bulk of the power now went into an arc and not into the resistive heating. In this way, it was possible to generate larger quantities of buckminsterfullerene and related compounds such as a new spheroidal hydrocarbon, $C_{60}H_{36}$. Smalley settled his score in the title of his paper, "*Efficient* production of C_{60} ..." (italics added), where he wrote of his intent

> to provide a cookbook-level recipe for the production of C60 in sufficient quantities for general chemical experimentation. The method is sufficiently simple that most laboratories should be able to set up rapidly to generate their own samples of this important new material.[37]

But the fact was that other laboratories were already making buckminsterfullerene with relative ease. The account read like a desperate attempt to recapture the crown lost to Krätschmer and Huffman.

BOSTON, MASSACHUSETTS

28 NOVEMBER–1 DECEMBER 1990

The announcement that Krätschmer and Huffman had made buckminsterfullerene and had done it, what's more, with the most basic of apparatus, had the effect of the starting gun for a city marathon. There erupted a prodigious outburst of Kuhnian normal science, the delayed reaction to Kroto and Smalley's paradigm shift of five years earlier.[38] At last, it became more productive to do real experiments than to make mathematical stabs in the dark. In addition to the ever-present trickle of theoretical papers predicting this or that property for buckminsterfullerene and its kin, there was now a flood of spectroscopic analysis, notable in its midst a blockbuster of a paper from Don Cox and his team at Exxon, describing the application of eight separate analytical techniques to C_{60} and C_{70}, as full an expiation of as anybody could desire for his former want of faith.[39] There were the beginnings of reports of the chemistry of buckminsterfullerene.

For their part, Krätschmer and Huffman now took something of a back seat. Their principal contribution at this time was a spectroscopic report, a tinge regretful in tone, announcing that the infrared emission spectrum of buckminsterfullerene in the gas phase did not fit the observed anomalous interstellar spectral emissions.[40]

For everybody else, it was in fact more of a sprint than a marathon. A tape was drawn across the track just two months down the road in the shape of the annual fall meeting of the Materials Research Society. Assembling this time in Boston, the week after Thanksgiving, the scientists who had acted on the stimulus of the Konstanz meeting or the papers published in its wake now had an opportunity to demonstrate their progress. It was a veritable fullerene *Fest*. Huffman and Krätschmer and Kroto told war stories. Other contributions came from IBM, Bell Laboratories, SRI International, the Menlo Park research company, NEC Corporation in Japan, Argonne National Laboratory, the Naval Research Laboratory, the University of California at Los Angeles, the University of Arizona at Tucson, North Carolina State University and the University of Toronto. The Rice group sang the praises of their "efficient" form of fullerene production. Exxon was truly back in the fray, now making more fullerene than anyone (up to ten grams a week). Jim Heath, who had moved on to a postdoctoral position at the University of California at Berkeley where he had spent much of his time researching the still neglected small carbon clusters, summarized C_{60} research there, including the synthesis with Joel Hawkins of the first true buckminsterfullerene compound.[41]

In his conference address, Don Huffman announced, with an impish dig back at Smalley, that it was

> a source of considerable satisfaction to the authors to be able to see so many workers in this session who have been able to follow our instructions and to produce considerable quantities of C_{60}/C_{70} within the past few months, after about five years of earnest searching for such a production pathway by various people....
>
> The discovery of the synthesis technique was serendipitous but not accidental. We had been trying for almost 20 years to make very small particles of graphitic carbon having as nearly a monodispersed size distribution as possible. With the help of nature, we succeeded beyond our wildest expectations.[42]

9

THE PEAK OF PERFECTION

Amid the frenzy of effort to characterize buckminsterfullerene, there was one prize above all that shimmered on the scientific horizon. Much of the gossip at the Konstanz conference concerned the hunt for one particular spectrum. Don Bethune, Robert Whetten, and Rick Smalley were united in agreement that it would prove the icosahedral structure of buckminsterfullerene with more elegance than anything else. One analytical procedure and one only would yield the most beautiful proof of the nature of the most beautiful molecule.[1]

That procedure was nuclear magnetic resonance spectroscopy. We have seen that nuclear magnetic resonance occurs when certain atomic nuclei are placed in a magnetic field and excited by radio-frequency radiation. The resonance happens when the characteristic rate of spin of the affected nuclei in the magnetic field coincides with the frequency of the radiation. The incoming burst of radiation barges into the already spinning nuclei, transferring its energy to them, in effect knocking them off kilter. The nuclei gradually shed the energy they have absorbed, passing it on to other nuclei or the lattice in which they are bound, and recover their equilibrium in a process called relaxation. Nuclear magnetic resonance (NMR) spectroscopy, which records the precise frequencies or magnetic field strengths at which this absorption occurs, is now regarded by most chemists as the best means of identifying molecules. Indeed, the very fact of obtaining an NMR spectrum for buckminsterfullerene would establish it once and for all as a regular molecule and not just an unusually tenacious cluster.

At Konstanz, there were rumours that someone already had the NMR spectrum. The chemists worried that Huffman and Krätschmer might have got it, but after Krätschmer's talk, with no mention of any NMR spectrum, their worry turned to bemusement. In fact, the NMR spectrum was not something that greatly interested Krätschmer or Huffman. They felt the direct physical evidence they were planning to present was more compelling. Besides, they did not have an instrument to hand and they were not inclined to mess around with the necessary preparatory work creating the solutions required to run a conventional NMR spectrum: such is the solid-state physicist's distaste for wet science. NMR spectroscopy was in any case something of a black art to them, at least compared to optical spectroscopy. At base, they simply did not know what all organic chemists know, that this form of spectroscopy was quite so powerful. Nor did they know that for this molecule of all molecules the NMR spectrum would be a wonder to behold.

Because all the carbon atoms in a molecule of buckminsterfullerene occupy equivalent chemical and magnetic environments, they share the same chemical shift relative to a given standard. The chemical shift, it will be recalled, provides a quantitative description of the chemical and magnetic environment of nuclei. Assuming that we have made, at some expense and difficulty, a molecular sample entirely composed of the carbon-13 isotope, the chemical shift for each nucleus will be the same. All sixty little absorption peaks coincide: buckminsterfullerene would possess as its carbon-13 NMR spectrum a single peak.

The conclusive proof it would offer of the structure of buckminsterfullerene was of course the primary stimulus in seeking the NMR spectrum, but few would deny that the aesthetic satisfaction of seeing this peak was a powerful secondary motivation. But for some, there was another reason. As early as 1986, Veit Elser at Cornell University and Robert Haddon at AT&T Bell Laboratories had picked up the gauntlet thrown down by Bob Curl in his and Kroto and Smalley's first report of buckminsterfullerene. Curl had proposed that the ring currents circulating on the surface of the carbon framework might give rise to an unusual chemical shift in the NMR spectrum of an atom enclosed within it.[2] Large ring currents circulating upon the surface of the molecule, induced by the applied magnetic field, would shield a caged nucleus to some extent from the field. But Elser and Haddon figured that the shielding would actually be very small.[3] Using calculations based on Hückel molecular orbital theory and assuming, for the sake of temporary convenience, that all the bonds on the carbon framework were of equal length and in character exactly like those in benzene, they estimated the ability of the molecule to generate ring currents in a magnetic field – its magnetic susceptibility – to be no more than one fifth of the maximum value for benzene, and of the opposite, negative, sign. In other words, buckminsterfullerene would be capable of being magnetized to a modest degree like some metals rather than repelled by a magnetic field like the vast majority of substances. When Elser and Haddon took account of the disparity in bond lengths and strengths thought from previous theoretical calculations to exist on the surface of the buckminsterfullerene molecule, they came up with a very different result. Whatever the actual result might turn out to be, it seemed that the magnetic behaviour was hypersensitive to the relative strength of the two sets of bonds. On the strength of these calculations, they could only conclude that although planar aromatic hydrocarbons and graphite had been employed for the past five years as worthy analogues in predicting its properties, buckminsterfullerene would possess magnetic behaviour quite "unlike that of any other molecule yet encountered".

Haddon's agenda was shaped to some extent by an early interest in the characterization of nonplanar organic molecules with chains or rings of alternating double and single bonds. How much did the fact that the pi bonds characteristic of such *conjugated* molecules would fail to overlap as fully in a contorted molecule as they do in a planar one alter its electronic behaviour and, in the case of a molecule with benzene-like rings, draw it away from aromatic behaviour? Other chemists had considered this problem, but Haddon's calculations forced him to a realization that the conventional wisdom regarding the orientation of the pi bonds and the extent of their "conjugation" was woefully inadequate. Haddon's reassessment of the degree of alignment required from the atomic orbitals on neighbouring atoms in order to form a strongly bonded conjugated system revealed that it was possible to have a molecule that was highly conjugated and yet not pancake flat – in effect, that electrons could turn corners.

At the core of his thesis was a parameter called the angle of pyramidalization. As a planar aromatic system is bent out of shape, so the carbon atomic orbitals that overlap to form its molecular bonds undergo distortion too. A completely flat conjugated system like graphite has sigma bonds between neighbouring carbons atoms formed by the overlap of sp^2 hybrid atomic orbitals. The conjugation comes from the formation of pi bonds from the weaker overlap between the remaining p atomic orbitals. At the opposite extreme is the tetrahedral bonding of methane or diamond based upon the overlap of sp^3 hybrid orbitals. On occasion, a molecule will form because there is a still a net energy benefit to be had by bringing together two cumbersome groups which form a strong bond and achieve a low-energy molecule despite the fact that it is necessary to introduce some distortion (and energy cost) to make them hold together. In such cases, the distorted parts of the conjugated structure will exhibit intermediate degrees of hybridization, and the pi bonds will gain some s orbital character. The pyramidalization angle, which is simply the angle between the bond direction and the unhybridized p orbital less the ninety degrees between these two in the standard case of graphite, is a convenient measure of this phenomenon. For buckminsterfullerene, this angle is 11.6 degrees.

When Haddon learned of the synthesis (coincidentally, by Leo

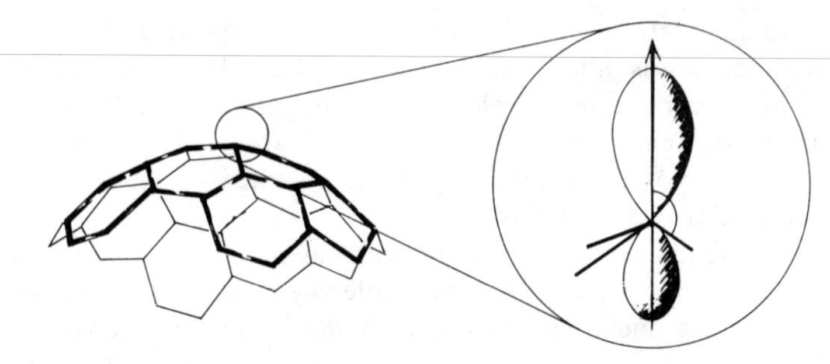

The angle of pyramidalization is a measure of the distortion in molecules that would prefer to be planar

Paquette, the architect of dodecahedrane) of a new organic molecule with a moderate angle of pyramidalization at certain points in its structure, it prompted him to go in search of literature describing the occurrence or synthesis of molecules with higher angles. He unearthed a description of the most distorted conjugated organic molecule ever to be isolated – with an angle of pyramidalization of 13 degrees – and brandished his find to confound those who thought that the truncated icosahedral structure proposed for buckminsterfullerene would be too strained for it to exist as a stable molecule. With its lower angle of pyramidalization, he wrote in a paper submitted just a year before, that isolation would be achieved:[4] "No particular problems are anticipated in the isolation of icosahedral C_{60}."

At about the same time, Patrick Fowler, a theoretical chemist at the University of Exeter in England, performed a different series of calculations in an attempt to predict the magnetic susceptibility of the still unproven buckminsterfullerene icosahedron.[5] He employed the *ab initio* mathematical methods that theoreticians typically use on far smaller molecules together with a supercomputer that could cope with the magnification of the problem to a sixty-atom species. He got a very different answer, predicting conventional magnetic properties pretty much in line with more normal aromatic species.

Fowler's challenge prompted a feisty rebuttal from Haddon who compared the two sets of results and decided that it still was not reasonable to describe buckminsterfullerene as being like a normal aromatic molecule.[6] Thomas Schmalz, a theoretical chemist at Texas A&M University in Galveston, weighed in on Fowler's side.[7] Only experimental measurement of the magnetic susceptibility could produce a winner of this debate. But equally, such a real result might show all disputants to be wrong and reveal only that the theoretical chemists had overreached themselves once again.

Such a debate might seem akin to theologists' arguments over how many angels can dance upon the head of a pin. But at its root the question remained: how aromatic was buckminsterfullerene? And at stake lay the entire character of its as yet virgin chemistry.

It was Roger Taylor's separation of the fullerenes C_{60} and C_{70} by the use of standard solvents that brought these molecules within reach of the conventional analytical techniques of organic chemistry for the first time. Attempts by the Sussex chemists during the early summer of 1990 to obtain the solid-state NMR spectrum of the small quantities of buckminsterfullerene they had made at that stage had failed. Now that they knew they could build up substantial concentrations of the stuff in solvents, however, NMR spectroscopy in solution was an altogether more attractive proposition.

It took eighteen hours' accumulation of scans for a signal to emerge. The spectrum showed a towering single peak – but a towering single peak, it very soon transpired, for the tiny residue of benzene retained by the buckminsterfullerene from an earlier stage in the preparation (the benzene ring with six equivalent carbon atoms bonded to six equivalent hydrogen atoms produces single peaks in both proton and carbon-13 NMR spectroscopy). The solvent, deuterated chloroform, produced another large peak for its own unique carbon nucleus. But there, near the benzene peak, was a third, Lilliputian, signal. This single line, occurring at about 143 parts per million (the shift is quoted as a fraction of the applied magnetic field strength), was what they were after.[8]

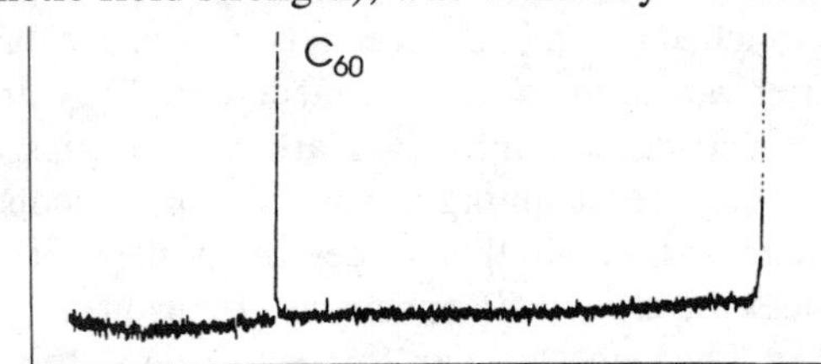

The chemists' proof of the C_{60} structure. The feature on the extreme right is due to the solvent used

At the Almaden Research Center, Don Bethune and his team's first inclination, having made their first tiny quantities of buckminsterfullerene in June of 1990, was also to try to record its NMR spectrum. Egged on by their colleague Heinrich Hunziker, who had more of a chemist's perspective than the rest, the IBM group debated their prospects. Costantino Yannoni felt that it would require the best part of a gram of the stuff to do the job, the sort of quantities that were not to become available until after people had learned of Krätschmer and Huffman's prolific fullerene generator. Since they had only a few hundred nanograms – about a millionth of this amount – they did not rate their chances highly. Yannoni sent Bethune and Gerard Meijer off in search of Bob Johnson in the laboratory's polymer science department. He was an expert in high-sensitivity NMR of solutions. Even he was sceptical that it could be done. He thought to send them packing when he said they would need several hundred micrograms – still three or four orders of magnitude more material than they had yet managed to scrape together. Instead, they began to think how they might begin to make up this vast shortfall. Hard graft would get them part of the way. Then, it would require an accumulation of many scans which could then be averaged to produce a clean signal – 50,000 scans was Johnson's guess. With a relaxation time after each resonance expected to be perhaps two minutes, the process could take weeks. (In the event, they found a way to compress the relaxation time to just one second, enabling a considerable acceleration.)

By the time of the Konstanz conference, Meijer was making soot as fast as he could. Johnson, working at the very limit of the equipment, thought of how best to obtain the clearest spectrum without undue background noise or the confusion of spurious peaks from chemical contaminants. They were getting spectra. But they had to be absolutely sure of what they had. One check was to re-run the spectra in different solvents; if what was thought to be the buckminsterfullerene peak shifted in a way predicted because of the change of solvent, then it made it less likely that the signal was a rogue.

When Bethune touched down in San Francisco after the day-long flight from Europe, Meijer was at the airport to meet him. They drove the fifty miles to San Jose and rushed into the dark laboratories. There on a computer screen were the scanning tunnelling microscope images of buckminsterfullerene and, on another screen in another darkened laboratory, the single peak of the NMR spectrum. Johnson had finished his checks. "Gentlemen," he said, "we can stop referring to this as 'the resonance', and can now call it C_{60}." They hastily wrote their paper in the hope that Kroto and his colleagues might still be checking their results. Having sent it off, Bethune called Kroto to swap information. He learned to his dismay that the Sussex paper had been submitted just the week before. Kroto faxed back a preprint together with the comment: "Congratulations on the one line. It is nice that C_{60} can make a lot of people happy. There's lots of it to go around."[9]

The Los Angeles chemists, Robert Whetten and François Diederich, had profited greatly by their visit to Krätschmer's laboratory in Heidelberg. They had learned how buckminsterfullerene could be made. But also, thanks to sight of the as yet unpublished Sussex paper, they no longer had reason to disbelieve the rumours of the past week that someone really had recorded its NMR spectrum. Two weeks after the IBM group, they sent off their own NMR results for publication.[10] The day after Whetten's paper arrived at the offices of the *Journal of Physical Chemistry* came Rick Smalley's paper describing the "efficient" production of buckminsterfullerene into which he had casually dropped a single sentence noting his contribution to the growing library of C_{60} NMR spectra.[11] The two papers were published back to back in the same issue of the journal. In another few weeks, Don Cox's group at Exxon appended their own confirmation.[12] By the time of these last two additions to the literature, it had already become redundant to publish the spectrum itself. It was sufficient merely to say that it had been obtained.

If buckminsterfullerene was a unicorn among molecules, then here surely was the evidence – one straight and slender horn. The capture of this trophy was the supreme moment of catharsis for the chemists. At last they felt they knew for sure that the magenta solution really was what they had all believed and hoped it was.

As Johnson, Meijer, and Bethune pointed out in their report, the NMR result bought buckminsterfullerene membership of the very exclusive club of icosahedral molecules, the only others being an icosahedral boron hydride ion with twelve vertices and Paquette's dodecahedrane with twenty vertices (the dodecahedron is the "dual" of the icosahedron and shares the same symmetry).[13]

In the event, its chemical shift was much like those of various plausible chemical analogues such as certain planar molecules combining hexagonal and pentagonal rings of carbon atoms. The peak lay "downfield" of the peaks observed in some planar analogues, the effect of the strained structure's slight "deshielding" of the carbon nuclei from the protection of their veil of molecular orbital electrons.

The NMR result also quelled the dispute between Patrick Fowler and Robert Haddon. The "shielding" of the carbon nuclei by the pi electron ring currents in buckminsterfullerene turns out from the spectrum to be about three-quarters of that felt by the carbon nuclei in benzene, confirming Fowler's view that the molecule would be only modestly aromatic, certainly no superaromatic wonder.[14] At the same time, the Bell Laboratories group, having finally obtained a sufficient sample of the molecule, performed an experiment to measure its magnetic susceptibility and confirmed this fact.[15]

Following the lead of the Sussex team, all these groups also attempted to record the NMR spectrum of C_{70}, although with varying degrees of success. Taylor's chromatographic separation of C_{60} and C_{70} made it equally easy for the Sussex chemists to record the spectra of both new molecules. The spectrum of C_{70} is less beautiful but more instructive than the single peak of its more perfect companion. Its greatest merit is perhaps that it confirms that a second fullerene also possesses a spheroidal geometry, thereby supporting the idea that an entire family of fullerenes so shaped might be unusually stable. The spectrum has five peaks, reflecting five distinct chemical environments of the molecule's seventy carbon atoms. These peaks are weighted in the proportion 10:20:10:20:10 as one moves downfield; their positions are immediately grasped by visualizing the elongated spheroidal "rugby ball" shape that the spectrum confirms. The ten carbon atoms that form the midriff of the molecule, in effect slotted in between two buckminsterfullerene hemispheres, are the least strained because the spheroid is most nearly flat around this equator. Thus, these nuclei have the lowest chemical shift and are to be found nearest the benzene peak. By contrast, another ten carbon atoms sharing identical circumstances, the two sets of five in the pentagons at each comparatively sharply curved apex of the rugby ball, are the most strained. Their chemical shift is greater than that of the twelve uniform pentagons of buckminsterfullerene. Between this equator and these poles lie the remaining fifty carbon atoms, ten of them along lines of latitude near the poles, twenty of them along lines of latitude near the equator, and the final twenty along a line in between these two. In these places, the curvature on the molecular shell is closest to that of a perfect sphere. Thus, these three remaining NMR peaks are grouped close together with chemical shifts close to that of buckminsterfullerene.[16]

The chemists' day was done. The physicists now took the stage, moving the game rapidly on to their favoured phase of matter. Their goal was to obtain the NMR spectra of the two principal fullerenes in the solid state. The spectra of solid substances are typically rather different from those of the same materials in solution. In general, a molecule locked into a solid lattice is no longer free to move and must endure the company of its neighbour molecules. These molecules may exert an influence that distorts the spectrum. The interpretation of the distortion might reveal something of the nature of the mutual positioning of individual molecules in the solid. In the case of buckminsterfullerene, it might shed some light on how the molecules are stacked in their crystalline arrays, a subject of some confusion and uncertainty.

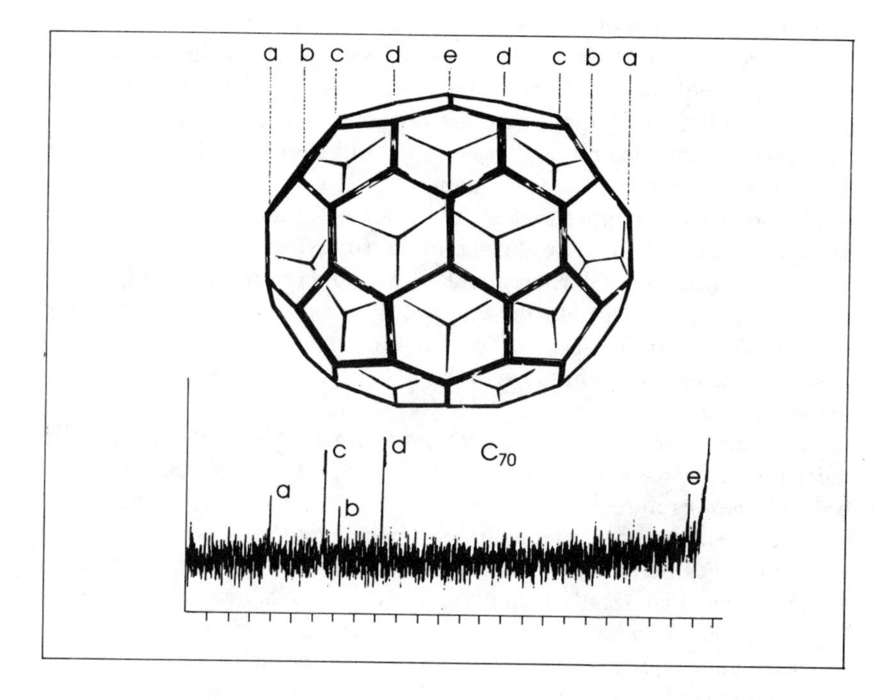

The NMR spectrum for C_{70} confirms the fact that there are five distinct environments for the carbon atoms

But the first report, from Rob Tyckò and others at Bell Laboratories, on room-temperature powder samples of buckminsterfullerene found the same sharp peak seen for the solutions.[17] Competition was still as sharp as ever, and the IBM group reported similar results a few days later.[18] What this curious repetition meant was that buckminsterfullerene molecules in the solid state behave in some respects like those in the solution. In particular, it showed that at room temperature they rotate rapidly on the spot within the crystal lattice. There is no preferred orientation for the axis of rotation; the molecules simply slip and spin around as they please up to a hundred million times a second. The surprising discovery of this rapid spinning provided yet more evidence – if it were needed by this stage – in support of buckminsterfullerene's being spherical. It also suggested that there was no preferred alignment of the axes of symmetry of the individual molecules with the long-range symmetry of the crystal array. The sharp peak was the giveaway. Were the molecules clamped in stationary positions within the crystal array, not all the carbon atoms on the surface of any given molecule would be in equivalent environments in the magnetic field of the spectrometer (as they are on average in solution where they are free to rotate and in only brief contact with other molecules). In this case, the fixed buckminsterfullerene molecules would be expected to show a range of quite different signals, some from those carbon atoms shielded by relatively strong ring currents induced in some directions by the magnetic field, and others from carbon atoms on parts of the molecular structure that are not oriented in the right direction for strong ring currents to arise. Furthermore, some carbon nuclei would be pressed up against the carbon atoms of a neighbouring molecule, while others would look out onto the comparative void of the octahedral and tetrahedral gaps in the lattice. The range of signals from all the carbon atoms would produce a broad spectral hump rather than a sharp peak for the molecular material as a whole. Repeating the NMR experiment at lower temperatures in order to freeze out the rotation of the molecules did indeed produce spectra with these broader features. From free rotation at ordinary temperatures, the molecules subside into a ratchet action at about 25 degrees centigrade below the freezing point of water, reorienting themselves periodically into new positions with the same symmetry as the old, and then finally lock into position at much lower temperatures.

The question of what was the preferred orientation of the buckminsterfullerene molecules relative to one another had not gone away, although it was now only relevant to ask it of the material in the low-temperature lattice. Do they share a common alignment such that a particular surface detail of hexagons and pentagons on one molecule faces a similar arrangement on the next and so on? Or are they oriented in no particular way, forming instead what Haddon describes as an "orientational glass" after the amorphous structure of glasses? Performing further solid state NMR spectroscopy on buckminsterfullerene at a range of temperatures, the Bell Laboratories group was able to confirm that the material experiences a transition between two distinct crystal structures.[19] Knowledge of these bulk structures is sure to be important in assessing the prospects for commercial applications of the fullerenes.

One last question remained to be cleared up. The undoubted beauty of the single peak of the NMR spectrum was unequivocal evidence for the buckminsterfullerene molecule having a structure in which all of its sixty carbon atoms are exactly equivalent. But which equivalent structure was it? It is possible to imagine a number of structures that comply. Some of these were discussed by the theoreticians. Others were so improbable that they had been barely mentioned. It is worth exhuming them here, if only in order to give them a decent burial. The easiest structure to imagine of sixty equivalent bonded atoms is not in fact the now familiar truncated icosahedron but simply a large ring. Carbon, as we have seen, readily forms chains and rings of various sorts. Why not a ring of sixty atoms? The answer lies in another question: Why sixty? Why not sixty-two (or sixty-one) or any other number? Although such a structure was theoretically possible, there was nothing sufficiently special – either geometrically or electronically (or, let us admit it, aesthetically) – to account for the prevalence of the sixty-atom species above all others. By these lights, C_{70} might equally be a ring with all its carbon atoms equivalent, but its NMR spectrum had shown otherwise.

Instead, as we have seen, the NMR spectrum was taken as conclusive evidence that buckminsterfullerene possesses icosahedral symmetry. But the truncated icosahedron with its surface detail of hexagons and pentagons is not the only such figure with the requisite sixty equivalent vertices. There is also its dual, the truncated dodecahedron. Three edges meet at each of the vertices of this polyhedron, so it too is in principle a suitable template for an all-carbon conjugated molecule. However, its surface is composed of regular decagons and equilateral triangles which if traced out in carbon bonds would form a highly strained structure. Whether because of the arrangement of its molecular orbital energy levels or because of simple structural strain, theoretical predictions had long ago shown this to be an uncomfortable alternative.[20] But it was worth disproving conclusively.

In addition, although the truncated icosahedron shape was now to all intents and purposes beyond question, precise information about the electronic and geometric structure remained to be filled in. X-ray diffraction and scanning tunnelling microscope images gave reasonable estimates of intermolecular spacing but were too fuzzy to provide a measure of interatomic bond lengths.

In conventional organic chemistry the "coupling" of nearby spinning nuclei is used to provide more detailed information about molecular structure. Hydrogen atoms attached to the same carbon atom have the same chemical shift, but it is possible to tell how many of them there are by examining the splitting of their NMR signal produced because of the different ways of summing the individual nuclear spins to yield values for the total spin of the nuclei in the external magnetic field. The splitting is proportional to the inverse cube of the distance between the coupled nuclei. Thus, it is possible to observe the splitting that arises as the result of spin-spin coupling between proton nuclei attached to the same carbon atom and occasionally the secondary splitting produced by the more distant coupling of two protons each bonded to neighbouring atoms.

It is not normally possible to study the spin-spin coupling in carbon-13 NMR spectroscopy, simply because, with only one atom in a hundred being the heavier isotope with the spinning nucleus, the chances that two of them are bonded together is vanishingly small. Nino Yannoni was nevertheless able to exploit this effect using buckminsterfullerene specially made with a far higher proportion of carbon-13. Working on the stilled molecules at low temperature, they observed the coupling of neighbouring carbon-13 nuclei and from these signals were able to deduce the carbon-carbon bond lengths of buckminsterfullerene for the first time. Because the splitting is proportional to the inverse cube of the internuclear distance, a very slight increase in bond length produces quite a significantly reduced splitting, thus providing a sensitive measure of the bond length. The values of the bond lengths came out to be 1.45 ångströms for the "single" bonds linking the hexagonal rings and 1.40 ångströms for the "double" bonds linking the pentagonal rings. The result unequivocally ruled out the ring or any kind of "fluxional" species – a catch-all phrase that had been used to describe less pretty structures that people might not have thought of. Yannoni and his colleagues also ruled out the truncated dodecahedron, calculating the likely NMR spectrum splitting for what would be the optimum bond lengths for the mesh of decagons and triangles, and finding it in poor correspondence with what they had observed directly.[21]

This series of experiments, carried out with such alacrity in the weeks and months after buckminsterfullerene had become available, provides a rich demonstration of the enormous utility of spectroscopic analysis and of NMR spectroscopy in particular. We have now seen this technique prove the equivalence of all the molecule's carbon atoms; we have seen it furthermore provide powerful evidence for the icosahedral structure in favour of other structures that could contain sixty equivalent carbon atoms; we have seen it reveal the spinning on the spot of the spherical molecules within their crystal lattice, and the stutter and cessation of that spinning at low temperatures. Further exploration of this phenomenon has uncovered valuable information about the various crystal structures that the bulk fullerene material chooses to adopt. Finally, we have seen it provide the best measure of the length of the molecule's bonds. Armed with this new information – and with final reassurance that buckminsterfullerene really was a molecular football – scientists around the world were at last ready to explore what it might do.

10

"MY LORDS, WHAT DOES IT DO?"

The old saw that nothing succeeds like success can be nowhere more true than in science. What scientist upon completing a fruitful project has ever said as he or she wrote up a report for publication in the literature or for the eyes of some funding agency: this area of research is now exhausted? The standard corollary is: more research is required. Saying this is partly a professional survival tactic, of course, but it is also said because it is true. Any discovery raises new questions. An important discovery raises many questions.

There could be no more dramatic demonstration of this truism than to witness the explosion of activity that has taken place since buckminsterfullerene was finally made in usable quantities in the latter half of 1990. The original work of Kroto and of Smalley and Curl and their students had stimulated much thought during the previous five years, but it was not until the announcement of the Krätschmer–Huffman technique and its rapid emulation by many other groups that the way was really open to all comers.

The journal *Science Watch* logs the citations of influential papers as they appear in subsequent papers in the various branches of science. In chemistry, fullerenes were the subject of nine of the ten most frequently cited papers by 1991. In the following year, all ten most cited papers were devoted to what was rapidly becoming an entire new branch of the subject. In the beginning, some hundreds of papers were given over to the characterization, by the means described in earlier chapters and others, of the new molecules. The emphasis soon shifted, however, to the making of fullerene derivatives and to the exploration of the likelihood that buckminsterfullerene, the other fullerenes, and their compounds might be of some practical use to humankind. By 1994, it was thought that perhaps as many as ten papers might be appearing every day on the science and technology of the fullerenes.

Small wonder, then, that buckminsterfullerene and its progeny began to attract wider notice. After 1990, the molecule enjoyed sustained coverage in the popular science media and beyond. It became a suitable subject for television quizzes. In Britain, it even came to the notice of the House of Lords.

It is not often that a new molecule is the subject of political debate. What these politicians have to say about it, as well as the quaint way they say it, reveals much about what society at large understands and expects of science. The following is the complete text of four minutes of debate (if this lordly banter earns the name) that took place in Britain's unelected upper legislative chamber on the afternoon of 10 December 1991. The dialogue appears under the heading "Buckminsterfullerene: Research Support":[1]

Lord Erroll of Hale asked Her Majesty's Government:

What steps they are taking to encourage the use of Buckminsterfullerene in science and industry.

The Parliamentary Under-Secretary of State, Department of Trade and Industry (Lord Reay): My Lords, the Government have been following with interest the emergence of Buckminsterfullerene and support research currently being undertaken at Sussex University through the Science and Engineering Research Centre [*sic*; it is in fact a Council]. However, it must be left to the judgment of firms whether they wish to pursue research into commercial applications of Buckminsterfullerene and other fullerenes.

Lord Erroll of Hale: My Lords, I thank my noble friend for his Answer, which is good so far as it goes. Can he not offer more substantial support in this country for the development of this exciting new form of carbon? It is already being manufactured in no fewer than three factories in the United States.

Lord Reay: My Lords, as I said, the Government continue to fund academic research into Buckminsterfullerenes at Sussex University. Many grants have been made available since 1986 which have gone towards that research. SERC also supports a number of researchers investigating the theoretical aspects of chemical bonding relating to fullerenes. The Government funding for collaborative research between industry and the academic world into the commercial application of Buckminsterfullerenes may be available also under the Link [through which the state matches private investment in specific research topics] or other schemes.

Baroness Seear: My Lords, forgive my ignorance, but can the noble Lord say whether this thing is animal, vegetable or mineral?

Lord Reay: My Lords, I am glad the noble Baroness asked that question. I can say that Buckminsterfullerene is a molecule composed of 60 carbon atoms known to chemists as C60. Those atoms form a closed cage made up of 12 pentagons and 20 hexagons that fit together like the surface of a football.

Lord Williams of Elvel: My Lords, is the noble Lord aware, in supplementing his Answer, that the football-shaped carbon molecule is also known, for some extraordinary reason as "Bucky ball"? It created a considerable stir within the scientific community. As the British Technology Group either has been, or is shortly to be, privatised, is this not a case that should be taken up by the privatised BTG and promoted as a British invention?

Lord Reay: My Lords, the privatised BTG will be free to take that decision. We do not feel that it is for the Government to say whether or not Buckminsterfullerenes have commercial usages, nor whether companies should become involved. It must be up to them.

Lord Renton: My Lords, is it the shape of a rugger football or a soccer football?

Lord Reay: My Lords, I believe it is the shape of a soccer football. Professor Kroto, whose group played a significant part in the development of Buckminsterfullerenes, described it as bearing the same relationship to a football as a football does to the earth. In other words, it is an extremely small molecule.

Lord Campbell of Alloway: My Lords, what does it do?

Lord Reay: My Lords, it is thought that it may have several possible uses; for batteries, as a lubricant or as a semi-conductor [*sic*. Does he mean superconductor?]. All that is speculation. It may turn out to have no uses at all.

Earl Russell: My Lords, can one say that it does nothing in particular and does it very well?

Lord Reay: My Lords, that may well be the case.

Lord Callaghan of Cardiff: My Lords, where does the name come from?

Lord Reay: My Lords, it is named after the American engineer and architect, Buckminster Fuller, who developed the geodesic dome, which bears a close resemblance to the structure of the molecule.

The debaters speak for various political parties and with diverse expertise. Only one, Lord Erroll of Hale, has any declared scientific background. He is an engineer by training, and was a member of the House of Lords Select Committee on Science and Technology. Lord Reay is enjoying a brief tenure as the Conservative government spokesman in the House of Lords on matters of trade and industry. Lord Williams is the deputy leader in the House of Lords of the Labour opposition party and opposes Reay as its spokesman on trade and industry and defence. Lord Callaghan is a former Labour prime minister. The others are by and large lawyers or historians.

The nature of their discussion provides confirmation, if confirmation were needed, of the persistence of the "two cultures" of sciences and humanities, the one quite unable to communicate with the other, that C.P. Snow feared a generation ago. This is a far cry indeed from Snow's call for "politicians, administrators, an entire community, who know enough about science to have a sense of what the scientists are talking about".[2]

Or course, it is amusing to find Lord Reay deducing that buckminsterfullerene is an extremely small molecule rather than that molecules in general are comparatively small. And one wonders if Baroness Seear does not really have a pretty good idea whether C_{60} is animal, vegetable, or mineral – although this is, perhaps, rather a good philosophical question to ask in regard to molecules of a certain size; buckminsterfullerene, however, is definitely mineral.[3]

Two serious points remain. The first is: what *does* it do? The second concerns the assumption underlying this question that "it must do something". This assumption is widespread among the non-scientific community. It is of significance because it influences policy-makers as they give shape to the scientific endeavour, dividing research into the basic and the applied, and deciding how to apportion funds between them.

I shall shortly describe some of the things buckminsterfullerene does do, some of the things it might have done but does not, and some of the things it might yet be shown to do. But first, we return to the vexed question of the supposed link between discovery and application. Some fundamental scientists find that the question recurs with such tiresome regularity that they have devised stock rejoinders ready for their challengers. Observers of the scientific scene, meanwhile, have devoted some energy to exploring the nature of the connection.

A biologist, Marston Bates, is the predictably irritated party in John Passmore's study of science in society, complaining of ordinary people that: "they immediately ask 'What good is it?' and mean 'In what respects is its existence helpful to human beings?' They assume that is, that whatever is must be for human use."[4] This secondary question is helpful in seeking to understand the motivation of non-scientists when they press their naive enquiry. Yet, it is often not clear what may or may not have a human use. Knowledge of planetary orbits or the discovery of a new species might seem at first to offer little in the way of practical benefits for humankind. Such benefits can accrue, even if sometimes only after the passage of years or centuries. But then again, as Passmore shows, there is no certitude of this:

> Considering, even, the science of our century, the knowledge that organic molecules exist in interstellar space or that the oldest moon rocks are of such and such an age does not by itself offer us any greater control over nature. Rutherford was wrong when he asserted, to the very day of his death, that his investigations into the structure of the atom were, and would remain, technologically useless. But he could have been right....[5]

It may be folly to rule out the technological potential of a discovery in this way. But on the other hand, it is not possible to proceed in basic research with a narrow, specific goal in mind for the solving of a practical problem of humankind. Nor can one know the potential harm or benefit of a discovery before it is made. A scientist engaged in basic research cannot say with much accuracy even to the body that supplies his or her funds whether such and such a project will prove useful or useless. It is thus unreasonable to blame the fundamental scientists if their discoveries are later put to destructive use. But by the same token, these scientists cannot expect to take all the credit for work that has a good outcome. Applied research is very different. Here, the scientists and technologists involved do deserve a share of the credit – or blame – for an application made possible through their work. However, it is fair to say that no small part of the enduring confusion between basic and applied research in the minds of the public and of scientists themselves stems from the fact that, being only human, scientists in both camps are wont to try to take credit and avoid blame out of proportion to their cause for doing so.[6]

Of course, this division between basic research and applied research is not hard and fast. The grey area between the two leaves considerable scope for cunning exploitation. In seeking funding, the fundamental scientist will often find it advantageous to hold up the promise of utility, not because it is of driving personal interest, but because this is what will interest the funding body:

> The scientist may not know whether his work will have technological applications; there may just be a chance that it will. He can ease his conscience as a scientist by appealing to the fact he does not know how his work will be applied, as a fund applicant on the ground that he does not know that his work will be useless.[7]

At the moment of its discovery, buckminsterfullerene – "or a derivative", to use Kroto and Smalley's careful phrase in *Nature* – could have been a cause of cancer or an agent in its cure. It could have contributed to weapons of war or to instruments of pleasure and prosperity. It could have polluted or helped to clean up pollution, or any number of other things. But such thoughts would have been remote at that instant. For scientists are not beholden to forge this link.

The appeal to scientists of apparently useless discoveries (or at least their disinterestedness in exploring their uses) has a long pedigree. Lewis Wolpert characterizes Archimedes in this vein.[8] But it can only really be since the Industrial Revolution that scientists have come to be regarded as capricious if they show no inclination to explore the possibility that their work might have a direct use.

Michael Faraday is the most conspicuous example of such a scientist:

> One question they asked him repeatedly, when they visited the Royal institution and saw him perform some experiment: 'But what's the use of it?'

> Faraday usually made the same answer that Benjamin Franklin made under similar circumstances: 'What's the use of a baby? Some day it will grow up!' On one occasion, however, he replied differently. Mr. Gladstone, then Chancellor of the Exchequer, had interrupted him in a description of his work on electricity to put the impatient inquiry: 'But, after all, what use is it?' Like a flash of lightning came the response: 'Why, sir, there is every probability that you will soon be able to tax it!'[9]

Faraday's pre-eminent achievement, of course, was his discovery of the interrelation of electricity with magnetism. The circumstances in which Faraday made this supreme addition to the storehouse of pure scientific knowledge are familiar today, not least to Kroto and Smalley, Huffman and Krätschmer, and others who played their part in the discovery of buckminsterfullerene.

Faraday spent much of his time in research on optical glasses and steel alloys, work funded by the defence agencies of the day including the British Admiralty. Despite its obvious practical purpose, this research proved as fruitless for immediate application as any piece of basic science. It was only when he could afford to set it aside for a time that Faraday was able to perform his famous experiment demonstrating the phenomenon of electromagnetic induction.[10]

Jacob Bronowski acknowledges that Faraday was not fussed about finding practical uses for his discoveries. He did not seek patents, perhaps because of his religious beliefs and a distaste for the accumulation of wealth.[11] But, Bronowski believes, there was a more subtle motivation that drove the great scientist, one that did stem from the ultimate promise of utility of the new science: "... Faraday worked all his life to link electricity with magnetism because this was the glittering problem of his day; and it was so because his society, like ours, was on the lookout for new sources of power."[12] But Faraday stated his own position, making it clear that he was happy to continue worrying away at aspects of the problem rather than turn his attention to the provision of electrical power to humankind:

> I have rather, however, been desirous of discovering new facts and relations dependent on magneto-electric induction, than of exalting the force of those already obtained; being assured that the latter would find their full development hereafter.[13]

In the event, it was fifty years until the electric motor was sufficiently developed for practical use. If Faraday and his contemporaries such as Joseph Henry had been minded to alter the course of their labours, this development might have been greatly accelerated. Such a shift would have demanded not only the dedication of the scientists but also the cooperation of their patrons. Even then, there could be no guarantee that application would follow on the heels of discovery.

Yet the wish that it should remains strong. For as J.D. Bernal wrote:

> Most scientists as well as laymen are content with the official myth that that part of the work of the pure scientists which may have human utility is immediately taken up by enterprising inventors and business men, and thus in the cheapest and most commodious way possible put at the disposal of the public. Any serious acquaintance with the past or present state of science and industry will show that this myth is untrue at every point, but just what is the truth is something much more difficult to find out.[14]

In subsequent paragraphs, Bernal attempts to discover this truth, but with little success. He is finally driven to refer to "The Time Lag in the Application of Science" as if it were somehow automatic.[15] But such a time lag is not a general condition. Nor is it in some way "necessary". There is no "mathematical" relationship that can describe this lag. It is not possible to say, for example, that the magnitude of a discovery is correlated in a particular way with the rapidity – or torpidity – with which it is put to general use. Nor is there any discernible correlation either between the provision of funding and the making of a discovery, or between the provision of further funding and learning whether that discovery can be put to the service of humankind, and if it can, how speedily this may be done.

If anything, there may be a kind of inverse relation at work. The discovery that is devoutly wished may be unattainable for a long time, during which its use is imagined in all manner of things. Its eventual arrival is thus discounted in the way that financial markets have often taken account of a widely anticipated interest rate change so that when the change actually comes its effect on the market is negligible. In this climate, the serendipitous discovery comes to possess an intrinsic importance; or at least it seems to because it exerts a larger spontaneous effect in the market of scientific ideas. This perhaps encourages the tacit assumption that every highly novel discovery *must* have some use. There is no particular justification for this view, but as the dialogue in the House of Lords shows, it persists, and, in the absence of a more rational scheme, serves as a basis for apportioning funds among research and development organizations. Thus, no funds were specifically made available for Kroto and Smalley's discovery of buckminsterfullerene. They obtained some funding in order to try to find how to scale up production of the new molecule. But Krätschmer and Huffman, who eventually discovered how to do this, had no funding for this purpose. These crucial discoveries made, there are suddenly abundant resources being poured into further research on the fullerenes and their applications.

A quick review of some other scientific landmarks will serve to illustrate the apparent randomness of the connection between discovery and application. In the 1860s, it had taken just four years for August Kekulé's solving of the ring structure of benzene to begin to inform the manufacture of new organic dyes. But neither Kekulé nor anyone else at the time could have envisioned the nature or breadth of uses to which benzene and its relations would one day be put.

Wilhelm Röntgen, on the other hand, immediately saw the potential for application of his accidental discovery of X-rays when he published his paper in the 1890s. (And this despite the fact that, as his enigmatic name for them implies, he did not fully understand the provenance of his rays.) Within a year, X-rays were put to use in medicine as he had foreseen, and they were used very shortly afterwards by archaeologists to examine mummies exhumed from tombs in Egypt. Röntgen's discovery unleashed a frenzy of scientific activity much like that seen with the fullerenes, with perhaps a thousand papers appearing within a year of his report.[16]

In this century, the interval between discovery and application has been equally hard to fathom. Nylon was on the market just three years after Du Pont's Wallace Carothers had formulated it. It was more than a decade before Fritz Haber's process for the conversion of atmospheric nitrogen to ammonia, the key to a lot of vital chemistry, was scaled up for industry. The transistor took well over a decade to become widespread and the laser even longer.[17] In the years after its invention, the laser was commonly described as an invention in search of an application. The label can fairly be said to apply still to various other innovations from memory metals to superconductivity.

Whatever else it was, the discovery of a new form of carbon certainly was not the answer to the glittering problem of the day. But the fact that buckminsterfullerene did not represent the desired culmination of a carefully planned and executed research programme did nothing to diminish people's interest in what it might be good for. Indeed, the very serendipity of the discovery stimulated speculation. Everybody from *Time* magazine to the Central Intelligence Agency wanted to know what it might do.

The earliest speculation as to what buckminsterfullerene might be good for came, it will be remembered, in the paper describing the original discovery of the molecule. Harry Kroto, Rick Smalley, and Bob Curl discussed the prospects for new classes of chemistry based on joining transition metal atoms to the carbon cages or on placing atoms inside them and on having a "topologically novel aromatic nucleus". They thought that the stable, symmetrical carbon structure might be a catalyst, its unusual surface providing a stage upon which some reactions might be led to occur with greater ease than elsewhere. Most specifically, they reasoned that the fluorinated buckminsterfullerene, $C_{60}F_{60}$, "might be a super-lubricant".[18] Their argument was based on the fact that polytetrafluoroethylene, the low-friction coating substance known as Teflon, also employs fluorine atoms as the sole element embellishing a plain carbon skeleton. Slippery balls of such stuff would presumably roll past one another better than ever.

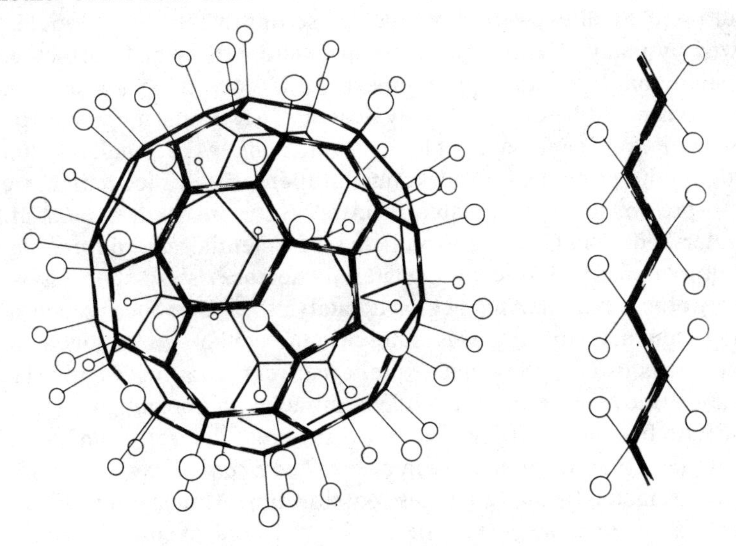

Hopes that fluorinated C_{60} might be a super-lubricant were based on its resemblance to Teflon (right)

The fact that Kroto and Smalley entertained such suggestions is testament to their keenness to persuade people of the importance of their discovery. In his cover letter to *Nature*'s Washington editorial office, Smalley had written: "We believe it is not only an intrinsically exciting discovery, but it is likely to have wide practical chemical applications."[19] It is also, perhaps, an unwitting illustration of the increasing pressure placed on fundamental scientists to do "relevant" work and even to play a part in the development and commercial exploitation of their new knowledge.

Some scientists are happy to make the connection between science and technology, between discovery and application. Others decide that their strength lies in pure, curiosity-driven research and that their talents would be poorly deployed in developing their discoveries. For their part, Smalley and Kroto did not shrink from general speculation, as we have seen, but they only pursued the possibilities that they raised to the extent that they overlapped with their own interests in the fullerenes' physics and chemistry. Smalley today has become interested in the field of nanotechnology, where fullerenes and related new carbon compounds seem likely to be important players. Kroto remains more preoccupied with the fundamental science – the rich chemistry of the fullerenes and whether they or related species are present in the interstellar medium, ready to explain the unknown spectral lines.

Kroto and Smalley were nevertheless setting a trend that would be followed by many scientists. In their speculation, they had been careful to stipulate that it would require a means of large-scale synthesis before applications could be explored or evaluated. Once Krätschmer and Huffman's method became available and scientists were finally able to investigate the properties of buckminsterfullerene, the speculation escalated to previously unimaginable levels. Writing just two months after this watershed, Don Cox of Exxon listed the potential for exploitation in "organic chemistry, lubrication, materials science, solid-state physics, energy storage, polymer science, and catalysis".[20] The sheer range was exciting, but his phrasing was safely vague. Other groups were less cautious. It is usually the popular press that runs away with enthusiasm for a new discovery, predicting unlikely miracles and opening up utopian vistas. With buckminsterfullerene, it was the press that felt bound to hold an excitable research community in check. The *Economist* wrote: "Scientists are fascinated by the fullerenes' possibilities. Almost too fascinated: in their excitement, some talk of far-fetched, and even silly, applications."[21] And *Science* reported how:

> When asked what 60 carbon atoms bonded into a soccerball-shaped molecules called buckminsterfullerene – alias buckyball, alias C_{60} – ultimately will bring to chemistry, most chemists start out by saying, "It's too early to speculate." Then they speculate, often at length and sometimes with reckless abandon.[22]

The appetite for the exploration of applications was sharpened in many cases by the chance to reap a commercial reward if patents could be obtained for them. The issue of the patentability of buckminsterfullerene, its derivatives, and the ways of fabricating and making use of all of them, is not simple.

Any patent application must jump a number of hurdles. In order to be patentable, an invention must demonstrate three things: it must be novel in the sense that it would not seem in retrospect like an obvious development to one who is expert in the field ("skilled in the art" to use the jargon); it must also be novel in the sense that it is not covered by previous patents (it must exceed "prior art"); finally, it must have a use or at least the strong likelihood of one.

There are two kinds of patent, both of which are relevant to the fullerene story, and both of which pose thorny problems of legal interpretation. One covers a novel thing; the other covers a novel way of making or employing that thing.

At the core of the problem for the first kind of patent – for the "composition of matter" – is the question of whether buckminsterfullerene is an invention or a discovery. The blurred distinction between the two, and the quandary it poses for patent protection, is already a fact of life in genetics. Because of this and for concomitant ethical reasons, the patentability or otherwise of biological material, and especially of human biological material, has become a topic of controversy. Biotechnology companies claim to make "biological machines", eligible for protection by patent like other artificial inventions. As patent law stands, genes or gene fragments whose function is unknown – and which therefore have no immediate "use" – should not be patentable, although some patents have already been granted. For ethical and political reasons rather than reasons to do with the logic of existing patent law, there are moves to ensure that human genes remain unpatentable. However, genes isolated outside the human body whose function and, therefore, use are known may be declared patentable. Conflict arises because "human genes" may be made artificially, copied from nature, rather than harvested from the body. The fault lies not so much with those who wish to patent genetic material but with an outdated legal framework designed to protect the physical inventions of the Industrial Revolution that is incapable of dealing equitably with invention at the molecular scale.

Thus, plants and animals have been the subject of patent applications. A few patents have been granted. The most celebrated is the Oncomouse, developed at Harvard University, a mouse treated by the addition of extra genes to make it unusually prone to cancer. A plant or animal patent cannot cover the form of life alone. Instead, it must describe some non-natural modification which it has undergone by human hands and/or the use to which it is put. The Oncomouse is a mouse that has been tampered with in such a way, and it has its use in cancer research. Another animal subject to a patent application is a nematode worm that attacks slugs. The worm is naturally occurring and cannot be patented. It naturally attacks the slugs and this biological occurrence cannot be patented either. But agricultural researchers hope to patent its use as a pesticide, nevertheless. The discovery that the worm attacks slugs is not the novelty. But adaptation of the worm as an element of a viable technology for agricultural use based on this discovery may be deemed novel and may be patentable. In other cases, the discovery that some species does something useful may not be patentable in its own right, but if it is necessary to go to some effort to isolate that species in order to exploit this property then the special process of doing so or the material so obtained may be patentable.

Chemistry is beginning to suffer some of the same confusion. Antibiotics are generally classified as discoveries, for example. More complex "designer drugs" created with the benefit of computerized techniques for planning the positioning of various chemical groups are more like inventions – one is setting out to design molecules that one hopes will have particular uses.

Many of these questions are quite as unresolved in the case of the fullerenes and their derivatives as they are for more complex molecules of biochemical interest. In 1985 and at various times thereafter, Harry Kroto and Rick Smalley considered pursuing patents in relation to their discovery. Partly out of distaste for the laborious procedures involved, they rejected the idea. Had they persevered, they could have sought a patent for buckminsterfullerene itself – a "composition of matter" patent. They could also have sought to obtain a patent to protect the way in which they made it, by laser vaporization of graphite. But at this time, though the novelty of the molecule was clear (at least, to most people) and though the process of making it was certainly not an obvious one, they were missing the third necessary ingredient – the use. This was something they could not hope to demonstrate without being able to make the new molecule in much larger quantities. The fact that they had not made it in the first place by design and assembly but had stumbled upon it serendipitously favours its being regarded as a discovery. When it was later found naturally occurring in a variety of circumstances, it could no longer rightfully be called anything but.

In 1990, Wolfgang Krätschmer and Donald Huffman sought a patent for their improved way of making buckminsterfullerene in visible quantities. In fact, they had first trodden this path in 1987 when they first suspected that they were making the molecule by the electrical heating of graphite rods, but they were forced to withdraw their patent applications when they could not get the process to work reliably. If their method had remained the only way of making buckminsterfullerene, then Krätschmer and Huffman might have stood to have benefited greatly from the grant of their patent. The many laboratories and corporations that dived into fullerene research would have had to obtain licenses to make their own fullerenes. Before long, however, research teams were making the molecule by a variety of methods. Once it proved so easy to make, the best hope for those ambitious of obtaining patents lay in finding particular uses for it.

Some of the early derivatives of buckminsterfullerene present a rather different picture. The metallofullerenes – carbon cages containing an atom of various metallic elements – fabricated by the Rice group in the years after the 1985 discovery do have the element of intent that was absent from the discovery of buckminsterfullerene itself. Smalley has applied for patents covering these molecules and the way of making them. Other groups have made other compounds and are seeking, and in some cases have obtained, patents for their composition.

Later, when other groups were able to make fullerenes, the patent traffic increased. If the individual buckminsterfullerene molecule was not patentable, then was "fullerite", the bulk crystalline form? And if not that, what about a special film of buckminsterfullerene constructed, as would later be undertaken, in order to conduct electricity in a certain way or to let pass certain gas molecules but not others?

A new property of a known substance – and such properties were being discovered practically weekly after 1990 – is not patentable. However, if it is then found that the property can be exploited for gain, then the corresponding invention may be patentable. Companies such as Du Pont and Exxon, AT&T, Xerox, and Hughes in the United States as well as Hoechst in Germany and NEC and Sumitomo in Japan have applied for patents to protect potential applications of the fullerenes and related novel carbon compounds. It is rumoured that no other single scientific discovery at present generates more activity in the world's patent offices.

The following survey of the properties of buckminsterfullerene and of the applications which might flow from them is not entirely in chronological order of their investigation. Nor is it in order of potential importance of the discovered property, simply because it is not yet possible to tell what this order is.

We begin simply with the most apparent physical characteristics – the ones that may be appreciated at a glance. It was because of its distinctive appearance that buckminsterfullerene quickly earned its nickname as the buckyball. The spherical molecular frame does indeed behave like the resilient surface of a ball. When Robert Whetten and his colleagues at the University of California at Los Angeles accelerated ionized buckminsterfullerene molecules to speeds of 17,000 miles per hour and fired them at walls of graphite and silicon, they bounced right back.[23]

One way of employing these molecular missiles might be as a rocket fuel. Researchers at the Jet Propulsion Laboratory in Pasadena, California, believe it may be possible to create beams of accelerated fullerene ions to drive spacecraft or steer satellites. The noble gas element, xenon, has traditionally been favoured for this task. The idea is to produce forward momentum for propulsion by jettisoning the accelerated particles. The requirements are therefore that the particles should have a high mass, be stable in storage before they are used, and that they be easy to ionize so that they may be accelerated by electrically charged plates. A molecule of sixty carbon atoms is more than five times heavier than an atom of xenon but twice as easy to ionize.[24]

The behaviour of individual buckyballs is in marked contrast to the bulk material. An individual molecule is extremely robust – in some respects harder than diamond. When stacked in a crystalline array with other identical molecules, however, the resulting material, the "fullerite", is soft. While the sixty carbon atoms in each molecule are held in position by strong chemical bonds, each carbon sphere as a whole is only bound to its neighbouring spheres by weak van der Waals bonds, the same bonds that hold flat layers of graphite together but cannot prevent them slipping past one another when a modest force is applied.

There were hopes for buckminsterfullerene as a lubricant on these grounds. The solid would perhaps be as slippery as graphite, not only in one plane like graphite, but in any direction that the crystalline material would allow planes of molecules to slip past one another. These hopes promptly evaporated along with the buckminsterfullerene itself at a mere 300 degrees centigrade, a temperature far exceeded by the friction generated in devices where lubricants are required. This was perhaps the first imagined application of buckminsterfullerene to fall at the hurdle of some other fundamental property of the molecule. It would not be the last.

Because of the forgivingness of the van der Waals bonds that keep the individual fullerene balls spaced their regulation three ångströms apart, the bulk fullerene material may be squeezed elastically up to a point where the molecules are essentially pushed hard up against each other. It is at this level of compression that the properties of the individual balls are felt and the material becomes harder than diamond. When Manuel Núñez Regueiro and a group at the Centre des Recherches sur les Très Basses Temperatures, Grenoble, squashed buckminsterfullerene even harder in a simple tool known as a diamond anvil, they found that it suddenly transforms to diamond, an occurrence of potential interest to industry. It is already possible to transform graphite in this way, but the fullerene transition took place at a lower pressure.[25] Fullerenes, especially C_{70}, have also been found by scientists at Northwestern University in Illinois to assist the growth of thin films of diamond which may be used to harden the edge of tools or to create electrically insulating barriers in electronic circuitry.[26]

The next obvious quality is the hollowness of the buckyball. Its nominal diameter from one carbon nucleus to its furthest opposite number is seven ångströms. After allowing for the "space" occupied by the electrons orbiting in molecular bonds across the outside and inside surfaces of the sphere, the interior has an effective diameter of about half this amount, large enough to trap an atom of more or less any element of the periodic table. It seemed likely that the properties of a trapped atom might be modified in useful ways by the cage and that the cage itself might provide a means of temporarily isolating such an atom from particular environments for storage or transport. Calculations confirmed this supposition, and also indicated likely new electrical conduction and semiconductor behaviour. The carbon cage might provide a new way to manipulate atoms of radioactive isotopes used in radiotherapy. Or it might provide an alternative to the use of ultra low-temperature matrices for the trapping of highly reactive ions.

At a second remove, a number of other potential uses become apparent from the physical nature of the fullerenes. The physical geography of a fullerene surface, with its cobbled pattern of carbon domes, might provide a fit for certain atomic or molecular species that impinge upon it, but not for others. It might thus be useful as a catalyst or as a filter. Douglas Loy and Roger Assink at Sandia National Laboratories in Albuquerque, New Mexico, found that certain gases would diffuse into pure crystals of buckminsterfullerene.[27] These gases can be cajoled free again by raising the temperature or by reducing the surrounding gas pressure. Molecules of hydrogen, nitrogen and oxygen nestle within the cavities between the fullerene molecules, but molecules of methane do not. This fullerene surface might therefore be useful in the purification of natural gas which is largely methane but frequently contaminated with other small molecules.

Arthur Hebard at AT&T Bell Laboratories in Murray Hill, New Jersey, has constructed a membrane a quarter of an inch square and just one hundred buckminsterfullerene molecules thick by depositing the molecules on layers of silicon nitride and silicon, then etching these away to leave just the carbon. Such a membrane, held together solely by van der Waals bonds between the balls, might also pass some gases and not others. But Hebard's aim is more general. He wants to show that fullerenes can be handled in the sorts of processing environments used to make silicon-based electronic components, a matter germane to almost any technological exploitation.[28]

Dissolved in suitable organic solvents, buckminsterfullerene exhibits a variety of curious optical characteristics. In some circumstances, the behaviour is "non-linear" – the light one shines into the solution is not the same as the light one gets out. This property, which these molecules share with certain other organic molecules such as polymerized acetylene chains, arises through interaction of the incident light with the delocalized electrons in the molecular orbitals. In some cases the fullerenes alter the frequency of the light; in others they transmit proportionally less light when exposed to brighter sources. Thin films of buckminsterfullerene could serve as "optical limiters" providing eye protection in laser or welding environments. The effect operates at lower levels of light than other optical limiters, making the material potentially useful in protecting sensitive light detection equipment. It could also be used as switches in optical circuitry where photons take the place of electrons. Alan Kost and Lee Tutt at Hughes Research Laboratories in Malibu, California, who first noticed this behaviour in fullerene solutions, are trying to produce it in solid films whose greater stability would be a help in realizing practical applications.[29]

Another concomitant of the unusual electronic and geometric structure of the molecule is that it is also photoconductive, passing electricity when irradiated by a light of a suitable frequency. Researchers at Xerox Corporation and Du Pont have separately demonstrated this effect at visible wavelengths. This property is useful in light detection and in the electrostatic imaging processes employed in photocopiers and printers. The photoconductive effect relies on the promotion of electrons in the fullerene molecular orbitals into higher energy unoccupied orbitals where they achieve the mobility necessary for conduction. In larger fullerenes, the successive orbitals are more closely spaced, making it possible for light of lower energy to promote the electrons. Thus, it is thought that fullerenes with more than the standard sixty atoms may exhibit photoconductivity at infrared wavelengths, something that is currently expensive or impossible to achieve using silicon-based films.

One property was soon discovered which was to put all others in the shade. The phenomenon of superconductivity is, of course, something of a holy grail in applied physics. It describes the condition in certain materials which, at temperatures very close to absolute zero, are able to conduct electricity with no losses due to resistance. If the effect can be produced at higher temperatures, it would have the potential virtually to eliminate losses in energy transmission. Much effort has been devoted to this end and a number of more or less esoteric materials have been found to exhibit superconductivity at higher (though still not exactly convenient) temperatures.

The discovery that buckminsterfullerene was another such material came from scientists at AT&T Bell Laboratories.[30] Ironically, no new property could have been less welcome at the renowned research centre. The ghost of high-temperature superconductivity already stalked the corridors there in the aftermath of a programme of massive investment that had yielded little of commercial merit.

The story of this discovery has humble beginnings in investigations of buckminsterfullerene's less exceptional electrical behaviour. Some linear and planar carbon compounds show improved electrical conductivity when "doped" with alkali metals. After earlier unsuccessful forays, it was in early 1991 that Robert Haddon and others at Bell Laboratories doped C_{60} and C_{70} by exposing them to the vapour of such alkali metals as lithium, sodium, and potassium and found that they too would conduct.[31] The metal atoms slot into the one octahedral and two tetrahedral voids per fullerene molecule in the crystal lattice, rather like tennis balls distributed among the gaps in a stack of soccer balls. As they do so, they release electrons to become ions. The fullerenes accept the electrons given up by the metal atoms, and the material, with its crystalline array of spherical molecules effectively in contact because of their delocalized electron populations, begins to conduct. In the experiments it was found that lithium and sodium ions were too small to occupy the octahedral gap satisfactorily; caesium was too big; and potassium and rubidium produced the highest conductivity.

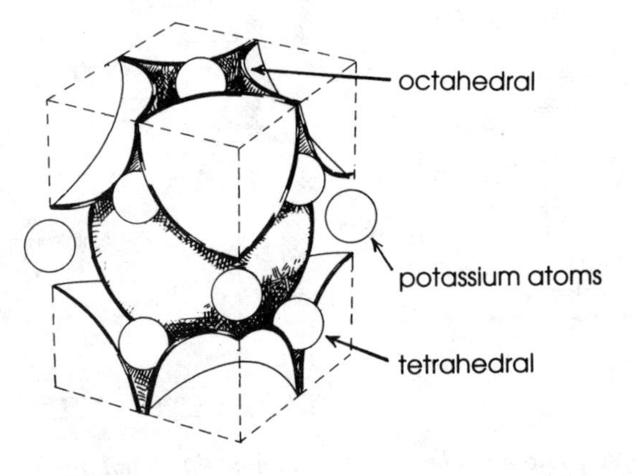

Alkali metal atoms fill the tetrahedral and octahedral interstices between the C_{60} stacked spheres

In order to understand what is happening, we must return to the molecular orbital diagram for buckminsterfullerene – the league table of energy levels of the molecule's pi bonds. It happens that the lowest unoccupied molecular orbital can accommodate three pairs of electrons in all.

It is initially empty and the material is insulating. As each alkali metal atom is added per fullerene molecule, it contributes one electron to this orbital. The conductivity rises in proportion to the number of metal ions added, reaching a maximum when there are three metal ions to each fullerene molecule (it is a happy coincidence that the crystal structure can accommodate three metal ions per fullerene molecule, although both threesomes stem from symmetry considerations); in this case, the three donated electrons are unpaired. Because these orbitals are only half filled, it is possible for the electrons in effect to migrate from one molecule's half-filled orbitals to its neighbour's. This mobility allows a current to flow, although the conductivity is far lower than that of a conventional metal. As the concentration of the alkali metal dopant is increased to a saturation level of six metal ions for each fullerene, the conductivity falls back to zero. Six electrons have been contributed, in three paired sets, the orbitals are full, and there are no vacancies at the same energy level between which these electrons may migrate, and thus no conduction takes place.

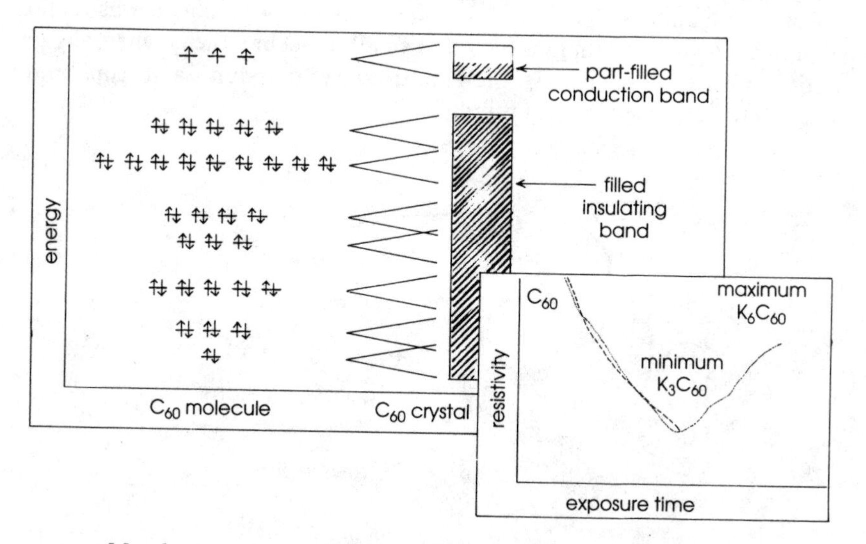

Metal atoms donate electrons to a conduction band producing first a rise and then a fall in conductivity

Haddon billed the doped buckminsterfullerene as the first three-dimensional organic conductor. Most previously known organic compounds that conduct electricity are one-dimensional or two-dimensional polymers. The conduction arises because delocalized electrons are free to pass along a molecular chain or across a molecular plane. Graphite doped with potassium, for example, is a good conductor in the line of the graphite planes, but a poor one perpendicular to these lines. A three-dimensional conductor such as the doped buckminsterfullerene may carry an electric current with equal ease in any direction and eliminates the very real practical problem of having to align the material to produce optimum conduction.

Haddon's earlier involvement in AT&T's high-temperature superconductivity programme together with the novelty of this three-dimensional conductivity dictated that he should investigate whether doped fullerenes would be superconductors. It is a rule of thumb in the field that a relatively poor metallic material, such as he had shown the doped fullerenes to be, is sometimes a suitable candidate for superconductivity. It was also known that many otherwise promising one- and two-dimensional superconductor materials fail to deliver because the move to low temperatures produces a transition to an insulating ground state. The three-dimensionality of the fullerenes might just side-step this difficulty.

The potassium-doped buckminsterfullerene, with the formula K_3C_{60}, had been the best conductor. So it was with this material that Haddon and his team set to work. They cooled specially sealed sample films to which electrodes were attached and found that electrical resistance fell to zero below a transition temperature of 18 kelvins (–255°C).[32] Although very low by everyday standards, it was an astonishingly high temperature for superconductivity. The material's two-dimensional analogue, potassium-doped graphite, becomes superconducting only when its temperature falls to within a fraction of one degree of absolute zero. This new substance exhibited superconductivity at the highest temperature yet observed for a molecular material as opposed to a metal alloy or a ceramic.

In another month, they had observed superconductivity in the rubidium-doped buckminsterfullerene at an amazing 28 kelvins.[33] Just days later, two other groups – one at Argonne National Laboratory in Illinois and Robert Whetten and François Diederich's group at the University of California at Los Angeles – reported similar results for the rubidium.[34]

Superconductivity is one of those important arenas of modern science where theoreticians can do little but scratch their heads. The effect is observed; but the reasons for it are not fully understood. Such theoretical models of superconductivity as do exist are notoriously inadequate. Superconductivity of the doped fullerenes squares with current theories, although since those theories are frustratingly incomplete, it is perhaps more noteworthy to comment that these new experimental results were valued for the contribution they might make towards refining the theory as much as for their own sake.

When isolated molecules with their ladders of molecular orbital energy levels are brought into proximity – as when they crystallize, for example – each line at a precise energy smears out to become an energy band. Bands fully populated with electrons are insulating. Partially filled ones are conducting. Metals happen to have such bands at a level where they are partially filled, hence their power of conduction. (Imagine a sealed tank of water placed upon a seesaw. Tipping the seesaw is analogous to applying an electric field or voltage. The water in a half-filled tank moves to the downhill side – a current flows. In a full tank, it cannot.) Semiconductors have electrons that may be easily promoted from a full band into an empty band at slightly higher energy where the electrons become mobile. More properly expressed, the Fermi level, that energy level at which the probability of finding an electron is exactly one half, lies within the conduction band for metals, within the valence band of bonding electrons for insulators, and between the two for semiconductors.

Superconductivity relies upon a mysterious pairing of electrons within the conduction band. The so-called Cooper pairs of electrons, perhaps mediated by quanta of vibrational energy arising from changes in the vibrational states within or between the fullerene molecules, provide a means for a current to flow without hindrance through the lattice. Cooper pairs form more readily in a narrow conduction band. When it is broad, there is less inducement for the electrons to pair up in this way.

The doped fullerenes provide an environment in which there are abundant opportunities for molecular vibration and also ample numbers of electrons of appropriate energies. The fullerenes doped with the larger alkali metal ions are found to be better superconductors – rubidium is better than potassium, and a compound with a few caesium atoms squeezed into the larger, octahedral, interstices in place of some of the rubidium, Rb_2CsC_{60}, is better than all of them. This is because the larger ions push the lattice apart slightly; thus the fullerene molecules become more nearly isolated one from another; their bands become narrower; and more electrons form the Cooper pairs needed for superconductivity.

The Bell Laboratories group and others, including Whetten and Diederich, Rick Smalley in collaboration with John Weaver and others at the University of Minnesota, and a group at the NEC Corporation in Japan, soon pushed the superconducting transition temperature up still further to greater than 30 kelvins. Each new result was a precious step closer to the 77 kelvins at which liquid nitrogen boils, the realistic threshold for many practical applications of superconductivity.

The high temperature at which the doped fullerenes become superconductors – higher than all but one of all known superconducting materials – is of course their principal attraction. But there are other advantages.

Fabrication of these materials for use in practical devices is potentially rather more straightforward than for some other superconductor materials. The fact that they conduct and superconduct equally well in all directions means that it would be unnecessary to consider orientation of the crystals of the material during device fabrication. Because the conduction properties change markedly depending on the proportion of alkali metal dopant, it might one day be possible to fabricate "heterostructures" with regions offering insulating, semiconducting, conducting, and superconducting properties exactly where they are required on a continuous stretch of fullerene substrate.

Here too, however, there is a powerful brake on the potential of such a technology. The doped fullerenes degrade rapidly in air. In its way, this was a repetition of the lesson learned at Bell Laboratories in its previous exploration of high-temperature superconductivity. Any application demands a material that not only demonstrates the required effect but which may also be manipulated in everyday circumstances. Superconductivity is a fine idea, but the picture currently conjured up of superconducting devices wasting no electricity yet yoked to massive refrigeration plants in order to make them work at all is hardly encouraging. Nevertheless, the fullerenes remain among the most favoured contenders for practical superconductivity. For Robert Haddon and many other scientists, the search continues for new combinations of metal dopants that are more stable while still superconducting.[35]

Doping the fullerenes may be regarded as a physical process or as a chemical one. Likewise, the trapping of atoms within a fullerene cage is a purely physical action; but what this action yields may in some ways be thought of as a new chemical compound. For many people, however, the priority was to explore the more conventional organic chemistry of the fullerenes. At the back of many minds must have been the knowledge of August Kekulé's discovery of the structure of benzene. Once this structural riddle had been solved, the way was opened for an entire new class of organic chemistry – one which, as it very quickly turned out, led directly to the synthesis of new types of compound with new uses for humankind.

It is worth recalling that, from the moment of its discovery, much had been made of the remarkable lack of chemical reactivity of buckminsterfullerene. These observations were not so much made as a comparison with other chemicals that one might find in a laboratory stock room, but with novel clusters that are made with the expenditure of much energy and exist but momentarily in elaborate apparatus. Nevertheless, pure, unadulterated fullerenes are indeed still rather unreactive by the normal standards of organic chemistry. At the heart of reactivity are two factors: electronic structure and shape. Buckminsterfullerene (and many of the other fullerenes) has an electronic structure that favours certain kinds of chemical combination. But its shape and symmetry leave it with no favoured handle by which an incoming reactant may grab hold of it. It has a structure that is slightly strained, which again would favour certain reactions, but the strain is spread with such evenness across the surface of each molecule that there is no weak point in its defences. In most cases, it would require a key to unlock the door to greater reactivity, but once this door was open paths would be found to lead out in many directions.

It would be impossible to provide a comprehensive discussion of the chemical derivatives of buckminsterfullerene, far less to divine their properties likely to prove useful to humankind. Any chemical reaction that can reasonably be imagined to occur might produce a viable compound of an entirely new class never seen before. Each compound might in turn have a range of properties. And each property might suggest a number of potential applications. These possibilities are, of course, multiplied for all the fullerenes with other than sixty carbon atoms.

It is convenient to group the fullerene derivatives into three kinds according to their topology. There are the endohedral complexes, the cage compounds. There are the exohedral compounds, fullerene molecules with one or more chemical groups added onto the outside of the spherical shell of carbon atoms, which offer the richest potential for new compounds. Finally, there is a third class of chemical variants where one or more of the carbon atoms on the shell of the molecule have been replaced by an atom or atoms of different elements.[36] These three kinds of derivative are not mutually exclusive, and permutations among them are sure to provide still greater interest.

Much of the groundwork on endohedral complexes – fullerenes with single atoms or possibly with very small molecules trapped within them – had been laid by Rick Smalley's group at Rice University in the months and years following the initial discovery of buckminsterfullerene. All the members of the original team – Smalley, Kroto, Robert Curl, Jim Heath, Sean O'Brien, Yuan Liu, and Qingling Zhang – played a part in this work. They had presented evidence to suggest that they had made quite a selection of fullerenes with carbon cages of different sizes and enclosing atoms of a variety of metallic elements.[37] The relation between the size of the atom and the size of cage required to enclose it strongly supported the hypothesis, then still unconfirmed, that fullerenes were indeed closed carbon structures. These "shrink-wrapping" experiments had served the useful purpose of banishing most people's scepticism that the fullerenes were actually spheroidal in shape as Kroto, Smalley, and Curl had postulated.

Invaluable though this work was, it had not of course yielded sufficient quantities of these "metallofullerenes" to embark upon an exploration of their chemistry. Nor did the advent of Krätschmer and Huffman's method greatly improve matters. With a vastly more efficient way to make empty buckminsterfullerene, it did not follow that it would become easier to make the endohedral metallofullerenes. While it was sufficient for Robert Haddon at Bell Laboratories to create potassium-doped buckminsterfullerene merely by placing a film of the carbon molecules in a potassium vapour, the procedure for efficiently enclosing an atom of potassium, or any other chosen element, clearly did not lie in starting with closed fullerenes. Logic called for some modification of the fullerene factory itself.

Such modification has proved comparatively simple to effect. In place of graphite electrodes, one employs hollow electrodes with a core filled with a mixture of carbon and the desired metal for encapsulation, or electrodes that have been dipped or otherwise suitably treated with this metal or an appropriate compound of it. The difficulty comes later. There is no guarantee in this method that each curling carbon shell will wrap itself around an atom of the chosen endohedral element on its way to forming a fullerene. In fact, only a modest proportion do so. Thereafter, it is a matter of collecting the soot that has formed, and separating not only the fullerenes from the carbon residue, but also of trying to separate the metallofullerenes from the empty fullerenes. By the end of 1993, groups at NEC and elsewhere in Japan as well as Smalley's group at Rice University and Don Bethune's team at IBM Almaden Research Center had encapsulated atoms of a variety of metallic elements, from the alkali metals, potassium and caesium, to transition metals, iron and cobalt, uranium, and more esoteric elements in the same group of the periodic table as lanthanum.[38] Separating the few of these metallofullerenes that are made in the chaos of a vapour of curling carbon network fragments and metal atoms from the mass of empty fullerenes remains a challenge.

It seems only fair that scientists should be experiencing this difficulty. After all, the principal novelty envisaged for these unusual compounds was that they would shield the enclosed atom's behaviour; it is clear that if the shielding is sufficient, then they will often be indistinguishable for the purposes of separation. Scientists hope to devise a chemical process that will unlock the carbon cages in a controlled manner, permit the insertion of desired atoms, and then lock them up again. This has been made to happen, albeit under somewhat gruelling conditions, with helium and neon by Martin Saunders and colleagues at Yale University and at the

University of Rochester. Buckminsterfullerene heated to 700 or 800 degrees centigrade in an atmosphere of helium or neon is thought to snap a bond to create a rupture in the carbon shell large enough to admit an atom of the noble gas. When cooled, a very few of the now restored fullerene molecules retain gas atoms. The discovery raises the prospect of enclosing the largest of the noble gases, radioactive radon, which might be transported through the human body encased in a suitable fullerene derivative for use in cancer therapy.

In the absence of experimental quantities, the properties of the endohedral fullerenes remain something of a mystery. There are two, somewhat contradictory, themes, depending upon whether one observes the carbon cage to be effectively "opaque" or relatively "transparent". If the cage is opaque, the enclosed atom will not exert much influence on the world outside the cage. The physics and chemistry of the fullerene will remain dominant. If this is the case, elements such as uranium or helium could be coaxed into organic solution by this means. Other, generally refractory, elements could be made volatile – the atom in its carbon balloon might evaporate where the naked atom would not. In the alternative scenario, the qualities of the trapped atom might shine through a "transparent" carbon cage. The Nobel prize-winner, Donald Cram, at the University of California at Los Angeles, has shown that this situation is more nearly the case for some other cage compounds. The cage would produce only subtle modifications of the standard behaviour of an enclosed atom which would become a swollen chemical shadow of its former self. It might be possible to conduct a new chemistry of interatomic reactions negotiated through a mesh of carbon with the genteel manners of dancers at a masked ball. The trapped atoms would exert a corresponding influence on their cages, perhaps introducing the useful ability to tune the basic physical and chemical properties of the fullerenes to particular needs.

Of course, it is in how the balance is struck between opacity and transparency that the potential for new technology lies. The chemistry of a completely shrouded atom would be of little interest. So would that of an atom whose chemistry remained apparently oblivious of the cage around it. The imagined applications for endohedral fullerenes, from electronic memories and switches to batteries and catalysts, will succeed or fail according to how well they marry the properties of the trapped atoms that are fundamental to the application with the protection offered by the carbon cage that may make it practicable.

The list of species that may be enclosed within a molecule of buckminsterfullerene is soon curtailed. There is no such limit for the formation of exohedral compounds. All the reactive groups in chemistry might try their hand as suitors for the carbon beauty. The exterior shell of the fullerenes presents an array of carbon-carbon double bonds like the "unsaturated" bonds in fats. The first matter for investigation is to see whether they might be saturated by the addition of atoms of hydrogen or of the halogen elements such as fluorine or larger chemical groups. Kroto, Smalley, and Curl had imagined breaking all thirty of the double bonds to add one fluorine atom to each of the sixty carbons in buckminsterfullerene thus obtaining a spherical analogue of Teflon that might be, in effect, a minute non-stick ball-bearing with super-lubricant properties.

The opportunity to put this idea to the test finally arose in 1991. The fluorination of C_{60} to form $C_{60}F_{60}$ proceeded well enough, if extremely slowly. A team that included Roger Taylor and David Walton together with chemists from the universities of Leicester and Southampton simply passed fluorine gas over solid buckminsterfullerene, observing a series of colour changes over a period of twelve days from brown to white.[39] They confirmed, by mass spectrometry and nuclear magnetic resonance spectroscopy, that white product was indeed the spherical Teflon analogue. The intermediate products signalled by the different colours seen over the days of the reaction were $C_{60}F_6$ and $C_{60}F_{42}$. It seemed that once the first double bond has been broken and the first two fluorine atoms added, things proceed smoothly to complete fluorine saturation, although the process becomes very slow as the final fluorine atoms squeeze in to find their place on the crowded fullerene surface. Exactly where the fluorine atoms add on to form the intermediate compounds – which double bonds they break – remained a matter for conjecture.

Once they had the fully fluorinated product, however, they noticed the vials were becoming etched by hydrofluoric acid made by the action of traces of water on the new compound. The implication was clear. Although this compound was stable in air, any lubricant property it might possess would be rendered useless by its susceptibility to attack by water to form this particularly corrosive acid.[40]

The unwanted side reaction closed one door, but it opened another. The attack by water was an example of nucleophilic substitution, the linchpin of much synthetic organic chemistry. The breaking of double bonds to add chemical groups (addition reactions) is all well and good, but when an organic molecule has no double bond where one wishes to add a new chemical group, then it is necessary to devise a "substitution reaction" by which an incoming chemical group does not simply add on but approaches in such a way that it can push aside a previously added chemical group and bond in its place. Nucleophilic substitution takes place when an attacking group, such as a halogen ion, has a negative charge and seeks out a site of relative positive charge. In the case of the breakdown of the fluorinated buckminsterfullerene, it appeared that the hydroxide ions of water were displacing fluorine ions.

Reactive fluorine is an ideal candidate for the initial addition reaction. Once this addition has been achieved, substitution reactions can be used to introduce a great variety of chemical groups in place of the fluorine. Organic chemists have assembled over the years a battery of standard nucleophilic reagents designed for this purpose in order to build up more complex organic molecules. Following the success of the fluorine reaction, it was soon found that other addition reactions also worked. Buckminsterfullerene would add hydrogen, bromine, hydroxide groups, the benzene ring units called phenyl groups, and other units by more or less direct methods. In all these cases though, it was hard to control the number and position of the chemical groups added. Before these additions and substitutions had been made, Joel Hawkins of the University of California at Berkeley had made the first pure fullerene derivative, a complex based on the tetroxide of the metal osmium which hooked onto the spherical molecules on a one-to-one basis. Paul Fagan at Du Pont has made other exohedral complexes with nickel, platinum, and other metals. Toshiyasu Suzuki and Fred Wudl at the University of California at Santa Barbara found organic reagents that will prise open one single bond on the fullerene surface to form a point of attachment for exohedral adducts.[41]

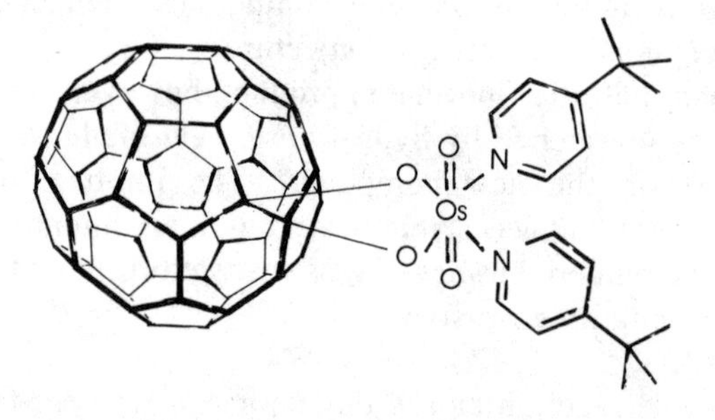

This complex with the metal osmium was the first cleanly made compound of buckminsterfullerene

The keys had been found. The door was open. Fears that the symmetrical molecules would lose too much of their stability if disrupted by an added chemical group had been shown to be unfounded. The aim now was to be able to add chemical groups in a controlled manner, in just those places and numbers desired. (The situation is superficially parallel with the substitution chemistry of the benzene ring, where many thousands of molecules can be made, with myriad uses, by substituting different chemical groups in appropriate permutations on the six carbon sites on the ring. Often, the sequence of reactions is critical in determining which end product is produced, and it is occasionally necessary to introduce temporary "protecting" groups on some sites while an otherwise incompatible reaction is encouraged to proceed elsewhere on the molecule. The "protecting" group is only removed at the last stage to be replaced by the more fragile group that could not have weathered the earlier stages of the synthetic sequence. With sixty bonding locations on the surface of buckminsterfullerene, the synthetic permutations are virtually endless.)

Chemists' interest is likely to centre on molecules that retain as much as possible of the spherical framework of carbon pi bonds while adding chemical groups of contrasting nature. Wudl wrote at the end of 1991: "Novel molecules with intricate, spherical architecture and as yet unknown properties await preparation."[42] One goal is to marry the fullerene spheres with organic chain molecules. The resulting ball-and-chain chemistry could have important uses, and would perhaps be analogous to that of soaps and detergents whose molecules incorporate both water-soluble and oil-soluble regions.

Wudl's group has linked C_{60} to polymer chains. Incorporated into familiar polyester and polyurethane, the fullerene molecules retain their characteristic electrochemical and non-linear optical properties. Wudl has also made a magnetic material by combining buckminsterfullerene with an organic electron-donor compound – creating an entirely non-metal magnet. The ability to encapsulate the fullerenes within a stable medium such as a polymer could prove instrumental in harnessing such properties to practical ends. Edward Samulski at the University of North Carolina has made what he calls "flagellanes" with multiple polystyrene chains flailing from the fullerene surface. The new molecules are soluble and readily form films and fibres. Wudl talks of linking fullerene molecules with short lengths of polymer chain in "pearl necklaces" or hanging them from these chains to form "charm bracelets".[43]

It is not easy to find an overarching logic for this vast miscellany of potential fullerene compounds. The exohedral chemistry activated by the opening of the surface double bonds is already coming to resemble a novel extension of the chemistry of the unsaturated organic compounds, the alkenes. Although early synthetic products are typically described as the fullerene with a chemical group or groups attached, it is likely that useful chemistry will increasingly be seen to come from conventional

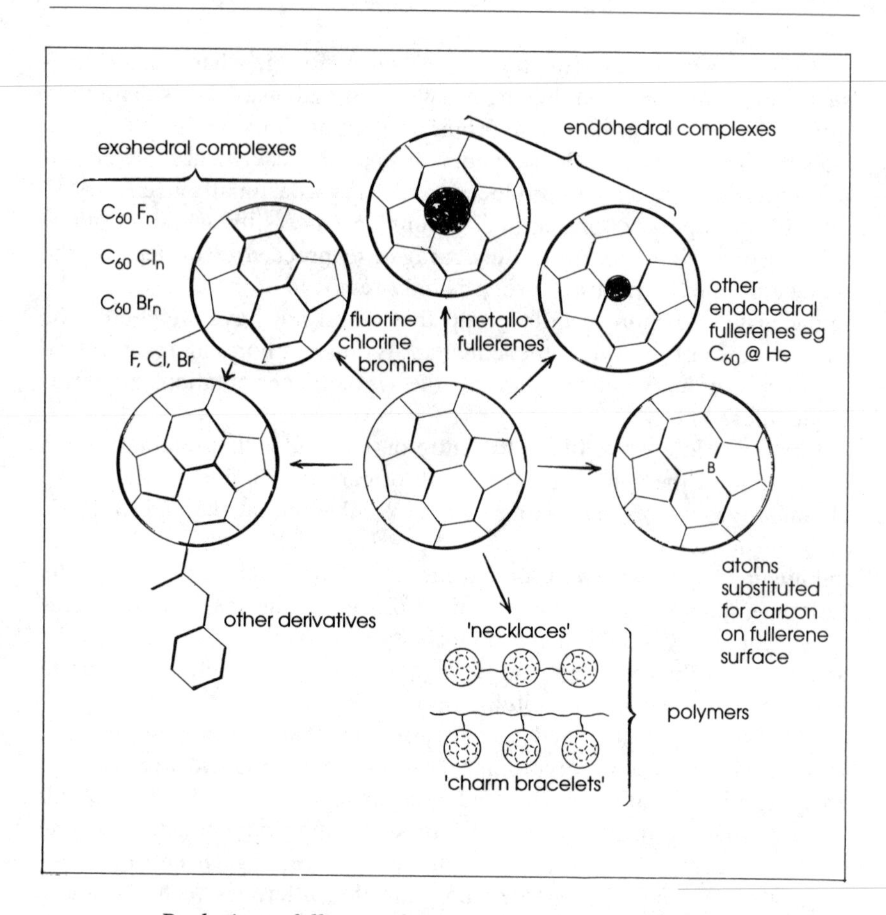

Buckminsterfullerene chemistry is seeing rapid growth, with many novel reaction pathways to be explored

organic molecules with a fullerene attached. The distinction may seem trivial but this transition would mark the assimilation of these novel molecules into the body of standard chemistry.

The endohedral fullerenes are more in a class of their own. It is possible to envision them as a kind of periodic table, with each element capable of being inserted into the cage appearing as if through a carbon mist. The periodic table is a useful metaphor also for the family of naked fullerenes, whose various sizes, roundish shape, and molecular orbital symmetries give them a kind of natural valency that entitles them to be regarded as analogues of the elements. This notion has already been realized to some extent in the superconducting films made at Bell Laboratories which are in effect "supersalts" where positive metal ions and negative fullerene ions form regular crystals with bond strengths somewhere between those of regular crystalline ionic bonds and the weak van der Waals bonds of the undoped solid fullerene. Kroto talks of the prospect of a "carbon periodic table" of lightweight fullerene superatoms of different sizes with different "pseudovalencies" and associated chemistries. This periodic table begins to build up by symmetry just like the table of the elements. Whereas the elements are grouped by the symmetry of their atomic orbitals, however, the fullerenes are grouped according to their symmetry at a molecular level. C_{60}, C_{120}, C_{180}, C_{240}, all with icosahedral symmetry, would form one group of the table, for example. Other fullerenes have various lower symmetries and could be grouped accordingly.

The "carbon periodic table" has a few nuances that are absent from Mendeleev's table. Some fullerenes exist in several forms or isomers with the same number of carbon atoms but different symmetries. Others of these carbon superatoms are already known to have bizarre properties. Most notable is perhaps C_{76}, which may exist in left- and right-handed versions according to the way the pentagons and hexagons spiral around its surface. This property could perhaps be exploited to twist polarized light into the correct orientation for use in fibre-optic communications. Most chemical chirality is based on a chiral centre; any carbon atom with four different groups bonded to it forms a chiral centre of the molecule of which it is part; most amino acids are chiral in this way, existing in two forms that are mirror images but which cannot be mapped onto one another. C_{76} has no such chiral centre and so could perhaps be used as a template to separate other chiral molecules. There are tantalizing links to biochemistry here since many biologically important molecules exist in handed variants.

We conclude this discussion of the potential uses of fullerenes with a parable. Quite aside from the intriguing matter of chirality, the unusual size and shape of buckminsterfullerene had already suggested that it might be "biologically active", playing a part in the lock-and-key biochemistry of proteins and enzymes, perhaps inhibiting viral attack. In early 1993, two separate groups of researchers used samples of a water-soluble fullerene derivative to test this proposition. Both groups selected the human immunodeficiency virus as the obvious target. Simon Friedman at the University of California at San Francisco designs drugs by testing their likely action using computer simulations. His theoretical studies suggested that a water-soluble buckminsterfullerene derivative would fit snugly into the active site for the protease specific to the virus, thereby inhibiting viral activity. His practical experiments mixing the derivative and the protease confirmed this.[44] At the same time, a team from Emory University in Atlanta led by Raymond Schinazi performed similar experiments using the same fullerene derivative. They confirmed the San Francisco result and also found evidence for direct inactivation of the virus itself.[45]

The episode illustrates a general rule that fullerene science has been advanced not so much by fundamental consideration of what avenues it might prove fruitful to explore, but by energetic researchers wheeling out their pet obsessions rather like hopeful breeders showing a stud dog to a prize bitch. It illustrates, too, the somewhat overheated atmosphere that had come to hang over the subject in the year or two after buckminsterfullerene had become widely available. Were these experiments the product of a sincere hope that it might provide a cure for AIDS? Or were they merely a clumsy attempt to grab headlines by combining chemistry's most popular molecule with nature's most infamous virus? It is significant that these results were submitted to the most famous journals, but were rejected by them, although later accepted by other respected publications.

The first demonstration of biological activity of a fullerene was surely newsworthy. But it seems it was the cruel misfortune of these experiments that it was the human immunodeficiency virus that was the prime candidate for study. This virus, of all viruses, just happened to cry out for an inhibitor the size and shape of the buckminsterfullerene molecule. However, this combination of beauty and the beast was perhaps too much for dispassionate evaluation. In any case, it is true that many compounds are biologically active against viruses, including some other novel carbon compounds.[46] Which brings us to a final observation: yes, a buckminsterfullerene derivative is biologically active, but not more so than some existing compounds, notably, of course, in the context of HIV, the drug AZT.

The fullerenes are the only intrinsically pure forms of carbon. (Diamond has a monolayer of hydrogen on its surface; graphite also has edge contaminants.) This purity, and the fact that it is the only form of pure carbon that may be sublimed and dissolved, will in time perhaps prove a convenience in scientific laboratories. The lasting impact of the discovery of the fullerenes will depend on whether this molecular form finds uses beyond those which prize it merely as a superior allotrope of carbon.

The question to ask is: "Does buckminsterfullerene yet do *anything* better than *something* else?" As of mid 1994, the answer has to be no.

But these are still early days. It is less than ten years since buckminsterfullerene was first described – less than five since people have been able to experiment with it. This is short compared to the lag that commonly separates discovery from application. This interval has seen the arrival of a number of fullerene start-up companies, not least in Houston and Tucson. The price of buckminsterfullerene has fallen from $2000 for a gram in 1992 to $500 in 1993. Smalley hopes to see it drop to around $100 per pound and expects that C_{60} may ultimately be no more expensive than aluminium.[47] The price has fallen not only because the molecule is easier to make, but also because of market forces. Demand for buckminsterfullerene is high because research activity is intense. Some applications that presently appear sound will be found to conceal flaws; others, not yet imagined, might well bear fruit. At the present pace of progress, it would be a disappointment if this chapter were not out of date before it was printed.

11

THE MOLECULAR ARCHITECTS

The opening chapter of this book described some of the challenges that chemists confront as they try to make new molecules. The molecules under discussion were, by and large, relatively small; by implication, they would be moderately easy to make provided they met the usual criteria of geometric and electronic feasibility. The main question was not whether they could be made, but how. Yet these molecules are mere specks. They extend no more than a few ångströms in any direction. They are, in effect, zero-dimensional. Polymer scientists have learned to build one-dimensional molecules – chains of polyethylene or similar. "But", as the Nobel laureate chemist, Roald Hoffmann, points out, "in two or three dimensions, it's a synthetic wasteland."[1]

In this scheme of things, buckminsterfullerene is just another zero-dimensional molecule, complacent in its perfection. Much effort has been devoted to extending the fullerene family. Soot prepared *à la méthode* Krätschmer–Huffman yields about three parts of C_{60} to one part of C_{70}, but also a smattering of larger fullerenes (presumably the ones responsible for the sequence of mass spectrum peaks into the hundreds of carbon atoms observed long ago by the Exxon group). Lamb and Huffman believe they have seen fullerenes with up to 900 carbon atoms apiece in their more recent experiments. Chemists have isolated C_{76}, C_{78}, C_{84} and other fullerenes in addition to C_{60} and C_{70}, thanks apparently to the extraordinary sensitivity of the chromatographic separation process to the slightly varied shapes of fullerenes with different numbers of atoms. Robert Whetten and François Diederich found a way to coalesce two buckminsterfullerene molecules to form C_{120}. The insoluble portion of the soot is believed to contain far larger fullerenes.[2] Some progress has also been made in the extraction and purification of endohedral metallofullerenes, but nothing like that made in the case of the empty fullerenes.[3]

Various ingenious ways have also been found to make larger quantities of buckminsterfullerene itself. By the end of 1993, it had been made not only in sooting flames but also by novel physical and chemical means – by its original discoverers among others. With one eye on the past success of his laser vaporization apparatus and the other on a rosy commercial future,

Rick Smalley reported that the molecule would form from the carbon vapour created above a graphite sample by the action of focused sunlight. The simple apparatus he described – a glass cell positioned at the focal point of a parabolic mirror – could clearly be scaled up to industrial proportions.[4] Harry Kroto's group made it chemically, although by the arduous chemistry of pyrolysis of naphthalene at 1000 degrees centigrade, rather than by the elegant means of assembly beloved of traditional organic chemists. The naphthalene molecules, with ten carbon atoms apiece, patch together by sixes and sevens to form C_{60} and C_{70}.[5]

These were but modest steps, however. Other scientists were pursuing their own avenues of enquiry that were greatly to broaden the canvas of the new carbon chemistry. Three variants of the fullerene unit are worth reviewing. They may be classified informally as elongations, magnifications, and inversions of the basic spherical geometry.[6]

The elongation of the fullerenes has its origin in a long-running research programme at the NEC Corporation's Fundamental Research Laboratories where Sumio Iijima had been making graphitized carbon particles since the 1970s by a technique similar to that of Krätschmer and Huffman. The material he made was mostly amorphous, but in 1987, Iijima looked back at some of the electron micrographs he had produced during this period of relatively spherical particles and found at the core of one of them a spherical unit of a diameter about the same as buckminsterfullerene. He announced his find by means of that rarity, a scientific paper with an exclamation mark – "The 60-carbon cluster has been revealed!" One may wonder why, having recorded this image as long ago as 1980, it is not Iijima whom we now credit with the discovery of buckminsterfullerene. In the context of his experiments, however, the phenomenon was a singular anomaly in a large set of micrographs, a far cry from the attention-seeking sixty-carbon peak in the mass spectra recorded five years later at Rice University.[7]

This trophy was no more than the prelude to a startling and significant discovery that Iijima made in the summer of 1991 when he reported the preparation of a new type of carbon structure consisting of "needle-like tubes" of graphite chicken-wire sheets wrapped around to form tiny cylinders. Working with an electric arc between two graphite electrodes as in fullerene synthesis, Iijima found that needles grew as coaxial graphite sheets at the negative end of the electrode, forming anything from two to fifty concentric tubes with spacing between them similar to that between planes of graphite.[8] The minimum possible tube diameter would simply be that of the C_{60} molecule at its equator as seen by Iijima in his earlier work.[9]

The fact that the carbon tubules or "nanotubes" are found in regions at the electrode (and not, like fullerenes, in the soot that condenses away from the electrodes) suggests that the electric field is key to their growth. The importance of the electric field is further underlined by the results of calculations which show that, for the sorts of numbers of carbon atoms present in a tube, the most stable arrangement is not in fact as a tube but as a fullerene. Whereas fullerenes form as the major constituent of a condensing carbon vapour, the major proportion of the carbon in an electric field condenses as tubes. The presence of the field must be preventing the formation of fullerenes and favouring the growth of the tubes. While the mechanism for the formation of fullerenes is still a matter of heated controversy, it comes as no surprise that the way nanotubes grow also remains a mystery. Whatever theory emerges, it will have to be a good one to explain the overwhelming bias towards one of the two carbon morphologies in favour of the other in each of these different physical regimes.[10]

The obvious means of growth would be to add carbon fragments to the jagged end of an open nanotube. But the almost complete absence of such open-ended tubes in the electron micrographs suggests that something else must happen. The tubes are for the most part terminated with hemispherical carbon caps of hexagons and pentagons like hemifullerenes. (Iijima illustrates this point in his lectures by showing a slide of traditional Japanese baskets whose immaculate hexagonal weave is compromised only at the corners where the weavers have long known to introduce pentagons in order to produce sufficient closure to make a usable container.) The fact that the nanotubes have this feature in common with the fullerenes strongly suggests that a mechanism for nanotube growth might have much in common with proposed mechanisms for the growth of fullerenes themselves. Although these nanotubes grow from the graphite electrode, it is easy to imagine molecular nanotubes with caps at both ends (hypothetical molecules that were dubbed capsulenes or, more humorously, zeppelenes). Indeed, the first in the series can be thought of as C_{70}; the ten extra carbons that form its waistband are simply the first slice in the growth of a nanotube with two hemi-buckminsterfullerene ends. Further extension rings could be added to form nanotubes of any length.

Unlike carbon fibres which are made by the controlled incineration of hydrocarbon polymers, the carbon nanotubes are in principle single molecules once they are complete with their end caps. Their structure is made up entirely of strong chemical bonds. Material strength is currently limited by faults that propagate through imperfect structures. The carbon bonds are strong enough to contain faults locally so overall strength

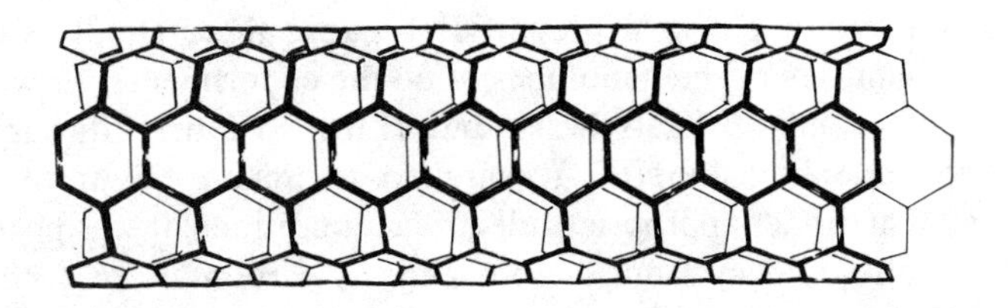

Carbon nanotubes are essentially unadulterated graphite sheets rolled around to close upon themselves

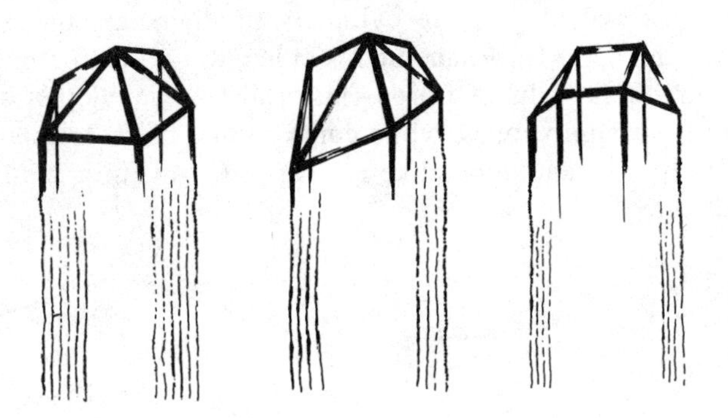

Nanotubes can grow in the form of many concentric cylinders while their leading edges are faceted like quartz

should be maintained even in slightly imperfect structures. As Iijima wrote in *Nature*: "The formation of these needles, ranging from a few to a few tens of nanometres in diameter, suggests that engineering of carbon structures should be possible on scales considerably greater than those relevant to the fullerenes."[11]

The negotiability of the length and diameter of the nanotubes is not their only claim on our interest. They prefer to grow in a screw-like fashion rather than evenly like a stretching telescope. (Although this might seem unlikely on first consideration, it is merely a reflection of the fact that there are several helical ways to initiate the growth of a nanotube, but only one square-cut cylindrical one.) This helical growth leaves some nanotubes with an inherent handedness. The exact arrangement of pentagons on the end cap where carbon is being added is thought to be important in determining how the tubes grow.[12] Theory indicated that the helicity as well as the diameter of the nanotubes would determine their electronic behaviour.

As with the fullerenes, it was not yet possible to test predictions of exotic electronic or physical properties because there simply were not sufficient quantities of the nanotubes to do the experiments. However, in a remarkable echo of Krätschmer and Huffman's breakthrough with buckminsterfullerene, the NEC researchers overcame this hurdle when they found that, under appropriate electrical conditions, it was possible to pull one electrode away from the other in the carbon arc in such a way that the build up of the carbon deposit on the stationary electrode keeps pace with the receding electrode. The bulk of the graphite in the sacrificial electrode is vaporized and reconstituted on the other electrode as nanotubes. The experiment yielded a host of nanotubes each with at least two concentric rolled graphite cylinders of diameters overall of 100 nanometres and up to 5000 nanometres in length.[13] Electron micrographs reveal luxuriant growths of fibrous material like bundles of asparagus. Some tubes taper into cones, while cones funnel down to smaller tubes, before finally tube and cone alike are capped with fullerene-like hemispheres of carbon.

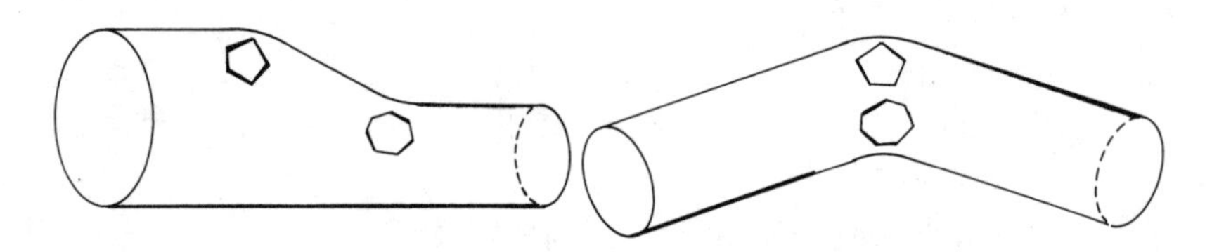

Pentagons or heptagons along their length enable nanotubes to taper, flare, or bend

Theoreticians at the Naval Research Laboratory in Washington DC predicted that these nanotubes should be metallic in character.[14] Other groups refined this forecast: one third of all nanotubes would be metallic, while the other two thirds would be semiconducting, they said.[15] These two discoveries raised the prospect that it might be possible to fabricate "nanowires" of two concentric nanotubes with the outer shell insulating and the inner one conducting like PVC-coated copper reduced to molecular scale.

Bringing his knowledge of endohedral metallofullerenes to bear on the new nanotubes, Rick Smalley predicted that metal-impregnated nanotubes would yield another kind of nanowire. Just as the fullerenes comfortably hold atoms of various elements, so nanotubes can in principle hold a number of atoms like peas in a pea-shooter. Such a composite material would constitute the finest possible metallic wire, combining the conductivity of the core with the high tensile strength of the strongly bonded sheath.[16] The puzzle was how to get metal atoms inside the capped tubes.

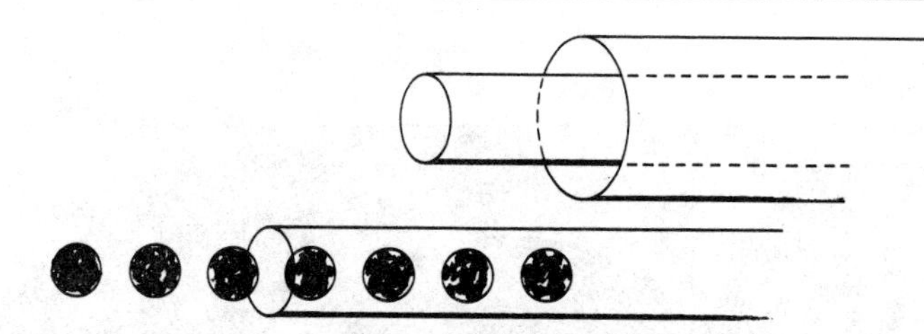

Concentric nanotubes and nanotubes with metal atoms sucked up into them show novel electrical properties

Once again, Iijima and his colleagues promptly came up with an answer, finding that the nanotubes would suck up atoms of lead by capillary action. Oxygen present from the air is thought to play a key role in opening up the tube ends giving the lead atoms the opportunity to slip in.[17]

The magnification of the fullerenes had been imagined from the very beginning. It is an idea that follows naturally from the spherical geometry and icosahedral symmetry of buckminsterfullerene itself. An infinite series of ever larger fullerenes can be constructed with this same shape and symmetry. Many more can be made without this perfect symmetry (and at larger sizes the symmetry becomes progressively less important in creating stable structures).

In September 1992, a year after Iijima's discovery of nanotubes, Daniel Ugarte, a physicist at the Ecole Polytechnique Fédérale in Lausanne, Switzerland, found that as many as several million carbon atoms would rearrange to form "onion skins" of fullerene shells, nested like Russian dolls, with each radius three ångströms or so larger than the last, again the spacing between layers of planar graphite.[18]

Such a discovery had not been on Ugarte's mind. He had been seeking to inspect clusters of gold trapped inside particles of carbon by means of an electron microscope. But Ugarte assailed his carbon soot with an electron beam more than ten times the intensity of that typically used in the instrument for producing electron micrographs. Even at normal intensities, the electron beam has the effect of "graphitizing" a certain amount of amorphous carbon. Ugarte found that at these much higher intensities the amorphous carbon soot would rearrange to form not graphite but vast nested fullerenes.[19] Irradiation of Iijima's nanotubes produced the same transformation. After twenty minutes of irradiation, almost all the carbon is present in the form of spherical particles or "onions" with up to seventy concentric fullerene shells. The larger shells are made up of thousands of carbon atoms bonded together in the familiar pattern of hexagons with the usual ration of regulation pentagons permitting closure of the shell.

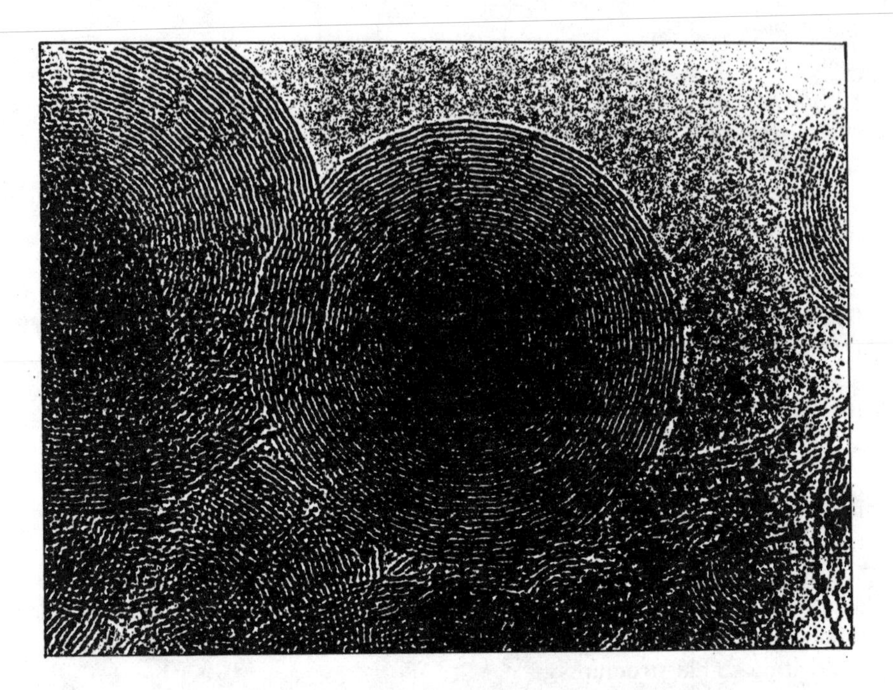

As many as several million carbon atoms may be made to rearrange to form giant nested fullerenes

Ugarte believes it may be possible to make much larger spherical carbon particles as well as hollow carbon "balloons" with only a few large fullerenes enclosing vast empty chambers that might have novel uses. As with the condensation of buckminsterfullerene from the carbon vapour in Smalley's apparatus, the question had to be asked: how do such ordered species take shape in this chaotic environment? And once again, the answer appeared to lie in overcoming the mental block that carbon always chooses to develop in the form of planar layers. As Harry Kroto wrote in an article accompanying Ugarte's paper in *Nature*, the discovery called into question the assumption that graphite was the most stable form of carbon, not only for groups of sixty or six hundred atoms, but at any scale.[20]

The most Alice-in-Wonderland of the variations that have been worked on the theme of the fullerenes is best described by recourse to the mental gymnastics of Kroto and Smalley and Jim Heath as they considered what would cause a hexagonal mesh to close in on itself and consume its own edges, and eventually realized that the insertion of pentagons in the hexagonal mesh would do the trick. One pentagon surrounded by hexagons (the carbon framework of the molecule corannulene) produces a shallow dish; the addition of five more pentagons produces a deeper, hemispherical dish; twelve pentagons generate a closed sphere. Now consider the inverse problem as a kind of thought experiment. If insertion of a number of pentagons into the hexagonal mesh produces a convex shape that leads ultimately to closure to form a complete three-dimensional bonded structure, what will the insertion of heptagons produce? The answer is that it does not produce a convex curve, nor yet a concave one, bending the plane in the opposite direction, but a saddle shape bending the carbon sheet upwards along one direction along its plane and downwards along a perpendicular direction. Further heptagons continue this curvature. This time the edges do not converge to produce an edgeless sphere. They diverge and carry on diverging as long as heptagons keep being introduced.

Iijima had observed that his growing nanotubes sometimes tapered off into cones before they were finally capped off. A pentagon or two in an otherwise regular hexagonal carbon mesh is all that is needed to start the conical descent like the shaft of a pencil as it reaches the shaved wood that tapers to the lead tip. In other cases, Iijima had noticed that the cones fail to converge to a point and instead flare out and continue growing as tubes. In the saddles, as cone turns to tube, there must be heptagons present. (In the absence of direct images of the molecular surface at these special points, Iijima illustrates the principle with examples of more intricate basket-weaving.)

No material has yet been isolated which is known to incorporate heptagons in a large expanse of hexagonally bonded carbon. However, Alan Mackay, a crystallographer at Birkbeck College in London (appropriately enough, he occupies the chair once filled by J.D. Bernal), has described in some detail a range of hypothetical "negatively curved" carbon networks.[21] The negative curvature arises from the regular inclusion of heptagons (or octagons) rather than pentagons in the otherwise hexagonal net. The resulting surfaces may be described mathematically as hyperbolic rather than ellipsoidal. The surfaces do not close in on themselves to form discrete molecules like the fullerenes, but in their eternal divergence produce instead beautiful repeating structures in three dimensions. The unit cells of these unusual "hyperfullerene" crystals are regions of the network surface that look like scaffolding joints. Repetition of the scaffolding units in all directions builds up a curious dual structure, a tight three-dimensional weave of two spatial labyrinths, the one forever separated from the other by the carbon membrane. In some of these hypothetical structures, all the carbon atoms lie on a surface of constant curvature (this through-the-looking-glass inside-out sphere is called the hyperbolic plane) and so occupy exactly equivalent environments just as in the positively curved buckminsterfullerene.

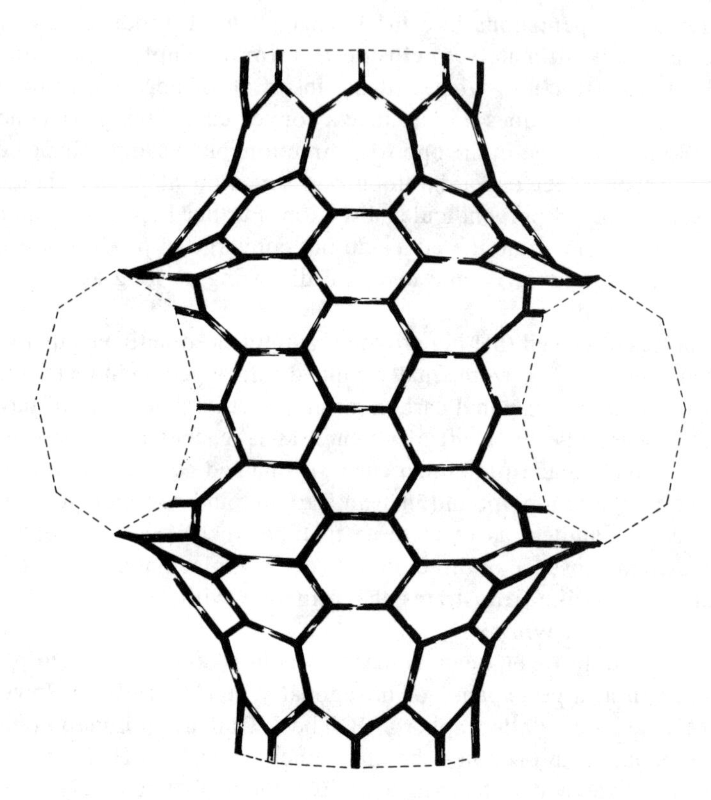

Adding heptagons rather than pentagons into the carbon network generates continuous "hyperfullerenes"

Veit Elser and his colleagues at Cornell University performed energy calculations which showed that these negative fullerene surfaces might be even more stable than buckminsterfullerene. They christened the imaginary material schwarzite, partly in anticipation of its likely colour (*schwarz* is German for black), but mostly in honour of the nineteenth century mathematician of this name who first explored these topological curiosities.[22] There is no requirement for the growth of schwarzites to be highly symmetrical; they could be randomly branched and interconnected rather like some seaweeds. The Cornell group reasoned that if fullerenes form in a comparatively cool, "annealing" environment in which positive curvature of the growing carbon shell quickly gobbles up the dangling bonds on the advancing front, then the reverse conditions – a very hot supersaturated carbon vapour in which dangling bonds proliferate and which surely also exists in the vicinity of a carbon arc – might generate the negatively curved forms of carbon. All told, it seems more than likely that at least some of the residue left after the solvent extraction of fullerenes might contain regions with this sort of structure.

As a coda to this catalogue of exotic carbon surfaces, it is worth adding that there is a very special case of a hypothetical structure that combines surfaces of positive and negative curvature in such a way that a closed topological unit – a complete molecule if it were to exist – is formed. This unique shape in effect bends a nanotube around to unite its ends to form a toroid or doughnut shape (a nanoscale smoke ring?), and had been envisioned by Patrick Fowler in 1986.[23]

Where does this ragbag of novelties take us? None of these carbon morphologies is entirely new, after all. Combustion scientists have for decades observed the growth of carbon filaments, the creation of twisted sheets in flames, and the nucleation of spherules of graphite in pitch and other hydrocarbons. However, they have in general lacked either the wherewithal to describe these things fully or, where it was called for, the nerve to reject the paradigm of planarity. Accordingly, they forgot about the resilience of the carbon-carbon bond. Rather than extrapolate the logic used to describe relatively small molecules, they chose a Procrustean view of things in which graphite layers give rise to all kinds of shapes, no matter how unsuitable their initial shape or orientation seems for the task.[24]

The discoveries of Iijima and Ugarte show that there is little to justify the ubiquity of this model. They argue, on the contrary, that certain extremely large aggregations of carbon atoms may still be regarded to all intents and purposes as molecules. This is no mere semantic quibble. Such definitions play a part in determining which scientific disciplines are likely to pursue a topic and which are not. We have described how the chemists are discovering how to add functional groups to the small fullerenes in order to make them do potentially useful things. It might be expected that materials scientists, who work on the macroscopic scale, would claim the limitless larger carbon architectures for themselves. Yet materials scientists seeking to fabricate superconductors, or semiconductor devices, or sensors of one kind or another often will not stop to consider chemical manipulation as an option. There is a world of difference – in scale, but also in outlook – between those who "fabricate" and those who "synthesize". It is not yet possible to tell whether uses for the fullerenes and these other carbon structures will depend upon the fabrication of devices or the synthesis of compounds. But the fact that they are formed entirely of carbon atoms joined by conventional chemical bonds (not by van der Waals and other vague long-range forces) gives the chemists a legitimate call on them, no matter what their size. It may be said at a stroke that much of the chemistry discussed in the previous chapter with regard to buckminsterfullerene is potentially applicable to giant fullerenes, nanotubes, and other carbon networks as well.

This change of perspective perhaps asks a lot of chemists, although Roald Hoffmann, for one, seems well prepared:

> Science is part discovery, part creation.... In describing their work most scientists will stress the discovery metaphor, while most artists will emphasize creation. Well, I think much of what we do in science is creation. Especially so in chemistry. The synthesis of molecules not present on earth before is clear evidence of this. Synthesis is a marvelous congeries of discovery and creation that brings chemistry close to the arts – and to engineering.[25]

In the new carbon chemistry, the discoveries (enough of them for the moment, anyway) have been made; let the creation begin. Quite how closely this creation is likely to resemble the arts, especially architecture, and engineering, perhaps even Hoffmann, writing in 1988, could not have imagined.

But both Kroto and Smalley were quick to see the potential. As Smalley was inspired to write on first looking down into this realm of carbon:

> Scanning tunneling microscopy has a way of conjuring up fanciful thoughts. While lecturing about C_{60} and the fullerenes recently, I visited one of the leading labs using this technique and looked at images of these hollow, nanoscopic geodesic domes of carbon, one at a time. Seeing such pictures it is easy to imagine a whole new world down there on the chemist's atom-by-atom length scale, and there are those who will view this world with the eyes and ambitions of a molecular architect, or a physicist, a materials scientist, or a chemical engineer. Are these the first elementary building blocks of a new carbon-based technology?[26]

Taking this family of carbon structures as a whole, the ultimate goal becomes the possibility of building custom carbon architectures, joining modules of spheres, tubes and other shapes according to a blueprint. We are already assured of the compatibility of these parts. C_{70} is an extension of C_{60}, with its waistband of ten extra carbon atoms. By adding further rings of ten atoms, one can imagine building longer and longer fullerenes. At some point, perhaps with hundreds or thousands of such rings added, it would become a nanotube. Nanotubes of larger diameter might likewise be imagined as if extruded from larger fullerenes. The larger fullerenes themselves bear the closest geometric kinship with buckminsterfullerene. Concentric spheres – Ugarte's "onions" – have their analogue with concentric tubes ("leeks"?). It might seem harder to relate the negatively curved (hyperbolic) carbon networks to the positively curved (ellipsoidal, cylindrical, and conical) structures. But they are presumably present in the small areas where cones and tubes join. From here, it is easy to imagine employing these saddle regions to negotiate the bends in joining nanotubes of different diameters and at different angles, or where a nanotube neck joins a fullerene chamber in a nanoscale version of a chemist's round-bottomed flask.

This family of carbon construction modules is part of a grander overall scheme. Rick Smalley believes that what is emerging for the first time from his own discovery with Harry Kroto and from the recent discovery of Iijima's is a complete picture of how carbon crystallizes from a vapour. The two familiar forms, diamond and graphite, are the three-dimensional and two-dimensional manifestations of the element. We now know that nanotubes are the one-dimensional form and that the fullerenes, as discrete molecules, are in effect zero-dimensional. (The hyperfullerenes might yet prove to be an alternative three-dimensional morphology, perhaps even interconvertible with diamond.)

Using this kit of parts, we could build a city of domes and tubes like a tiny scale model of a science fiction spaceship. But this is only part of the story. Such carbon edifices would be mere castles in the air if they were of no use. In practice, they would be the rugged framework for all kinds of functionality. The addition of bonded chemical compounds or metal "contaminants" at required points – inside the fullerenes, interspersed between them, sucked up into nanotubes, or in any number of other locations – could introduce a wide range of electromagnetic, physical and chemical properties to particular regions of the overall structure chosen by design.

Much of this we do already, of course, in silicon. But silicon-based semiconductor devices are large compared to fullerene cages and pea-shooter nanotubes. The element must be handled with great care, in part because it does not form such strong bonds as carbon. This makes device fabrication difficult and the devices themselves rather frail. In addition,

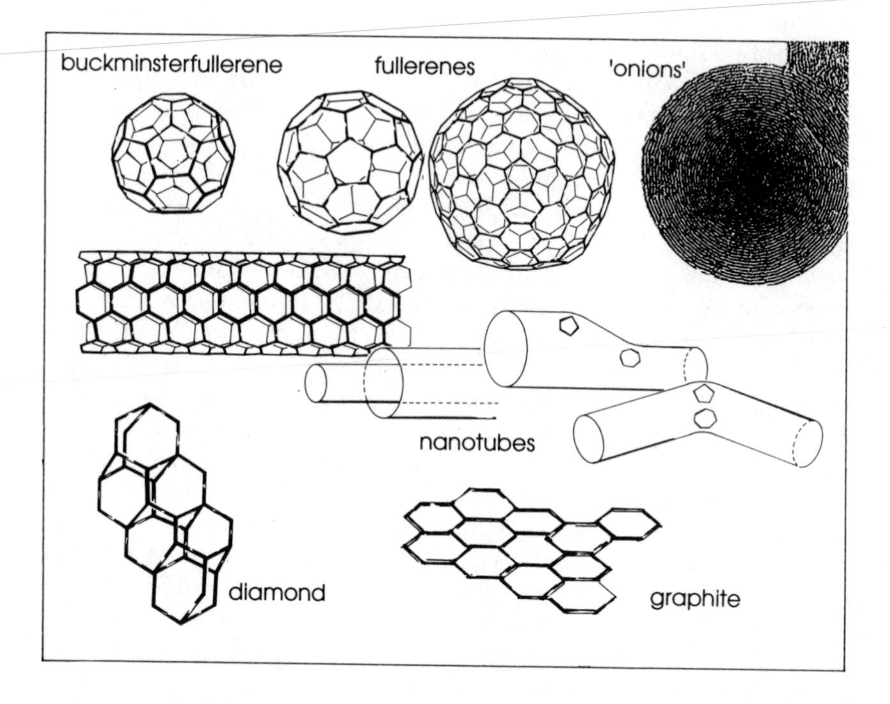

Carbon in its various shapes and sizes can be imagined as a kit of parts for future projects in nanotechnology

the versatility of silicon is rather limited; while it is an ideal semiconductor, it is not a good material for much else.

One merit of this new carbon science is to have such a range of properties in one basic material. The fact that this breadth arises not only from fundamental differences in chemical composition, for example when dopant metals are added, but from differences in geometry (seen most dramatically in the varied electrical properties of nanotubes of different sizes) may ultimately make it easier to fabricate electronic and optical devices that will be stronger and more chemically and physically stable than any available today. It is, of course, a great bonus that this basic material happens to be carbon whose propensity to form complex and intricate structures is unrivalled and whose behaviour we understand better than most elements' (although still not well as this entire tale of discovery testifies).

The prospect of carbon being exploited in these ways should be viewed in the knowledge of progress already made in the fabrication of carbon-based composite materials. The basic ingredient of these is carbon fibre made from rayon or other polymer fibres from which the hydrogen and other elements have been driven out by heating. These fibres are not graphitic but simple chains of carbon atoms linked by double bonds. They have fewer flaws along their length than many other fibre materials which makes them proportionately stronger. They may be woven in two dimensions to form carbon sheets far stronger than graphite and also in three dimensions to form solid blocks. These blocks are frequently modified in their turn by impregnation with further carbon-rich material to produce what are called "carbon-carbon composites". In these materials the bonds between the carbon fibre weave and the carbon-based matrix that suffuses it are strong chemical bonds, which is not always the case with other composite materials. This makes them suitable for use in aircraft brakes and engine components and as rocket motor linings and spacecraft re-entry heat shields.[27]

In theory at least, our new family of carbon materials is far superior. There is no need to weave fibres when we can design molecular nets to cover any eventuality in two and three dimensions. Our nanotubes are made up entirely of a chemically bonded mesh unlike carbon fibres which have strong bonds along the individual carbon chains but little cross-bonding between them. This not only makes nanotubes potentially even stronger than present-day carbon fibres, it also inhibits their infiltration by impurities and arrests what few defects there might be, preventing them propagating and hence weakening the overall structure.[28] Finally, they offer the bewildering array of magnetic, electronic and optical properties demonstrated in three short years. They are, in short, aristocrats among materials, suited for far better things than the brute uses of carbon-carbon composites.

Consider how a typical consumer product such as a calculator might take shape in this new scheme of things. The circuitry would be made of carbon chips, not silicon chips. With transistors of molecular size, it would be possible to incorporate more powerful processing in the hand-held unit. The machine would gain its energy from photovoltaic cells made of a fullerene derivative. It would store the energy in batteries made of fullerenes adapted to store charge. Its display would be made of light-emitting diodes based on yet another fullerene variant. Insulated nanowires with conducting cores would carry current between these parts. As now, the casing for the calculator would be made of hydrocarbon polymer and the transparent display and solar cell coverings would be of a transparent plastic such as polystyrene. Both materials would be entirely compatible with the other carbon components and neatly chemically bonded to them rather than clumsily glued or snapped into place.

Another powerful attraction of carbon-based molecular architecture is the element's compatibility with biological systems. Three-dimensional hyperfullerene networks might be rather like bone, combining extremely light weight with high strength and a degree of porosity that allows interaction with fluids in the body. Science-fiction writers have traditionally favoured carbon-based computers over silicon-based ones because they would be more like our own brains.

Needless to say, all of this lies a long way in the future. It is far off but not far-fetched. It is easy to imagine all these things, considerably less easy to see how they might be made. However, it is at least possible to foresee an emergent discipline under whose umbrella they might begin to pass into reality.

That discipline is nanotechnology – technology based upon the manipulation of matter atom by atom and molecule by molecule. It is commonly agreed that its remit runs from the scale of individual atoms up to about 100 nanometres, where it meets conventional electronics coming the other way but beginning to run into problems because of quantum effects such as "tunnelling" by which electrons no longer remain where they ought to remain according to classical theory. The range over which nanotechnology extends concurs nicely with the observed sizes of fullerenes and nanotubes. Whereas conventional microelectronics finds its aims thwarted by such quantum-mechanical side effects, nanotechnology will harness quantum effects to its own ends. Nanotechnology promises electronic memories and switches that operate by the action of single electrons located within "quantum wells" and currents that flow along "quantum wires". Nanotechnologists envision devices just dozens of nanometres in their dimensions, designed to tolerances on the order of single bond lengths – fractions of a nanometre – and operating at speeds measured in nanoseconds rather than the micrometres and microseconds of today's microelectronics. Naturally, much of the push to develop this new technology comes from the microelectronics industry as it seeks to cram ever more power into ever less volume. For their part, chemists (and biologists) are already working at the nanoscopic scale. Indeed, nanostructures are for chemists intimidatingly large things to contemplate making.

In his exciting, and at times disturbing, book, *Engines of creation: the coming era of nanotechnology*, Eric Drexler is wisely vague on what materials will comprise the new nanomachines and nanocomputers that will labour on our behalf, doing everything from repairing damaged body tissue to building extraterrestrial habitation by scavenging building materials in space. But implicit in every mention of "nanomachines of tougher stuff than protein" or the ability "to bond atoms together in virtually any stable pattern" is the candidacy of the fullerenes and these other carbon networks.[29]

Once again, the difficulty is not in imagining these things, but in describing how they might be constructed. It is all very well to have detected nanowires created in one chaotic environment and giant fullerenes in another, but to be of practical use it is likely that controlled means of fabrication will have to be found. Wires must be drawn; giant fullerenes must be grown like pearls or blown like glass. These are problems that have been solved to some extent for silicon and the other semiconductor elements for which we have developed sophisticated sequences of deposition and etching which, for all their drawbacks, have a thirty-year head start on any carbon-based rival.

As ever, there are two routes of attack, the chemical and the physical. The chemical route branches into two: there is the sloppy but successful path taken in the manufacture of nature's own nanomachines – the proteins and enzymes; and there is the tidier-seeming possibility of doing chemistry in the gas and solid phases without intervening solvents.

Much of the discussion in Drexler's book is built around the assumption that nanoscale manufacture will be based on biological mechanisms of replication, the cell being the archetypal nanofactory. The milieu for this chemistry is invariably the aqueous solution, muddy waters swimming with a cocktail of chemical compounds. Biological systems exhibit some fine prototypical nanomachines, from bacteria that are propelled along by nanoscale biological motors to viruses that do their dirty deeds by micromechanical means. But the real lessons are to be learned at the strategic level. Biological self-assembly is the process by which molecules form spontaneously from smaller precursors under ambient conditions, with no poking and prodding from us and our clumsy tools. Emulation of the principles of self-assembly might help us to make molecules and molecular architectures that are currently too big – and hence too difficult or too tedious – to make by conventional synthetic techniques.

The other fork in the chemical route is the dry path, the logical extension of Kroto and Smalley's early experiments in making clusters and arcane molecules. At the end of this rainbow lies the promise of a new kind of chemistry. Molecules could be built by design, atom by atom, using just the right amount of each ingredient. Precisely calculated bursts of energy, sufficient to initiate the desired connection but no other, would snap each new atom or chemical group into place on the molecule under construction. The only reaction product would be the desired product. There would be no by-products and no waste, a production line in place of the murky old world of will-it-or-won't-it bucket chemistry.

Finally, there is the purely physical route which parallels the process of architecture at the human scale. Bricks are moved and positioned and affixed one by one in the building of a wall. In similar fashion, the scanning tunnelling microscope, whose potential was so strikingly demonstrated by IBM scientists in 1990 when they spelled out their company's logotype by positioning single xenon atoms on a smooth metal surface, could be employed to position fullerenes and nanotubes. Electron beams could "weld" them together. The assembly of ambitious nanostructures by such means might seem doomed to be very slow. However, the parallel action of many such devices able to work on nanosecond timescales would soon speed things up. After all, biological nanomachines are able to construct enormous, intricate objects with great rapidity.[30]

Nanotechnology will undoubtedly advance on all these fronts. But could it be that the availability of these new carbon building blocks will hasten this advance? Already, great strides have been made towards bringing the small fullerenes into the domain of traditional wet organic chemistry, inducing them to combine with other chemical groups in the solution phase. Dry techniques continue to show promise too. If the electric field is the key to nanotube growth, for example, there may be a way of controlling that field to tease out individual tubes. Equally, since it is known that oxygen and other simple molecules are instrumental in opening, closing, and controlling the diameter of the nanotubes, it may be that a stream of these molecules is what is needed to create growth to order. The recent detection of nanotubes coiled like telephone cords and in other bizarre patterns suggests that such physical or chemical guidance systems are there to be understood and harnessed in the manufacture of tubes that follow a planned course rather than the whim of circumstance.[31]

To date, however, of all this diverse class of carbon materials, only buckminsterfullerene itself may be produced with any degree of efficiency. A few other small fullerenes may be made in small quantities. The creation to order of a fullerene of a given diameter or of a job lot of fullerenes nested or stacked in such and such a way is still well beyond the scope of present techniques. There is as yet no reliable guide to the sets of conditions that will encourage the production of fullerenes or nanotubes to specification. Nor do we yet have a clear idea whether properties useful in one context (conductor or semiconductor behaviour in a nanowire, for example) are compatible with those useful in another (the tensile strength of that wire). The assembly of two- and three-dimensional carbon structures – designed planes of graphite or networks of diamond or hyperfullerene – pose still greater problems.

The challenges are vocational as much as technical. Since most physicists have never learned about molecules and most chemists are poorly equipped to understand larger-scale phenomena, there are limits on how much any one discipline is likely to achieve. Real progress is likely to demand sustained collaboration between many disciplines. It calls not only for materials scientists and physicists but also chemical engineers, biochemists, molecular biologists, and even computer scientists, structural engineers and architects. When in 1993 Rick Smalley convened an interdepartmental meeting at Rice University in order to formulate a strategic policy on nanotechnology, he summoned representatives from seven faculties. Chemists must be central to the collaboration, says Smalley, because "atoms have to hold hands". The prospect of such teamwork – very much in the spirit of the cross-fertilization that led to the original discovery – would have warmed the heart of Buckminster Fuller.

There is one last obstacle that must be overcome, perhaps the toughest of all. Any technological or nanotechnological wonder must find a market. If it does something no one needs done or only does it in inconvenient circumstances or at an exorbitant price, then failure will be the consequence, regardless of the ingenuity of the idea. Recall the prospect of a carbon calculator. A conventional calculator costs, say, ten dollars. It requires replacement batteries and eventually falls apart, but no one minds. Our superior machine will have to compete on these terms, and that is certainly something that will not be possible any time soon.

If it seems deflating to conclude this adventure in chemistry by dangling such banal fruit as the distant prospect of "improved" calculators – or more realistically, merely the components for such things, or, still worse, a commodity as dull as a new photocopier toner – then that is no more than a reflection of the state of mind that expects miraculous uses for every discovery. It is the destiny of most scientific discoveries to find application – if they ever find it – as incremental improvements in mundane objects and everyday processes. No fullerene-derived materials, fascinating and wonderful though they and their properties may be, are yet close to providing a practical superconductor, a drug to cure any disease, or anything else that is sure to produce even an infinitesimal improvement in our welfare. While we wait for these things with varying degrees of anticipation, it is worth remembering that there is already a gain to humankind, although one that is more spiritual than material. It is the beauty of the molecule and the serendipity of its discovery that remain its chief seduction.

EPILOGUE

SPOT THE BALL

The discovery of buckminsterfullerene and its cousins is a gift to science as well as the gift of science. While it is naive to expect these new molecules to be put to commercial use in the very near future, they will surely find their *métier* in due course. They could have been made and identified some decades before 1985, but their belated fabrication and identification is welcome, nonetheless, as a spectacular success in the physical sciences, especially after recent spurious breakthroughs such as the "discovery" of cold fusion. And yet, for all this, the puzzle that stimulated this find remains as perplexing as ever.

For, as Wolfgang Krätschmer and Donald Huffman were reluctantly obliged to point out, the spectra of buckminsterfullerene recorded in the laboratory do not coincide with any of the unexplained astronomical spectra that motivated both Harry Kroto and Krätschmer and Huffman in their research.

Surely, such a molecule is too beautiful *not* to occur in nature. The appeal of such a perfect object emerging from chemical chaos is not only aesthetic, it is also logical: what neater way to explain the occurrence of some peculiarly dominant species throughout the interstellar medium? It is made with such ease (we can now afford to say) in flames and carbon arcs that it must be present somewhere in the universe. Mustn't it? If the spectral evidence is not there for buckminsterfullerene itself, are there variants of it that might do the trick? The molecule ionized in various ways? Those ions that have been made and spectrally measured do not match the astrophysical observations either. Particular ratios of carbon isotopes in the fullerenes? Endohedral or exohedral fullerene complexes with various elements? These have so far proved too difficult to make, so their spectra cannot be compared. Among this class of compounds, fullerenes with a few hydrogen atoms added onto their surfaces are perhaps the most plausible candidates. Hydrogen, after all, is the most abundant element in space, and various hydrogenated fullerenes have been successfully made on earth, although there is not the one dominant compound that would make the idea really compelling. The spanner in the works here is that the elegance of this solution diminishes in proportion to the number of hydrogen atoms one adds. With one or two hydrogen atoms added, the spectrum of the molecule might still be reasonably clean. But with more, the number of spectral lines would rapidly mount up.

Historically, one of the intriguing aspects of the diffuse interstellar bands was that they were thought to be rather few in number, suggesting the presence of one or at most a few species. However, as the resolution of astronomical observation has grown, more lines have become visible. A new clump of them was detected in 1994. There are now so many that it is unlikely that a single buckminsterfullerene derivative (or any other single molecule) is responsible for them. A family of molecules would provide a better explanation. There is nothing to rule out a family of fullerene derivatives. This leaves the sequence of hydrogenated fullerenes in with a chance. Another possibility is a range of scalloped carbon shell fragments of fullerenes. But equally, it could be something as prosaic and disappointing as a cocktail of run-of-the-mill polycyclic aromatic hydrocarbons (naphthalene, anthracene, and their like).[1] In the best tradition of astrophysics, it is easy to reject individual suitors if there is the evidence available, but virtually impossible to find conclusively in favour of a particular candidate or class of candidates, especially while it remains difficult to make them in terrestrial laboratories.

If variants or fragments of buckminsterfullerene itself are not the answer, then perhaps the carbon "onions" may fit the bill. Calculations suggest that the 220 nanometre absorption line may be graphite of some kind, but it has not been possible to check this experimentally. In order to produce a match with the observed interstellar absorption, it is thought that the particles must be spherical and small, but not so small as typical single molecules. It is possible to juggle particle diameters and the optical constants for graphite in such a way as to produce a match for the absorption band. This much has been known for some time, but has been ignored because people could not reconcile the two key provisions – the graphitic nature of the particles and their sphericity. Nested fullerene "onions" like those made by Daniel Ugarte square the circle. It might at first seem unlikely that carbon onions are widespread in space while the small fullerenes and buckminsterfullerene itself are not. However, the onions gain stability from their structure of concentric shells and can endure even more severe conditions than naked buckminsterfullerene. Their spectra, however, remain to be recorded.[2]

The reason that the diffuse interstellar bands and the other unidentified astrophysical spectra are so puzzling is that scientists have been unable to reproduce them in their laboratories. The spectral transitions occur among chemical species that either have not been studied or – more likely since there exists an exhaustive bibliography of recorded spectra – are not present on earth. The fullerenes seemed like good candidates to explain these features because they too were unknown on earth.

In 1992, however, Peter Buseck and Semeon Tsipursky, mineralogists at Arizona State University in Tempe, were sent samples of shungite, a rare carbonaceous Precambrian rock found in Karelia on the Russian-Finnish border. Their aim was to study the carbon microstructure using an electron microscope as they had done for many other carbon-rich minerals. Examining the lustrous black rock, they found a small vein with the distinctive "fullerite" array of stacked spheres (familiar to them from an earlier examination in their laboratory of a manmade buckminsterfullerene sample). Subsequent analysis confirmed the presence of both C_{60} and C_{70}.[3]

The discovery of the first naturally occurring fullerenes raised some interesting questions. How, in this unexpected context, did the fullerenes form? Were they formed at the same time as the bulk of the rock – between four billion and 500 million years ago – or by some more recent trauma? Either way, how had they survived for many millions of years the oxidation and degradation known to afflict laboratory fullerenes?

One hypothesis that might explain the presence of fullerenes here and their scarcity elsewhere in nature invokes their creation by lightning strikes, one of few natural conditions that can compare with the conditions required to produce fullerenes in the laboratory. Buseck and Daly followed up this speculation by examining a specific class of rock, fulgurite, that forms when lightning strikes the ground, fusing rock and surrounding soil into a glassy solid. One sample among a large number taken for assessment duly revealed the presence of fullerenes here too.[4] The fulgurite source minerals are essentially free of carbon, and it is thought that the source of the fullerene carbon is organic matter, perhaps pine needles, that becomes locked into the fused rock at the time of the lightning strike. Trapped inside the rock in this way, the fullerenes are protected from oxidation. This second find does not necessarily imply that the fullerenes in the shungite were formed by the same means, and indeed fullerenes have subsequently been found in other geological environments, including in a meteorite sample[5] and in rock formed at the boundary between the Cretaceous and Tertiary eras 65 million years ago.[6] In the first instance, it is thought that the fullerenes were formed from terrestrial carbon at the time of impact, not that the fullerenes rode pillion through space on the meteorite. In the second, the age of the rock invites speculation that the fullerenes may have formed as a consequence of wildfire associated with the cataclysm that befell the dinosaurs.

It is, of course, a disappointment to Huffman and Krätschmer as well as to Kroto, Curl and Smalley that they have not explained the astrophysical feature that played such a crucial part in stimulating their research. But they may take a perverse solace from the fact that the diffuse interstellar bands remain as mysterious as ever: they – and everyone else – still have this problem to ponder as well as a whole set of new ones to tackle in the burgeoning field of the fullerenes.

It is the paradox of serendipitous discovery that while science fails in meeting its stated objective, it can succeed wildly in surpassing some other objective that no one has thought to express. Everyone has failed to resolve the mystery of what lies between the stars. Yet, what glorious failure! For there is still the chance to succeed. There is still the chance, too, that further attempts to solve the problem will lead not to its solution but to further serendipities.

NOTES AND REFERENCES

CHAPTER 1

1 Atkins, P.W., *Atoms, electrons, and change* (New York: Scientific American Library, 1991), p 11.

2 Wolpert, L., *The unnatural nature of science* (London: Faber and Faber, 1992), p 121.

3 Schwartz, J., *The creative moment* (London: Jonathan Cape, 1992), pp 30-31.

4 Although some "soft" sciences have come to seem rather foolish in their effort to appear "hard", it is not true that the unquantified – perhaps unquantifiable – sciences are worthless. Sigmund Freud, for example, gives the lie. Passmore, J., *Science and its critics* (Duckworth: London, 1978), p 53. Passmore adds that Freud was at his worst when trying to be most scientific.

5 Gaston, J., *The reward system in British and American science* (New York: John Wiley, 1978), p 45; Gaston, J., *Originality and competition in science* (Chicago: University of Chicago Press, 1973), p 60.

6 Medawar, P., *Pluto's republic* (Oxford: Oxford University Press, 1982), p 235. The comment comes in the context of an extended review of D'Arcy Thompson's famous essay, *On growth and form*, a work that will have an unexpected bearing on events in later chapters.

7 Hoffmann, R., "Under the surface of the chemical article", *Angewandte Chemie* 27 (1988): 1593-1602.

8 *The creative moment*, p 16.

9 "The division of scientific labour into innumerable subject specialties gives rise to a naive conception of scientists as people who devote their whole lives to the study of some very obscure and very detailed question, until, as the old joke has it, they know 'everything about nothing'." Ziman, J., *Knowing everything about nothing*, (Cambridge: Cambridge University Press, 1987), p 18. Ziman confirms that there is an element of truth in this, but adds that a larger number than might be imagined work "on diverse problems or in a broadening way".

10 Ibid., p 78.

11 Dietz, T.G., M.A. Duncan, D.E. Powers, and R.E. Smalley, "Laser production of supersonic metal cluster beams," *Journal of Chemical Physics* 74 (1981): 6511-6512.

12 Michalopoulos, D.L., M.E. Geusic, S.G. Hansen, D.E. Powers, and R.E. Smalley, "The bond length of Cr_2," *Journal of Physical Chemistry* 86 (1982): 3914-3915.

13 Of the fifteen million chemical compounds that humankind has added to nature's total the great majority are organic. Hoffmann, R., "How should chemists think?," *Scientific American,* February 1993: 66-73. The pace at which new organic molecules have been added is prodigious. Less than forty years ago, the inventory stood at no more than half a million. Wells, A.F., *The third dimension in chemistry* (Oxford: Clarendon Press, 1956), p 120.

14 Faraday, M., *The chemical history of a candle* (London: Scientific Book Guild, 1960), p 1.

15 *Atoms, electrons, and change*, pp 18-19.

16 Experiments to measure the vapour pressure of graphite as late as the 1960s show results in disagreement by a full order of magnitude with "no obvious explanation" for the variation. Palmer, H.B., and M. Shelef, "Vaporization of carbon," *Chemistry and physics of carbon*, ed. P.L. Walker (New York: Marcel Dekker, 1968) 4: 85-135.

17 Mass spectrometer studies during the Second World War indicated the presence of carbon species of up to fifteen atoms. These experiments were repeated twenty years later producing evidence for up to 33 carbon atoms in a single cluster. Structural and other details remained elusive, however. Ibid.

18 Smalley, R.E., "The third form of carbon," *Naval Research Reviews* December (1991): 3-14 .

19 Deming, H.G., *In the realm of carbon* (New York: John Wiley and Sons, 1930), p 66. Deming's source is an account given at the Kekulé Memorial Lecture in 1898 by which time Kekulé's dream had been burnished into legend by many tellings.

20 See, for example, Roberts, R.M., *Serendipity: accidental discoveries in science* (New York: John Wiley and Sons, 1989) , pp 75-81. Roberts concedes that Kekulé did not refer to the dream in his contemporary accounts but takes the myth at face value nevertheless.

21 Anschütz, R., *August Kekulé* (Berlin: Verlag Chemie, 1929) 1: p 304.

22 For a more detailed exposition see Murrell, J.N., S.F.A. Kettle, and J.M. Tedder, *Valence theory* (London: John Wiley and Sons, 1970), Chapter 15.

23 Many examples are gathered in Vögtle, F., *Fascinating molecules in organic chemistry* (Chichester: John Wiley and Sons, 1992).

24 The International Union of Pure and Applied Chemistry has devised a logical system of names for organic compounds. The system works well enough for simple molecules and is much favoured in high schools. It becomes increasingly useless for more complex molecules where first history and then poetry take over in inspiring names.

Consider the simple family of molecules that consist of two carbon atoms and a number of hydrogens. The saturated form has the maximum number of hydrogens with only a single bond joining the two carbons. This is ethane. If we remove two hydrogens and upgrade the carbon-carbon bond from single to double we obtain the unsaturated ethylene or, in the IUPAC nomenclature, ethene. Lose two more hydrogens and join the carbons now with a triple bond and we get ethyne which is more attractively but less systematically known as acetylene. The vowel sequence -ane, -ene, -yne that signifies the nature of the carbon bond is logical enough. In ordinary alcohol, one hydrogen atom is replaced by a hydroxide group comprising one oxygen and one hydrogen atom. The old chemical name is ethyl alcohol, the new one, ethanol. So far, so good. Matters become complicated very rapidly with larger molecules, branched chains, "foreign" atoms. Benzene is cyclohexatriene but no one ever calls it that. Many new molecules made in modern syntheses are too complicated to make a systematic name worth attempting.

25 Paquette, L.A., R.J. Ternansky, D.W. Balogh, and G. Kentgen, "Total synthesis of dodecahedrane," *Journal of the American Chemical Society* 105 (1983): 5446-5450.

26 There are five Platonic solids in all. I have described the hydrocarbon analogues of the cube and the dodecahedron. Of those remaining, tetrahedrane is too structurally strained to be isolated; octahedrane and icosahedrane could not exist because carbon's tetravalence would be more than satisfied by the number of edges (bonds) that meet at each vertex (carbon atom).

27 Kroto, H.W., "Semistable molecules in the laboratory and in space," *Chemical Society Reviews* 11 (1982): 435-491.

28 Kroto's experiments owe much to the work of George Porter who was awarded the Nobel Prize for chemistry for devising the technique known as flash photolysis. Lest it seem that molecules created in such ways, and which cannot be stored or poured from a beaker, are of only academic interest, it is worth remembering that one of Porter's new molecules was chlorine monoxide now known to be intimately involved in the depletion of the atmospheric ozone layer by manmade chlorofluorocarbons, Porter, G., "Flash photolysis and spectroscopy. A new method for the study of free-radical reactions," *Proceedings of the Royal Society* 200 (1950): 284-300.

29 For a brief popular account of Miller and Urey's experiment see Hoyle, F., *The intelligent universe* (London: Michael Joseph, 1983), pp 19-21. Hoyle recounts the experiment in the context of his argument that life did not emerge from a primordial soup, an assertion he bases on the fact that none of the many further necessary stages of biomolecular assembly occurs from the soup without the assistance of an added enzyme. So, asks Hoyle, whence the first enzyme? These arguments are set forth in more – but ultimately unconvincing

– detail in Hoyle, F., and C. Wickramasinghe, *Lifecloud* (London: J M Dent and Sons, 1978).

30 Kroto, H.W., "C_{60}: Buckminsterfullerene, the celestial sphere that fell to earth," *Angewandte Chemie* 31 (1992): 111-129.

31 Duley, W.W., and D.A. Williams, *Interstellar chemistry* (London: Academic Press, 1984) summarizes the rapid progress in this new science.

32 Allamandola, L.J., A.G.G.M. Tielens, and J.R. Barker, "Polycyclic aromatic hydrocarbons and the unidentified infrared emission bands: auto exhaust along the Milky Way!," *Astrophysical Journal* 290 (1985): L25-L28. Others had earlier found a good fit for these molecules, but used more measured language, Leger, A., and J.L. Puget, "Identification of the unidentified infrared emissions of interstellar dust," *Astronomical and Astrophysical Letters* 137 (1984): L5-L8 .

33 Douglas, A.E., "Origin of the diffuse interstellar lines," *Nature* 269 (1977): 130-132.

34 "C_{60}: Buckminsterfullerene, the celestial sphere that fell to earth".

35 The British science budget was growing by 13 per cent a year in 1966-1967, Wilkie, T., *British science and politics since 1945* (Oxford: Blackwell, 1991), p 77.

36 "C_{60}: Buckminsterfullerene, the celestial sphere that fell to earth".

37 One of Kroto's best pieces was the cover of the 1977 Sussex undergraduate and graduate chemistry prospectus. This showed a map of Britain with each university city's position denoted by a schematic drawing of a squat Bunsen burner. The only burner with a flame issuing from it is the one for Brighton. The message: Only the Sussex chemistry course will light your fire. The joke is complete with the knowledge that Brighton's best known landmark is John Nash's Regency extravaganza, the Royal Pavilion. Its centrepiece is an onion dome inspired by Moghul architecture and in outline just like the chimney and flame of the Bunsen burner.

38 Dale, J., "Structure and physical properties of acetylenic compounds; the nature of the triple bond," *Chemistry of acetylenes*, ed. H.G. Viehe (New York: Marcel Dekker, 1969), p 62.

39 Ibid., pp 76-77.

40 Although these episodes form a logical sequence alternating between laboratory synthesis and astrophysical observation, it will be noted that the interval between them is a matter of a year or more. This pace may be considered fairly typical of scientific events, given the need to book time to use equipment, delays in the publication of results, and the usual commitments to

teaching and administration. Later chapters will reveal just how much faster things can move when circumstances warrant.

41 The sequence of discoveries is reviewed by Lorne Avery, one of the astronomers involved in the collaboration at what is now called the Herzberg Institute of Astrophysics, named for the spectroscopist Gerhard Herzberg whose tables of spectral data of many atoms and molecules are the bibles of chemical physics. Avery, L.W., "Long chain carbon molecules in the interstellar medium," *Interstellar molecules*, ed. B.H. Andrew (Mont Tremblant, Québec, Canada: D. Reidel Publishing Company, 1979), pp 47-55. Other researchers added the eleven-carbon cyanopolyÿne to the group in 1982. With a far weaker spectral signal, the first cyclic molecule, the three-carbon ring cyclopropenylidene, was only detected in 1985.

42 Winnewisser, G., F. Toelle, H. Ungerechts, and C.M. Walmsley, "Long chain carbon molecules in the laboratory and in space", *Interstellar molecules*, pp 59-64.

43 Sakata, A., "On the formation of interstellar linear molecules," *Interstellar molecules*, pp 325-328.

44 *Interstellar chemistry*, p 198.

45 Kroto, H.W., "Chemistry between the stars," *New Scientist* 10 August 1978: 400-403; "Semistable molecules in the laboratory and in space".

CHAPTER 2

1 Schwartz, J., *The creative moment* (London: Jonathan Cape, 1992), p xix.

2 "... what a scientist thinks of as his 'data' are not only corrigible in principle but for the most part theory-dependent." Passmore, J., *Science and its critics*, (Duckworth: London, 1978), p 82.

3 "AP2 chronology 1985," (Rice University: unpublished, 1992) . This lengthy chronology was faxed to me with some alacrity when I announced my plan to write this book.

4 Perhaps the most widely read of the reviews to appear during this period were the two that appeared in consecutive issues of *Science* in November 1988. Curl, R.F., and R.E. Smalley, "Probing C_{60}," *Science* 242 (1988): 1017-1022; and Kroto, H., "Space, stars, C_{60}, and soot," *Science* 242 (1988): 1139-1145.

5 Baum, R.M., "Laser vaporization of graphite gives stable 60-carbon molecules," *Chemical and Engineering News* 23 December 1985; Browne, M., "Molecule is shaped like a soccer ball," *New York Times* 3 December 1985.

6 Taubes, G., "Great balls of carbon," *Discover* September 1990: 52-59;

Baggott, J., "How the molecular football was lost," *New Scientist* 6 July 1991: 56; Baggott, J., "Great balls of carbon," *New Scientist* 6 July 1991: 34-38.

7 Smalley, R.E., "Great balls of carbon, the story of buckminsterfullerene," *The Sciences* March-April 1991: 22-28.

8 Kroto, H.W., "C_{60}: Buckminsterfullerene, the celestial sphere that fell to earth," *Angewandte Chemie* 31 (1992): 111-129.

9 Taubes, G., "The disputed birth of buckyballs," *Science* 253 (1991): 1476-1479.

10 There are two principal sources of disagreement between Smalley and Kroto. One concerns the name of the molecule, the other its structure. It seems certain that Buckminster Fuller was mentioned and discussed by Smalley and Kroto and also Bob Curl during the first full week of Kroto's visit. Smalley maintains that he and Kroto jointly thought of the name "buckminsterfullerene" in the course of a Monty Pythonesque banter at this time. This seems unlikely for a number of reasons: 1) Invoking Fuller and his geodesic domes implies that a spheroid closed structure was known, which it was not at this stage; 2) The -ene suffix implies that the molecule was known to have unsaturated double bonds like benzene or anthracene, which it also was not at this time; 3) In subsequent discussion and literature beginning with the announcement of the new molecule in *Nature*, Smalley is at best equivocal about the merits of the name. Based on this circumstantial evidence and in the absence of material evidence or independent witnesses' testimony to the contrary, I have assumed that the molecule was not named until after its structure was revealed to the group on Tuesday 10 September.

The second controversy concerns the forcefulness with which structural ideas related to Fuller's domes and Kroto's stardome were advanced, when, and by whom. Smalley and Curl, as stated above, maintain that Fuller came up in the first week. Whether he did or not, his influence was not powerful at this stage, for it was not until Monday 9 September that Smalley went and found a book on his architecture from the university library. Kroto does not remember mentioning the stardome until that day. While Kroto must have mentioned the probability that pentagons were present on the surface of the stardome in order that Smalley would recall them during his model-making, it seems unlikely that he was particularly insistent about the role of pentagons since: 1) Smalley was persistently reluctant to try pentagons when considering possible structures; and 2) Heath, in his model-making efforts, did not try pentagons at all; 3) If Kroto had been sure of this, it seems probable that he would have made his own models or at least confirmed the stardome structure by telephoning home. In the end, the fact is that it was Smalley who made the model that turned out to be correct and that, in order to do so, he recalled Kroto's mention of pentagons.

Whatever the truth of these two episodes, arising from more or less private conversation that took place (or not) between the two parties who now contest it, we shall probably never know, just as we shall never know whether August Kekulé really discovered the structure of benzene while dreaming of snakes chasing their tails or whether Archimedes ever took his famous bath.

11 The five principal members of the team that discovered buckminsterfullerene were interviewed by the author between January and March 1993.

12 Anonymous referees' comments on "C_{60}: Buckminsterfullerene", September 1985.

13 Kroto, H.W., J.R. Heath, S.C. O'Brien, R.F. Curl, and R.E. Smalley, "C_{60}: Buckminsterfullerene," *Nature* 318 (1985): 162-163.

14 Scientists at Exxon Research and Engineering Company then in Linden, New Jersey, were interested in the question of how graphite builds up on catalysts. Neither the Exxon group nor any of those who saw their presentations at various scientific meetings during 1984 and 1985 realized the significance of the spectra they showed in which a sixty-atom carbon fragment was the largest by a small margin in a range of cluster peaks. Rohlfing, E.A., D.M. Cox, and A. Kaldor, "Production and characterization of supersonic carbon cluster beams," *Journal of Chemical Physics* 81 (1984): 3322-3330. Researchers at Bell Laboratories in nearby Murray Hill, New Jersey, also recorded spectra with a prominent peak at sixty carbon atoms, but did not investigate further. Bloomfield, L.A., M.E. Geusic, R.R. Freeman, and W.L. Brown, "Negative and positive cluster ions of carbon and silicon," *Chemical Physics Letters* 121 (1985): 33-37.

15 Sources for the description of AP2 are: Dietz, T.G., M.A. Duncan, D.E. Powers, and R.E. Smalley, "Laser production of supersonic metal cluster beams," *Journal of Chemical Physics* 74 (1981): 6511-6512; Powers, D.E., S.G. Hansen, M.E. Geusic, A.C. Pulu, J.B. Hopkins, T.G. Dietz, M.A. Duncan, P.R.R. Langridge-Smith, and R.E. Smalley, "Supersonic metal cluster beams: laser photoionization studies of Cu_2," *Journal of Physical Chemistry* 86 (1982): 2556-2560; Smalley, R.E., "Mass-selective laser photoionization," *Journal of Chemical Education* 59 (1982): 934-938; "C_{60}: Buckminsterfullerene"; Smalley, R.E., "Down-to-earth studies of carbon clusters," (NASA, 1987), 199-244; "Probing C_{60}"; Smalley, R.E., "Supersonic carbon cluster beams," *Atomic and molecular clusters*, ed. E.R. Bernstein. (Amsterdam: Elsevier, 1990) 1-68; "Great balls of carbon, the story of buckminsterfullerene"; the briefing document, "Some facts related to the discovery of the fullerenes," (Rice University: unpublished, no date); and interviews.

16 "Production and characterization of supersonic carbon cluster beams".

17 Murrell, J.N., S.F.A. Kettle, and J.M. Tedder, *Valence theory*, (London: John Wiley and Sons, 1970), pp 270-273 leads through the analogous mathematics for double-bonded ethylene and butadiene.

18 They did this experiment at a later date using ammonia and methyl cyanide, both species containing nitrogen and hydrogen and both known to be present in interstellar space. Both experiments worked but the findings were not published until 1987. Heath, J.R., Q. Zhang, S.C. O'Brien, R.F. Curl, H.W. Kroto, and R.E. Smalley, "The formation of long carbon chain molecules during laser vaporization of graphite," *Journal of the American Chemical Society* 109 (1987): 359-363. The fact that they had now made cyanopolyÿnes in hot carbon vapour using ingredients available in space lent powerful support to Kroto's theory that these molecules are made in the atmosphere around stars and not in the cold clouds where they may be detected today. Though this was terrestrial chemistry, the implications were significant enough to warrant another paper, in the *Astrophysical Journal* – Kroto, H.W., J.R. Heath, S.C. O'Brien, R.F. Curl, and R.E. Smalley, "Long carbon chain molecules in circumstellar shells," *Astrophysical Journal* 314 (1987): 352-355.

19 The structure of DNA had been discovered famously by the model-making efforts of James Watson and Francis Crick as had the structure of the α-helix by Linus Pauling. Watson, J.D., *The double helix* (Harmondsworth, Middlesex: Penguin Books, 1970), pp 47-48.

20 Unknown to Rick at this time, this carbon framework had already been synthesized. The dangling bonds on the rim of the dish were completed with hydrogen atoms to form a stable compound called corannulene, the name derived from the large carbon ring compound, annulene, now with a smaller ring at its core.

21 "Great balls of carbon, the story of buckminsterfullerene".

22 Kroto, Smalley, and Curl soon learned that their geodesic dome of carbon had been predicted with varying degrees of detail. In 1970, Eiji Osawa suggested that the soccer ball structure of C_{60} might be stable. Osawa, E., "Superaromaticity," *Kagaku (Kyoto)* 25 (1970): 854-863. His paper was in Japanese but it clearly showed both icosahedral and soccer ball structures. A few years later, scientists in the then Soviet Union published (in Russian and in an English translation edition) Hückel calculations confirming that this structure was favoured. Bochvar, D.A., and E.G. Gal'pern, "Carbododecahedron, s-icosahedron, and carbo-s-icosahedron hypothetical systems," *Proceedings of the Academy of Sciences of the USSR* 209 (1973): 239-241 . It is notable that there are few significant cultural barriers in science today, but the large volume of literature published in the Japanese and Russian languages must surely constitute one of those that remains for Western scientists.

The Rice group's fears that they would be beaten came closest to being realized with the knowledge that Tony Haymet, a theoretician at the University of California at Berkeley, had been working independently but concurrently with them and in ignorance of the Japanese and Russian precedents. In his paper, which appeared only a month after the *Nature* paper, he predicted that C_{60}, which he called footballene, "will have remarkable properties" and pointed out that metal ions could sit within the carbon cage. Haymet, A.D.J., "Footballene: a theoretical prediction for the stable, truncated icosahedral molecule C_{60}," *Journal of the American Chemical Society* 108 (1985): 319-321.

By far the most entertaining prophecy came from David Jones who writes under the pseudonym of Daedalus, then in *New Scientist*, and still today, in *Nature*. Daedalus's fevered imagination saw a discontinuity in density between the gases and the liquids and solids, and proposed to plug the gap with large hollow molecules made of graphite sheets. Jones, D.E.H., "Hollow molecules," *New Scientist* 32 (1966): 245. On its first publication, Jones fudged the problem of how to close the hexagonal net, proposing to introduce "ill-fitting foreign atoms ... to warp them". He proposed a growth mechanism for the curving graphite sheets that was remarkably similar to the one that Kroto and Smalley would eventually publish. He even shared their prediction of lubricant properties for the molecules. By the time he came to collect his inspired inventions in a book, Daedalus had learned that it required pentagons to close the net into a sphere requirement. He commented of progress made in "real" science during the intervening sixteen years that: "Hollow-molecule chemistry has not come very far since I published this idea. The current record in hydrocarbon chemistry is a mere pentagonal dodecahedron, described by L.A. Paquette and co-workers in *Science* (Vol. 211, 1981, p.575). Keep at it, fellows!" Jones, D.E.H., *The inventions of Daedalus* (Oxford: Freeman, 1982) , pp 118-119.

23 See note 10.

24 Smalley, R.E., correspondence to *Nature*, 12 September 1985.

25 Heath, J.R., S.C. O'Brien, Q. Zhang, Y. Liu, R.F. Curl, H.W. Kroto, F.K. Tittel, and R.E. Smalley, "Lanthanum complexes of spheroidal carbon shells," *Journal of the American Chemical Society* 107 (1985): 7779-7780. Coming so soon after the discovery of buckminsterfullerene itself, the lanthanum paper, submitted less than two weeks later, had the useful effect of providing leverage enabling Smalley to attempt to hasten the *Nature* publication by warning its editors that it would look unfortunate if the latter were to appear after the *Journal of the American Chemical Society* paper rather than before it. Smalley, R.E., correspondence to *Nature*, 9 October 1985.

26 One referee wrote: "... in the spirit of stimulating scientific debate it certainly is a fun paper. In terms of substantial content I am not sure exactly what it contains other than the ability to emphasize the 'production' of the specific carbon cluster size, C_{60}." The other commented that much of the letter was "highly speculative, but much of the speculation is very interesting." One of the referees drew the authors' attention to a recently published paper (Krätschmer, W., N. Sorg, and D.R. Huffman, "Spectroscopy of matrix-isolated carbon cluster molecules between 200 and 850 nm wavelength,"

Surface Science 156 (1985): 814-821) which was to prove an eerie omen of events to come.

27 "C_{60}: Buckminsterfullerene".

28 Wolpert, L., *The unnatural nature of science* (London: Faber and Faber, 1992), p 57.

29 Hoffmann, R., "Under the surface of the chemical article," *Angewandte Chemie* 27 (1988): 1593-1602.

30 Ibid.

31 Terry Eagleton produces various definitions of what constitutes literature, most of which do not exclude scientific writing. Eagleton, T., *Literary theory*, (Oxford: Blackwell Publishers, 1983) pp 1-16. Roland Barthes admits that: "As far as science is concerned language is simply an instrument, which it profits it to make as transparent and neutral as possible: it is subordinate to the matter of science (workings, hypotheses, results) which, so it is said, exists outside language and precedes it." He is further sceptical of structuralism's claim to be the science of literature, but for all this says nothing to preclude its application to scientific texts. Barthes, R., "Science versus literature," *Structuralism: a reader*, ed. M. Lane. (London: Jonathan Cape, 1970) pp 410-416. For a more detailed discussion of the applicability of techniques of literary theory to scientific texts, see Locke, D., *Science as writing*, (New Haven: Yale University Press, 1992).

32 It is amusing to note that there was at least one attempt to name the molecule according to the convention of the International Union of Pure and Applied Chemistry. The prelude for this exercise in fact predated the discovery and concerned a putative saturated molecule, $C_{60}H_{60}$, nicknamed footballane or soccerane. Castells, J., and F. Serratosa, "Goal! An exercise in IUPAC nomenclature," *Journal of Chemical Education* 60 (1983): 941. The systematic name occupied a line and a half of close-spaced type. The authors, Josep Castells and Felix Serratosa from the University of Barcelona, added a postscript in the light of the recent synthesis by Leo Paquette of dodecahedrane, $C_{20}H_{20}$, noting that "the synthesis of footballane appears as the next 'goal.' Good luck!" Later, an error was pointed out in their naming methodology and the same authors published a pithy apology which they hoped would cost them no more than a single "penalty"! Castells, J., and F. Serratosa, "Replaying the ball: soccerane revisited," *Journal of Chemical Education* 63 (1986): 630.

CHAPTER 3

1 This chapter draws once again upon interviews with the five principal members of the team that discovered buckminsterfullerene and upon interviews with Donald Cox at the Exxon Research and Engineering Company,

Annandale, New Jersey; Orville Chapman and Robert Whetten at the University of California at Los Angeles; Doug Klein and Thomas Schmalz at Texas A&M University at Galveston; Eric Rohlfing at the Sandia National Laboratory, Livermore, California; and Fred Wudl at the University of California at Santa Barbara.

2 Paquette, L.A., R.J. Ternansky, D.W. Balogh, and G. Kentgen, "Total synthesis of dodecahedrane," *Journal of the American Chemical Society* 105 (1983): 5446-5450.

3 O'Brien, S.C., J.R. Heath, R.F. Curl, and R.E. Smalley, "Photophysics of buckminsterfullerene and other carbon cluster ions," *Journal of Chemical Physics* 88 (1988): 220-230.

4 Heath, J.R., Q. Zhang, S.C. O'Brien, R.F. Curl, H.W. Kroto, and R.E. Smalley, "The formation of long carbon chain molecules during laser vaporization of graphite," *Journal of the American Chemical Society* 109 (1987): 359-363.

5 Casti, J.L., *Paradigms lost* (London: Scribners, 1990), p 40.

6 Heath, J.R., S.C. O'Brien, Q. Zhang, Y. Liu, R.F. Curl, H.W. Kroto, F.K. Tittel, and R.E. Smalley, "Lanthanum complexes of spheroidal carbon shells," *Journal of the American Chemical Society* 107 (1985): 7779-7780. The entire cluster team at Rice was now committed to the exploration of the new carbon clusters. Kroto's name was added in proof at his protestation. Kroto felt that all those involved in the original discovery should be named on this secondary paper; Smalley that only those who physically did the experiment reported. Their difference of opinion perhaps reflects a difference between British and American protocol. This incident sowed the seeds of Kroto and Smalley's gradual falling out over the next few years, a rift that widened dramatically after Kroto had given what Smalley saw as a biased account of their work together to the Space Sciences Department at Rice University in the spring of 1986.

7 Rohlfing, E.A., D.M. Cox, and A. Kaldor, "Production and characterization of supersonic carbon cluster beams," *Journal of Chemical Physics* 81 (1984): 3322-3330.

8 Popper writes that a hypothesis strengthens a theory if it *increases* its degree of falsifiability. Popper, K.R., *The logic of scientific discovery*, (London: Hutchinson, 1959) , p 83. The hypothesis that a lanthanum atom may be enclosed inside a sixty-atom carbon cage thus strengthens the theory that carbon forms such cages at all. This theory was certainly stronger at this time than theories favouring polyÿne chains or graphite flakes, not least because it was the most open to rigorous testing. Although surprising and revolutionary, it was a simple idea. "Simple statements, if knowledge is our object, are to be prized more highly than less simple ones *because they tell us more; because*

their empirical content is greater; and because they are better testable." (p 142).

9 Cox, D.M., D.J. Trevor, K.C. Reichmann, and A. Kaldor, "C_{60}La: a deflated soccer ball?," *Journal of the American Chemical Society* 108 (1986): 2457-2458.

10 "Production and characterization of supersonic carbon cluster beams."

11 Haymet, A.D.J., "C_{120} and C_{60}: Archimedean solids constructed from sp^2 hybridized carbon atoms," *Chemical Physics Letters* 122 (1985): 421-424; Haymet, A.D.J., "Footballene: a theoretical prediction for the stable, truncated icosahedral molecule C_{60}," *Journal of the American Chemical Society* 108 (1986): 319-321. Other theoreticians soon confirmed that this geometry was likely to be the most stable of the possible isomers of C_{60}. Fowler, P.W., and J. Woolrich, "π-systems in three dimensions," *Chemical Physics Letters* 127 (1986): 78-83; Klein, D.J., T.G. Schmatz, G.E. Hite and W.A. Seitz, "Resonance in C_{60} buckminsterfullerene", *Journal of the American Chemical Society* 108 (1986): 1301; Shibuya, T., and M. Yoshitani, "Two icosahedral structures for the C_{60} cluster," *Chemical Physics Letters* 137 (1987): 13-16.

12 Sean O'Brien had done the Hückel calculation for buckminsterfullerene in the days after the Rice group discovery, but Smalley vetoed its inclusion in the papers being submitted at this time, perhaps because he did not see it as his group's job to throw theoretical conjecture into the arena when they could produce real experimental results, or perhaps simply for fear that the calculations might be proved wrong. (It was later learned that the calculation had also been performed several times over by various researchers around the world long before the molecule had actually been made.) The calculation – not a trivial one with such a large molecule, but manageable because of simplification stemming from its high symmetry – yielded values for energy levels of the molecular orbitals from which it was possible to draw a qualitative portrait of many of the molecule's likely electronic and chemical properties.

13 Letters to the editor, *New Scientist*, 24 August 1991, 21 September 1991.

14 Q. Zhang, S.C. O'Brien, J.R. Heath, Y. Liu, R.F. Curl, H.W. Kroto, and R.E. Smalley, "Reactivity of large carbon clusters: spheroidal carbon shells and their possible relevance to the formation and morphology of soot," *Journal of Physical Chemistry* 90 (1986): 525-528.

15 Cox, D.M., K.C. Reichmann, and A. Kaldor, "Carbon clusters revisited: the 'special' behavior of C_{60} and large carbon clusters," *Journal of Chemical Physics* 88 (1988): 1588-1597.

16 The language adopted by the Rice group in their reports was, of course, sometimes correspondingly loaded in their own favour. For example, Curl and

Smalley write in their major review during this period of the noteworthiness of "the increasing prominence of one cluster, C_{60} (and to a lesser extent, C_{70}), as the nozzle conditions are changed." We are not reminded that whether the cluster's prominence increases or decreases depends on *how* the nozzle conditions are changed. Curl, R.F., and R.E. Smalley, "Probing C_{60}," *Science* 242 (1988): 1017-1022.

17 Liu, Y., S.C. O'Brien, Q. Zhang, J.R. Heath, F.K. Tittel, R.F. Curl, H.W. Kroto, and R.E. Smalley, "Negative carbon cluster ion beams: new evidence for the special nature of C_{60}," *Chemical Physics Letters* 126 (1986): 215-217.

18 Hahn, M.Y., E.C. Honea, A.J. Paguia, K.E. Schriver, A.M. Camarena, and R.L. Whetten, "Magic numbers in C_N+ and C_N- abundance distributions," *Chemical Physics Letters* 130 (1986): 12-16.

19 O'Brien, S.C., J.R. Heath, H.W. Kroto, R.F. Curl, and R.E. Smalley, "A reply to 'Magic numbers in C_n+ and C_n- abundance distributions' based on experimental observations," *Chemical Physics Letters* 99-102 (1986).

20 "Photophysics of buckminsterfullerene and other carbon cluster ions." This was the first work on the fullerenes published by the Rice group that did not carry Kroto's name as a co-author. Even as they carried out the series of experiments to prove the special nature of buckminsterfullerene and keep the Exxon group at bay, Kroto and Smalley had drifted apart as each felt the need to present and defend his part in the discovery – one who knew them both well at this time describes them as behaving like rivals for the same lover. Upon his return to Britain after the last of his return visits to Rice University, Kroto saw no other option than to build his own cluster beam apparatus. With Tony Stace, a cluster specialist at Sussex, Kroto obtained funding to build a souped-up machine that it was hoped would allow them to accrue quantities of buckminsterfullerene and do new experiments on it. The apparatus was designed to pump large volumes of gas and operate with a continuous laser beam rather than a pulsed beam, measures to compensate for the presumed relative scarcity of fullerenes in the soot and to enable the separation of the various fullerenes as they were produced. The machine was redundant, at least as a fullerene factory, from the moment of its commissioning, because by then the secret of how to make quantities of buckminsterfullerene had been found...

21 "Carbon clusters revisited: the 'special' behavior of C_{60} and large carbon clusters." It is notable that the paper from the Rice group describing the new photophysics of buckminsterfullerene (cited in the previous note) was both received later and published sooner by the same journal.

22 "The capacity for self-delusion, even among scientists, should never be under-estimated: conviction can have profound effects on observation." Wolpert, L., *The unnatural nature of science* (London: Faber and Faber, 1992), p 141.

23 Weiss, F.D., J.L. Elkind, S.C. O'Brien, R.F. Curl, and R.E. Smalley, "Photophysics of metal complexes of spheroidal carbon shells," *Journal of the American Chemical Society* 110 (1988): 4464-4465.

24 Heath, J.R., R.F. Curl, and R.E. Smalley, "The UV absorption spectrum of C_{60} (buckminsterfullerene): a narrow band at 3860Å," *Journal of Chemical Physics* 87 (1987): 4236-4238. Puzzlingly, the recorded spectrum did not coincide with the camel spectrum that Wolfgang Krätschmer and Donald Huffman had become convinced was something to do with buckminsterfullerene. The spectral picture was confused for two reasons. The Rice group recorded their result, assigning the peak not to what from a casual glance at the Hückel diagram of orbital energy levels was apparently the first transition (of an electron from the highest occupied to the lowest unoccupied orbital of the molecule) which is disallowed by quantum-mechanical symmetry rules, but to the first allowed transition (from the highest occupied to the *second-lowest* unoccupied orbital). In the end, it transpired that it was a forbidden transition that was taking place, the usual rules suspended because of complications due to molecular vibrations. Haufler, R.E., Y. Chai, L.P.F. Chibante, M.R. Fraelich, R.B. Weisman, R.F. Curl, and R.E. Smalley, "Cold molecular beam electronic spectrum of C_{60} and C_{70}," *Journal of Chemical Physics* 95 (1991): 2197-2199.

25 Yang, S.H., C.L. Pettiette, J. Conceicao, O. Cheshnovsky, and R.E. Smalley, "UPS of buckminsterfullerene and other large clusters of carbon," *Chemical Physics Letters* 139 (1987): 233-238.

26 "Reactivity of large carbon clusters: spheroidal carbon shells and their possible relevance to the formation and morphology of soot."

27 Gould's encomium comes in the preface to a recent paperback edition of Thompson's great work, which was first published in 1917 and extensively revised by Thompson in 1942. In his discussion of the spherical exoskeleton structure of some radiolarians (an order of marine protozoans), Thompson writes: "But here a strange thing comes to light. *No system of hexagons can enclose space*; whether the hexagons be equal or unequal, regular or irregular, it is still under all circumstances mathematically impossible. So we learn from Euler: the array of hexagons may be extended as far as you please, and over a surface either plane or curved but *it never closes in*." This, of course, was the realization that had escaped Smalley and Heath as they had sought to form a closed graphite-like structure for their sixty carbon atoms. Thompson, D.W., *On growth and form*, ed. J.T. Bonner (Cambridge: Cambridge University Press, 1961), pp 157-158.

28 Versions of the spiralling carbon framework occur in a number of papers including "Probing C_{60}," and Kroto, H., "Space, stars, C_{60}, and soot," *Science* 242 (1988): 1139-1145.

29 In three successive papers, these authors use the phrase "uniquely elegant"

to describe the proposed geometry of buckminsterfullerene, Klein, D.J., W.A. Seitz, and T.G. Schmalz, "Icosahedral symmetry carbon cage molecules," *Nature* 323 (1986): 703-706; Schmalz, T.G., W.A. Seitz, D.J. Klein, and G.E. Hite, "C_{60} carbon cages," *Chemical Physics Letters* 130 (1986): 203-207; Schmalz, T.G., W.A. Seitz, D.J. Klein, and G.E. Hite, "Elemental carbon cages," *Journal of the American Chemical Society* 110 (1988): 1113-1127.

30 "Elemental carbon cages."

31 Kroto, H.W., "The stability of the fullerenes C_n, with $n = 24, 28, 32, 36, 50, 60$ and 70," *Nature* 329 (1987): 529-531.

32 It was later learned that the fullerenes actually form in far greater concentrations than the Rice group had ever imagined. This new knowledge obliged Smalley to reconsider this model for the growth of carbon clusters and shells. The modified scheme, which had been dubbed the Pentagon Road, favours the curved structure of a network of hexagons *and* pentagons for open shells as well as for fullerenes of all sizes, and – overcoming once and for all the ingrained notion that carbon always tries to form flat graphite sheets – with as many of the latter as possible without their coming into mutual contact. The old "party line" had required merely the haphazard incorporation of the occasional pentagon in a growing carbon shell. In the Pentagon Road scheme, the Rice group finally acknowledged and assimilated into their thinking the annealing process (the "cooking") that they had always said was important in the condensation of carbon clusters from a vapour. Annealing gives time for the many pentagons in the new scheme to space themselves optimally within the carbon network to avoid geometric strain and unfavourable electronic configurations. "It is a mechanism of graphite-sheet self-assembly that leads uniquely to C_{60}." Smalley, R.E., "Self-assembly of the fullerenes," *Accounts of Chemical Research* 25 (1992): 98-105; Curl, R.F., and R.E. Smalley, "Fullerenes," *Scientific American* October (1991): 54-63.

Jim Heath proposed a different mechanism, a reversal of the known fragmentation process, whereby the fullerenes might grow by the addition of C_2 units, Heath, J.R. in Hammond, G.S., and V.J. Kuck, ed., *Fullerenes: synthesis, properties, and chemistry of large carbon clusters*, (Washington DC: American Chemical Society, 1992) 481, p 1-23. Both schemes were somewhat superseded in later work by Yves Rubin and François Diederich, Harold Kroto and David Walton, and Michael Bowers and colleagues at the University of California at Santa Barbara. The case for and against various mechanisms is concisely reviewed in Curl, R.F., "Collapse and growth," *Nature* 363 (1993): 14-15.

33 Kroto, H.W., and K. McKay, "The formation of quasi-icosahedral spiral shell carbon particles," *Nature* 331 (1988): 328-331.

34 "Space, stars, C_{60}, and soot."

35 Gerhardt, P., S. Löffler, and K.H. Homann, "Polyhedral carbon ions in hydrocarbon flames," *Chemical Physics Letters* 137 (1987): 306-310.

CHAPTER 4

1 Arnheim, R., *Entropy and art* (Berkeley: University of California Press, 1971), p 7.

2 In Kepes, G., *Structure in art and in science* (New York: Braziller, 1965), pp 20-28.

3 A.I. Miller, "Visualization lost and regained: the genesis of the quantum theory in the period 1913-1927" in Wechsler, J., ed., *On aesthetics in science* (Cambridge, Massachusetts: MIT Press, 1978), pp 73-102.

4 J. Bronowski, "The discovery of form", in *Structure in art and in science*, pp 55-65.

5 See, for example, the comments of many of the scientists interviewed in Wolpert, L., and A. Richards, *A passion for science* (Oxford: Oxford University Press, 1988).

6 Chandrasekhar, S., "Beauty and the quest for beauty in science," *Physics Today* 32.25 (1978): 59-73 , reprinted in Chandrasekhar, S., *Truth and beauty: aesthetics and motivation in science*, (Chicago: University of Chicago Press, 1987).

7 Stewart, I., and M. Golubitsky, *Fearful symmetry* (Oxford: Blackwell, 1992), p 88.

8 Taubes, G., *Nobel dreams* (Redmond, Washington: Tempus Books, 1986), p 111.

9 Popper, K.R., *The logic of scientific discovery* (London: Hutchinson, 1959), pp 137-142.

10 See, for example, McLeish, J., *Number* (London: Bloomsbury, 1991); Joseph, G.G., *The crest of the peacock*, (London: I B Tauris and Co., 1991).

11 In his valuable study of the non-Western origins of science, George Gheverghese Joseph writes:

> Different explanations have been offered for the origins of the sexagesimal system, which, unlike base 10, or even base 20, has no obviously anatomical basis. Theon of Alexandria, in the fourth century AD, pointed to the computational convenience of using the base 60. Since 60 is exactly divisible by 2, 3, 4, 5, 6, 10, 15, 20 and 30 [the author omits 12], it becomes possible to represent a number of common fractions by integers, thus simplifying calculations... Indeed, while base 10 may be more 'natural', since we have ten fingers, it is computationally more inefficient than base 60, or even base 12.

> However, this explanation for the use of base 60 is unconvincing because of its 'hindsight' character. It is highly unlikely that such considerations were taken into account when the base was chosen. A second explanation emphasizes the relationship that exists between base 60 and numbers that occur in important astronomical quantities.

Joseph goes on to say that either 30, the number of days in a lunar month, or 360, the Babylonians' consequent estimate of the number of days in the year, was used as the numerical base, before the advantages of calculation in base 60 were recognized. *The crest of the peacock*, p 100.

12 Lawlor, R., *Sacred geometry* (London: Thames and Hudson, 1982), p 16.

13 Arnheim, R., *Visual thinking* (London: Faber and Faber, 1970), pp 280-281.

14 Ibid., p 282.

15 Coxeter, H.S.M., *Regular polytopes* (London: Methuen, 1948) ; Wenninger, M.J., *Polyhedron models*, (Cambridge: Cambridge University Press, 1971).

16 Wells, A.F., *The third dimension in chemistry* (Oxford: Clarendon Press, 1956), p 33.

17 *Polyhedron models*, pp 19-24. As for the enduring symbolic merit of these forms, the crystallographer Alan Mackay offers this thought: "Visiting our present day society as a cultural anthropologist, what would Plato or Plutarch make of the World (Football) Cup ceremonies in which just such an object is kicked around in scenes of mass hysteria – clearly rites of cosmological significance?" Hargittai, I., ed., *Symmetry: unifying human understanding*, (New York: Pergamon Press, 1986), p 22.

18 See, for example, Wölfflin, H., *Classic art* (Oxford: Phaidon Press, 1980), pp 23-29.

19 *Symmetry: unifying human understanding*, p ix.

20 *Fearful symmetry*, p 15.

21 The mathematics writer, Ian Stewart, points this out entertainingly with an invitation to "Consider a spherical frog". *Fearful symmetry*, p 152. See also Lewis Wolpert's discussion in *Symmetry: unifying human understanding*, pp 413-429.

22 Browne, Sir T., *Urne buriall and The garden of Cyrus*, ed. J. Carter (Cambridge: Cambridge University Press, 1958).

23 See, for example, Weyl, H., *Symmetry* (Princeton, New Jersey: Princeton University Press, 1952) and *Fearful symmetry*.

24 There is a companion book to the Eameses' film, Morrison, P., P. Morrison,

and the office of C. and R. Eames, *Powers of ten* (New York: W H Freeman, 1982).

25 Kepler, J., *Harmonices Mundi, Book V, Great books of the Western world*, ed. M.J. Adler (Chicago: Encyclopedia Britannica, 1952) 15: Ptolemy, Copernicus, Kepler.

26 Field, J.V., *Kepler's geometrical cosmology* (London: Athlone Press, 1988), p 16.

27 The earth was omitted from Kepler's scheme. Football fans and buckminsterfullerene researchers might care to note that the form most suited to symbolizing our planet is surely that which is intermediate between the dodecahedron (Mars) and the icosahedron (Venus), namely the truncated icosahedron!

28 Thompson, D.W., *On growth and form*, ed. J.T. Bonner (Cambridge: Cambridge University Press, 1961), pp 148-169.

29 Pearce, P., *Structure in nature is a strategy for design* (Cambridge, Massachusetts: MIT Press, 1978), p 19.

30 Haldane, J.B.S., "The origin of life," *Rationalist Annual* (1929), quoted in Bernal, J.D., *The origin of life* (London: Weidenfeld and Nicolson, 1967), p 244.

31 Francis Crick makes a more subtle observation, admitting the virus is the simplest living thing, but pointing out that it can multiply only within a supporting environment that contains the mechanisms for protein synthesis. Since these mechanisms are not usually present outside a living cell, Crick nominates a unicellular organism, the bacterium *Eschericia coli*, as his candidate for the species on the cusp of life. Crick, F., *Of molecules and men*, (Seattle: University of Washington Press, 1966), pp 49-50.

32 Electron micrographs reveal the roughly spherical and often icosahedral shape of many viruses. Madeley, C.R., *Virus macromolecules* (Edinburgh: Churchill Livingstone, 1972). Coxeter has amusingly pointed out the likeness of particular geodesic domes to certain viruses – the Arctic Institute on Baffin Island and the turnip yellow mosaic virus, the geodesic sphere on Mount Washington and the herpes and chicken pox viruses, the United States pavilion in Kabul and one of the adenoviruses, and so on. Coxeter, H.S.M., "Virus macromolecules and geodesic domes," *A spectrum of mathematics*, ed. J.C. Butcher (Wellington: Auckland University Press, 1971) 98-107.

33 Scott, A., *Pirates of the cell* (Oxford: Blackwell, 1985), p 37; Lisanti, M.P., M. Flanagan, and S. Puskin, "Clathrin lattice reorganization: theoretical considerations," *Journal of Theoretical Biology* 108 (1984): 143-147.

34 In the last ten years, the science of crystallography has been thrown into

turmoil with reports of crystals displaying both regions of icosahedral symmetry and the long-range periodicity (over thousands of atomic diameters) sufficient to generate the ordered diffraction patterns beloved of crystallographers. At one point, it was thought that the solid bulk buckminsterfullerene might join this exotic class of "quasicrystals" when scientists at AT&T Bell Laboratories reported that they had grown ten-sided columnar fullerene crystals. Though superficially similar to a decagonal quasicrystal, these crystals were shown upon closer examination to have imperfections in which one of the angles between different planes within the fullerene crystal lattice (just over 35 degrees) was nearly confused with the ten-fold rotation angle (36 degrees). Fleming, R.M., et al., "Pseudotenfold symmetry in pentane-solvated C_{60} and C_{70}," *Physical Review B* 44 (1991): 888-891.

35 *Symmetry: unifying human understanding*, p 4.

36 Weeks, D.E., and W.G. Harter, "Rotation-vibration spectra of icosahedral molecules. II. Icosahedral symmetry, vibrational eigenfrequencies and normal modes of buckminsterfullerene," *Journal of Chemical Physics* 90 (1989): 4744.

37 Linus Pauling discusses a number of such molecules, but makes an unfortunate prediction that dodecahedrane would suffer "steric hindrance" from the crowding on hydrogen atoms on its surface: "This feature of the structure may explain why chemists have not yet succeeded in synthesizing this hydrocarbon." Pauling, L., and R. Hayward, *The architecture of molecules* (San Francisco: W H Freeman, 1964), pp 33-44.

38 *Regular polytopes*, p ix.

CHAPTER 5

1 Kenner, H., *Bucky: a guided tour of Buckminster Fuller* (New York: William Morrow, 1973), p 57.

2 Fuller, R.B., *Synergetics* (New York: Macmillan, 1975), pp 15-17.

3 *Bucky: a guided tour of Buckminster Fuller*, p 81.

4 Cited in Fuller, R.B., and R. Marks, *The Dymaxion world of Buckminster Fuller* (Garden City, New York: Anchor Books, 1973), pp 6-7.

5 *Synergetics*, p 22.

6 Ibid., p 109.

7 Ibid., p 286.

8 Ibid., p 702.

9 Meller, J., ed., *The Buckminster Fuller reader* (London: Jonathan Cape, 1970), pp 132-133.

10 Ibid., p 100.

11 Fuller, R.B., *Nine chains to the moon* (London: Jonathan Cape, 1973), p 15.

12 *The Buckminster Fuller reader*, p 133; Critical Path, p 132.

13 *Bucky: a guided tour of Buckminster Fuller*, p 208.

14 Ibid., p 247.

15 Wechsler, J., ed., *On aesthetics in science* (Cambridge, Massachusetts: MIT Press, 1978) , p 25.

16 *Bucky: a guided tour of Buckminster Fuller*, p 123.

17 The theory of structures is perhaps best explained in J.E. Gordon's perennials, *The new science of strong materials* (Harmondsworth, Middlesex: Penguin Books, 1968) and *Structures* (London: Penguin Books, 1978).

18 Applewhite, E.J., Letter to Joseph D. Clinton (Buckminster Fuller Institute: unpublished, 1991).

19 Ibid.

20 Ward, J., ed., *The artifacts of R. Buckminster Fuller* (New York: Garland Publishing, 1985) 4 : p 293.

21 *The Dymaxion world of Buckminster Fuller*, p 63; Kepes, G., *Structure in art and in science* (New York: Braziller, 1965), p 78.

22 *Bucky: a guided tour of Buckminster Fuller*, p 302. Ed Applewhite, on the other hand, has admitted (at a Royal Society conference on buckminsterfullerene in London, 1-2 October 1992) that the architectural integrity of this building was limited to its outer structural shell. The inner framework was fudged even to the extent of having the polygonal facets overlap like shingles rather than abut precisely.

23 Ibid., pp 62-62.

24 *The Dymaxion world of Buckminster Fuller*, p 3.

25 *Synergetics*, p 68.

26 Although his grip on chemistry was unsure, Fuller surely would not have done as the author has done of the only biography that postdates the discovery of buckminsterfullerene who describes this molecule as "a virtually indestructible carbon *atom*" and as "the smallest *atoms* of carbon in soot" (italics added). Pawley, M., *Buckminster Fuller* (London: Trefoil Publications, 1990), p 146. It is ironic that the author, who speaks with an ignorance of

science not untypical in his vocation, is one of a school of British critics who profess to favour an architecture based on contemporary technology.

27 *Synergetics*, pp 517-518.

28 Fuller, R.B., *Critical path* (New York: St Martin's Press, 1981), p 16.

29 *Nine chains to the moon*, pp 59-60.

30 *Synergetics*, pp 8-9.

31 Ibid., p 35.

32 Ibid., p 123. Protons and neutrons are presently thought to be in motion in a shell-like sequence of orbits. According to quantum-mechanical rules, such motion precludes an overall symmetric structure such as Fuller proposes.

33 *Structure in art and in science*, p 72.

34 *Synergetics dictionary*, 3: p 284 and 1: p 224 and p 421, where Fuller adds, without exactly clarifying matters, that the cube relates to chemistry while the tetrahedron, octahedron, and icosahedron relate to physics.

35 *Bucky: a guided tour of Buckminster Fuller*, p 227.

36 *Synergetics*, pp 728-729.

37 Elements containing "magic numbers" of 2, 8, 20, 28, 40, 50, 82 or 126 protons *or* neutrons are typically the most abundant isotopes of the relevant elements. "Doubly magic" isotopes, such as calcium, with twenty protons and twenty neutrons, or Lead-208, with 82 protons and 126 neutrons, are particularly abundant. The reasons for this are not fully understood. See Mason, S.F., *Chemical evolution* (Oxford: Clarendon Press, 1991), p 42. Fuller refers to species with magic numbers of neutrons *and* protons *totalling* 2, 8, 20, 50, 82 and 126. Fuller, R.B., "NASA speech," (Buckminster Fuller Institute: unpublished, 1966), p 104; *Synergetics*, pp 604-605. He later adds 28 to his list, correlating the numbers "with the atoms of highest structural stability." Fuller, R.B., and E.J. Applewhite, *Synergetics 2* (New York: Garland Publishing, 1979), p 356.

38 *Synergetics*, p 753.

39 Casti, J.L., *Paradigms lost* (London: Scribners, 1990), pp 57-59.

40 *Fads and fallacies in the name of science* (New York: Dover, 1957) quoted in Gratzer, W., ed., *The Longman literary companion to science* (Harlow, Essex: Longman, 1989), pp 439-442.

41 Applewhite, E.J., "An account by R. Buckminster Fuller of his relations with scientists during the development of his energetic geometry," (Buckminster Fuller Institute Archives, unpublished, 1969).

42 "Tracking the buckminsterfullerene," *Trimtab (Bulletin of the Buckminster Fuller Institute)* 6.2 (1991): 9-15.

43 Fuller, R.B., *Operating manual for spaceship earth* (Carbondale: Southern Illinois University Press, 1969), p 13.

44 Fuller, R.B., *No more secondhand god* (Carbondale: Southern Illinois University Press, 1963).

45 *The Buckminster Fuller reader*, p 318.

46 *Bucky: a guided tour of Buckminster Fuller*, p 125.

47 "An account by R. Buckminster Fuller of his relations with scientists during the development of his energetic geometry".

48 Fuller argued against the convention of commodities brokers who saw steel as an unvarying substance in favour of an appreciation of the different types of steel with their very different properties and uses. He would surely have argued likewise for the carbon allotropes, the fullerenes and their variants. *Nine chains to the moon*, p 179.

49 Applewhite, E.J., letter to the author, 1993.

50 Urner, K., "The invention behind the inventions: synergetics in the 1990s," *Synergetica* 1.1 (1991): 8-25.

CHAPTER 6

1 Bernal, J.D., *The social function of science* (London: Routledge, 1939), p 89.

2 Yang, S.H., C.L. Pettiette, J. Conceicao, O. Cheshnovsky, and R.E. Smalley, "UPS of buckminsterfullerene and other large clusters of carbon," *Chemical Physics Letters* 139 (1987): 233-238.

3 Williams and Fleming make the same exception, although they note that mass spectrometry is not true spectroscopy because the process is determined by the reactivity of the molecule and its fragments rather than by inherent molecular qualities. Williams, D.H., and I. Fleming, *Spectroscopic methods in organic chemistry*, Fourth, revised ed. (London: McGraw-Hill, 1989), p 150.

4 I have consulted a number of classic texts including Banwell, C.N., *Fundamentals of molecular spectroscopy* (London: McGraw-Hill, 1972), and Dixon, R.N., *Spectroscopy and structure* (London: Methuen, 1965) . Denney, R.C., *A dictionary of spectroscopy* (London: Macmillan, 1973) was my main source of definitions.

5 Hearnshaw, J.B., *The analysis of starlight* (Cambridge: Cambridge University Press, 1986).

6 Wynn-Williams, G., *The fullness of space* (Cambridge: Cambridge University Press, 1992), pp 9-10.

7 Ibid., p 25.

8 Baly, E.C.C., *Spectroscopy* (London: Longmans, Green and Co., 1924) 1, p 8.

9 *The analysis of starlight*, p 27.

10 Ibid., p 47.

11 *The Scientific Papers of Sir William Huggins, Publications of Sir Wm Huggins Observatory* (W Wesley and Son, 1909) 2, cited in *The analysis of starlight* , pp 69-70.

12 Mason, S.F., *Chemical evolution* (Oxford: Clarendon Press, 1991), p 9.

13 An amusing illustration of the former, ever present, difficulty came in the 1960s when astrophysicists at the Observatoire de Haute Provence thought they had observed potassium flare stars, a unique phenomenon. It was later shown that a match struck near the spectrograph slit was the most likely cause of this never repeated "observation". *The analysis of starlight*, p 278. The persistence of the latter difficulty is amply demonstrated by the continuing mystery of the diffuse interstellar bands.

14 Bennett, J.A., *The celebrated phaenomena of colours* (Cambridge: Whipple Museum, 1984), p 18. The author of this catalogue contrasts the enthusiastic acceptance of the spectroscope with its two major predecessors in optical instrumentation, the telescope and the microscope, both of which were resisted at the time of their invention during the seventeenth century.

15 McGucken, W., *Nineteenth-century spectroscopy* (Baltimore: Johns Hopkins Press, 1969), p 133.

16 *The analysis of starlight*, p 100.

17 *Nineteenth-century spectroscopy*, p 133.

18 *Spectroscopy*, p 217.

19 *Spectroscopic methods in organic chemistry*, p 62.

20 Levy, G.C., R.L. Lichter, and G.L. Nelson, *Carbon-13 nuclear magnetic resonance spectroscopy* (New York: John Wiley & Sons, 1980).

CHAPTER 7

1 Levi, P., *Opere* (Torino: Giulio Einaudi, 1975) 1: pp 642-643; translation by the author, but see also *The periodic table* (London: Michael Joseph, 1985), p 225.

2 This chapter draws on interviews with Donald Cox, Robert Curl, Jim Heath, Donald Huffman, Wolfgang Krätschmer, Harold Kroto, Sean O'Brien, Eric Rohlfing, Rick Smalley, and Daniel Ugarte of the Ecole Polytechnique Fédérale in Lausanne.

3 Donnet, J.-B., and A. Voet, *Carbon black* (New York: Marcel Dekker, 1976), p 1.

4 See, for example, Savage, G., *Carbon-carbon composites* (London: Chapman & Hall, 1993).

5 *Carbon black*; Walker, P.L., ed., *Chemistry and physics of carbon*, (New York: Marcel Dekker, 1965) 1; *Carbon-carbon composites*; Walker, P.L., ed., *Chemistry and physics of carbon* (New York: Marcel Dekker, 1968) 4: pp 243-286.

6 Pitzer, K.S., and E. Clementi, "Large molecules in carbon vapor," *Journal of the American Chemical Society* 81 (1959): 4477-4485 . The overall situation is reviewed in Palmer, H.B., and M. Shelef, "Vaporization of Carbon", in *Chemistry and physics of carbon*, 4: pp 85-135. It is notable that after leaving Rice University with his doctorate, Jim Heath went to work with Richard Saykally at the University of California at Berkeley on the small carbon clusters, producing C_5, seen in carbon stars, and C_9 in a cluster beam apparatus, Heath, J.R., "The physics and chemistry of carbon clusters," *Spectroscopy* 5 (1990): 36-43 ; Hammond, G.S., and V.J. Kuck, ed., *Fullerenes: synthesis, properties, and chemistry of large carbon clusters* (Washington DC: American Chemical Society, 1992) 481: p 1-23. Saykally's group has very recently produced evidence that C_{13} is linear.

7 "Vaporization of Carbon", p 131.

8 *Fullerenes: synthesis, properties, and chemistry of large carbon clusters*, p ix.

9 Rohlfing, E.A., and D.W. Chandler, "Two-color pyrometric imaging of laser-heated carbon particles in a supersonic flow," *Chemical Physics Letters* 170 (1992): 44-50.

10 "Special issue: Fullerenes," *Carbon* 30.8 (1992).

11 This feud made an appearance in the popular scientific press. Taubes, G., "Great balls of carbon," *Discover* (September 1990), 52-59.

12 Frenklach, M., and L.B. Ebert, "Comments on the proposed role of spheroidal carbon clusters in soot formation," *Journal of Physical Chemistry* 92 (1988): 561-563.

13 Kroto, H.W., and D.R.M. Walton, "Postfullerene organic chemistry," *Carbocyclic cage compounds: chemistry and applications*, ed. E. Osawa, and O. Yonemitsu. (New York: VCH Publishers, 1992) 91-100.

14 Gerhardt, P., S. Löffler, and K.H. Homann, "Polyhedral carbon ions in hydrocarbon flames," *Chemical Physics Letters* 137 (1987): 306-310. The context of Homann's work is reviewed in Rohlfing, E.A., "High resolution time-of-flight mass spectrometry of carbon and carbonaceous clusters," *Journal of Chemical Physics* 93 (1990): 7851-7862.

15 Howard, J.B., J.T. McKinnon, Y. Makarovsky, A.L. Lafleur, and M.E. Johnson, "Fullerenes C_{60} and C_{70} in flames," *Nature* 352 (1991): 139-141. More lately, matters have come round still further towards the carbon shell view of things. Howard and his colleagues have now found more complex morphologies involving curved sheets of carbon and even that "fullerene shells are far more prevalent than graphic [sic] flat sheets", Howard, J.B., K.D. Chowdhury, and J.B. VanderSande, "Carbon shells in flames," *Nature* 370 (1994): 603.

16 "Postfullerene organic chemistry" and "High resolution time-of-flight mass spectrometry of carbon and carbonaceous clusters".

CHAPTER 8

1 Orville Chapman, interview with the author.

2 Bernal, J.D., *The social function of science* (London: Routledge, 1939), p 97.

3 Lewis Wolpert gives a striking example of two molecular biologists, Mark Ptashne and Wally Gilbert, in 1965 working in intense rivalry on different floors of same building at Harvard, both trying to isolate the same repressor protein that would demonstrate the molecular basis of the control of gene activity. Wolpert, L., *The unnatural nature of science*, (London: Faber and Faber, 1992), pp 73-74.

4 This section draws primarily on interviews with Donald Huffman, Wolfgang Krätschmer, and Lowell Lamb and other cited sources, Huffman, D.R., "Solid C_{60}," *Physics Today* 44 (1991): 22-31; Huffman, D.R., and W. Krätschmer, "Solid C_{60} – how we found it," *Materials Research Society Proceedings* 206 (1990), 601-610; Krätschmer, W., "How we came to produce C_{60}-fullerite," *Zeitschrift für Physik D* 19 (1991): 405-408.

5 Weeks, D.E., and W.G. Harter, "Vibrational frequencies and normal modes of buckminsterfullerene," *Chemical Physics Letters* 144 (1988): 366-372.

6 Larsson, S., A. Volosov, and A. Rosén, "Optical spectrum of the icosahedral C_{60} – follene-60," *Chemical Physics Letters* 137 (1987): 501-503. These authors, incidentally, were one of several groups to take up the challenge issued by Kroto and Smalley in their 1985 paper to devise alternative names for buckminsterfullerene. Their suggestion, follene, derives from the Latin for football. It did not catch on although they continued to plug it in their papers.

7 Krätschmer, W., K. Fostiropoulos, and D.R. Huffman, "Search for the UV and

IR spectra of C_{60} in laboratory-produced carbon dust," in *Dusty objects in the universe*, E. Bussoletti, A.A. Vittone, eds (1989), 89-93.

8 Krätschmer, W., K. Fostiropoulos, and D.R. Huffman, "The infrared and ultraviolet absorption spectra of laboratory-produced carbon dust: evidence for the presence of the C_{60} molecule," *Chemical Physics Letters* 170 (1990): 167-170.

9 Krätschmer, W., L.D. Lamb, K. Fostiropoulos, and D.R. Huffman, "Solid C_{60}: a new form of carbon," *Nature* 347 (1990): 354-358.

10 Their unease was well founded. Old riddle: Q. How many sides has a sphere? A. Two, an outside and an inside. Jack Fischer at the University of Pennsylvania wrapped this mystery up when he found that the X-ray diffraction signals which had been causing Huffman and Krätschmer some anguish were missing because of an accidental cancellation occasioned by the unanticipated diffraction of X-rays off the *insides* of the arrayed buckminsterfullerene spheres. Heiney, P.A., J.E. Fischer, A.R. McGhie, W.J. Romanow, A.M. Denenstein, J.P. McCauley, and A.B. Smith, "Orientational ordering transition in solid C_{60}," *Physical Review Letters* 66 (1991): 2911-2914. The "odd" diffraction pattern was a composite of the pattern recorded from X-rays reflected from the outer surfaces of the fullerene molecules and from the inner surfaces. (Gold atoms, which can be considered as points in space without inside or outside, proved to be an unreliable guide in this case.) Had they known of this accidental cancellation, Huffman and Krätschmer would have been able to aver not only the spacing between the fullerene molecules but also their sphericity.

11 Hammond, G.S., and V.J. Kuck, ed., *Fullerenes: synthesis, properties, and chemistry of large carbon clusters* (Washington DC: American Chemical Society, 1992), p ix.

12 Willis, E., Rennie, G.K., Smart, C., Pethica, B.A. "'Anomalous' Water," *Nature* 222 (1969) 159-161 quoted in Franks, F., *Polywater* (Cambridge, MA: MIT Press, 1981), p 62.

13 Anon., "Polywater," *Scientific American* 221 (1969) 90-95, quoted in *Polywater*, p 75.

14 As Exxon rejoined the fullerene field following Krätschmer and Huffman's breakthrough, one manager demanded to know how buckminsterfullerene differed from cold fusion (which Exxon like many other commercial laboratories had gone to some effort to corroborate). The discovery of how to make solid C_{60} even mollified officials at the National Science Foundation who had been getting flack from other scientists for pouring in funding since 1985 with apparently little to show for it.

Huffman's comments recorded in an interview with the author show clearly his preoccupation:

> We were very conscious of that in the summer of 1990; Krätschmer and I talked about it; Lowell and I talked about it. I think perhaps we were a little more conscious of the cold fusion story on this side of the ocean than over there [in Heidelberg] because it happened right there in Utah. Utah is our neighbouring state after all. The whole thing was just enormously on my mind. We wanted very badly to avoid every semblance of what they did wrong.

15 Huizenga, J.R., *Cold fusion: the scientific fiasco of the century*, (Rochester, New York: University of Rochester Press, 1992), pp 4-5 and 242-248.

16 Ibid., p 191.

17 Ibid., pp 227-228.

18 Letter to the editor of the *Wall Street Journal*, 26 April 1989, cited in *Cold fusion: the scientific fiasco of the century*, p 59.

19 This section draws primarily on interviews with Jonathan Hare, Harry Kroto, Roger Taylor, and David Walton and other cited sources, including Hare, J.P., and H.W. Kroto, "A postbuckminsterfullerene view of carbon in the galaxy," *Accounts of Chemical Research* 25 (1992): 106-112; Hare, J.P., "A tale of two fullerenes," (unpublished, no date).

20 The struggle was not least to obtain funding. Kroto's application for funds from the Science and Engineering Research Council to be able to make fullerenes and investigate their chemistry to the as the race to make fullerenes was gathering steam in September 1990 elicited this response under the heading "Scientific merit":

> Essentially what the Applicants have done is to ask SERC for £140,606 on the basis of one page of notes, two spectra and one model.
>
> There is little supporting evidence for their claims (copy of unpublished work would have been very helpful). There is no programme of work. There is no account of how the Applicants will spend 10 hours of their time each week for three years and of how one [research assistant] and one Technician will be fully occupied for three years. There is no critical description of the materials to be used, of their origin, their purity, their structures. There is no reason given for the purchase of the [chromatography apparatus] and other equipments. No hint is given of the way "important new findings" may be exploited. Importance has to be identified and exploitation has to be specific.
>
> The application should be rejected on the grounds that the applicants have not prepared, at all, a case (a basis) to support their application for funding.

Behind the schoolmasterish tones, there is something more sinister in the refusal to see that the proposal might have any merit at all and the willingness to reject solely on procedural grounds. That said, the situation is not uncommon. Even investigations that appear highly promising, and which are proposed by groups with excellent track records, fail to gain support. The designation "alpha unfunded" has long been a feature of British science. Wilkie, T., *British science and politics since 1945* (Oxford: Blackwell, 1991), p 92.

21 Parts were supplied by Hare's studentship award supervisor, Steve Wood, then with British Gas, who had himself been a postgraduate at Sussex, and who was performing his duties considerably more attentively than was really required.

22 Letter from Wolfgang Krätschmer to Jonathan Hare, 28 February 1990, cited in "A tale of two fullerenes".

23 One of Hare's diversions at this time was to help in the analysis of mass spectra obtained by the Giotto space probe on its passage through the tail of Halley's comet. The regular spacing of peaks in the spectra had been used to support the theory that the comet contained polymers (with molecular weights rising in arithmetic progression) of formaldehyde. Hare and Kroto and a third researcher, Simon Balm, showed statistically that pretty much any gaggle of organic molecules could explain the data and thus that a regular feature in the spectrum did not necessarily indicate a particular compositional order within the comet. Cheekily, they demonstrated this by choosing the molecules produced in Urey and Miller's "primordial soup" experiment, and found that they too closely matched the observed spectrum. Such procedures provide a useful check whenever scientists are wont to invoke data in support a theory rather than to form a theory on the basis of the data, which is more often than might be supposed, especially in astrophysics. Balm, S.P., J.P. Hare, and H.W. Kroto, "The analysis of comet mass spectrometric data," *Space Science Reviews* 56 (1991): 185-189.

24 "A tale of two fullerenes".

25 Hoffmann, R., "How should chemists think?," *Scientific American*, February 1993 : 66-73.

26 Interviews with Donald Bethune, Mattanjah de Vries, Costantino Yannoni, Robert Johnson and Jesse Salem and other cited sources.

27 Bethune, D.S., "The origins and early course of fullerene research at IBM Almaden Research Center," (unpublished, 1992).

28 Meijer, G., and D.S. Bethune, "Laser deposition of carbon clusters on surfaces: a new approach to the study of fullerenes," *Journal of Chemical Physics* 93 (1990): 7800-7802.

29 Meijer, G., and D.S. Bethune, "Mass spectroscopic confirmation of the presence of C_{60} in laboratory-produced carbon dust," *Chemical Physics Letters* 175 (1990): 1-2.

30 Interview with Robert Whetten.

31 Hahn, M.Y., E.C. Honea, A.J. Paguia, K.E. Schriver, A.M. Camarena, and

R.L. Whetten, "Magic numbers in C_N+ and C_N- abundance distributions," *Chemical Physics Letters* 130 (1986): 12-16.

32 Hare describes this relationship as an attempt to give people who have no conception of molecular scale a sense of buckminsterfullerene's size. Hare's Universal Constant is only valid within physicists' limits of accuracy: the earth is approximately twelve thousand kilometres in diameter; buckminsterfullerene is about 0.7 nanometres across; this requires the soccer ball to be only thirteen centimetres in diameter, just over half its actual size.

There is an altogether more sophisticated reading – unintentional, he insists – of Hare's calculation as a satire on the anthropic principle that number and proportion within the universe have some hidden meaning. In their book on the subject, John Barrow, who is coincidentally the professor of astronomy at the University of Sussex, and Frank Tipler, claim that interest in the anthropic principle has grown among cosmologists in recent decades. Put briefly, there are two kinds of anthropic principle, the weak and the strong. The weak principle states in essence that we cannot know what we cannot see and is comparatively uncontroversial. The strong principle is teleological, holding that nature is directed to particular ends, and states that the universe is constructed such a way as to require the emergence of (human) observers. The periodic emergence of new scientific theories has caused some disquiet among proponents of these principles, each time forcing something of a reappraisal on their part. Growing understanding of entropy and its consequence in the heat death of the universe occasioned one set of revisions; the arrival of the quantum-mechanical view of nature another. The fact that the anthropic principle must be revised after each discovery surely discredits the whole notion, but anthropic cosmologists remain unabashed. Adherents of these principles find great significance in the apparent coincidence of certain physical constants, and it is this aspect that Hare unintentionally parodies. For a detailed review, see Barrow, J.D., and F.J. Tipler, *The anthropic cosmological principle*, (Oxford: Clarendon Press, 1986). For a more concise discussion, see Davies, P.C.W., *The accidental universe*, (Cambridge: Cambridge University Press, 1982), especially pp 110-130.

33 Taylor, R., J.P. Hare, A.K. Abdul-Sada, and H.W. Kroto, "Isolation, separation and characterisation of the fullerenes C_{60} and C_{70}: the third form of carbon," *Chemical Communications* (1990): 1423-1425.

34 Bethune, D.S., G. Meijer, W.C. Tang, and H.J. Rosen, "The vibrational Raman spectra of purified solid films of C_{60} and C_{70}," *Chemical Physics Letters* 174 (1990): 219-222.

35 Wilson, R.J., G. Meijer, D.S. Bethune, R.D. Johnson, D.D. Chambliss, M.S. de Vries, H.E. Hunziker, and H.R. Wendt, "Imaging C_{60} clusters on a surface using a scanning tunnelling microscope," *Nature* 348 (1990): 621-622 ; Wragg, J.L., J.E. Chamberlain, H.W. White, W. Krätschmer, and D.R. Huffman, "Scanning tunnelling microscopy of solid C_{60}/C_{70}," *Nature* 348 (1990):

623-624.

36 Ajie, H., et al., "Characterization of the soluble all-carbon molecules C_{60} and C_{70}," *Journal of Physical Chemistry* 94 (1990): 8630-8633.

37 Haufler, R.E., et al., "Efficient production of C_{60} (Buckminsterfullerene), $C_{60}H_{36}$, and the solvated buckide ion," *Journal of Physical Chemistry* 94 (1990): 8634-8636.

38 Thomas Kuhn is the philosopher who, in *The Structure of Scientific Revolutions,* gave the world the now much abused phrase "paradigm shift". He intended the term to describe the comparatively rare moments in science when we are forced to change our ideas about nature. The change brings about a sometimes painful transition between a period when "normal science" was pursued placidly according to an old paradigm and when the new paradigm is accepted and a new flow of "normal science" can ensue. "Few people who are not actually practitioners of a mature science realize how much mop-up work of this sort of paradigm leaves to be done or quite how fascinating such work can prove in the execution. And these points need to be understood. Mopping-up operations are what engage most scientists throughout their careers." Kuhn, T.S., *The structure of scientific revolutions* (Chicago: University of Chicago Press, 1962), p 24.

39 Cox, D.M., et al., "Characterization of C_{60} and C_{70} clusters," *Journal of the American Chemical Society* 113 (1991): 2940-2944.

40 Frum, C.I., R. Engleman, H.G. Hedderich, P.F. Bernath, L.D. Lamb, and D.R. Huffman, "The infrared emission spectrum of gas-phase C_{60} (buckminsterfullerene)," *Chemical Physics Letters* 176 (1990): 504-508.

41 The compound, an osmylate complex, has four oxygen atoms and two organic groups branching from a central atom of the metal osmium. Two of the oxygen atoms bridge to the fullerene. Computer graphics show a molecule that is no longer spherical but rather resembles a freshly plucked apple with two leaves protruding from the stem. To others the addition looked like rabbit ears, and Hawkins's molecule quickly earned the nickname, bunnyball. The synthesis was highly significant as the first indication that buckminsterfullerene could play a fairly conventional role in organic synthesis, opening up the prospect that innumerable novel compounds might be made, but largely by means of established synthetic routes. Hawkins, J.M., T.A. Lewis, S.D. Loren, A. Meyer, J.R. Heath, Y. Shibato, and R.J. Saykally, "Organic chemistry of C_{60} (buckminsterfullerene): chromatography and osmylation," *Journal of Organic Chemistry* 55 (1990): 6250-6252.

42 "Solid C_{60} – how we found it".

CHAPTER 9

1 This chapter draws on cited sources and interviews with the following: Don Bethune, Mattanjah de Vries, Robert Johnson, and Costantino Yannoni at IBM Almaden Research Center; Robert Whetten at the University of California at Los Angeles; Rick Smalley at Rice University; Wolfgang Krätschmer at the Max-Planck-Institut für Kernphysik and Donald Huffman at the University of Arizona at Tucson; Robert Haddon at AT&T Bell Laboratories; Jonathan Hare, Harry Kroto, and Roger Taylor at Sussex University.

2 Kroto, H.W., J.R. Heath, S.C. O'Brien, R.F. Curl, and R.E. Smalley, "C_{60}: Buckminsterfullerene," *Nature* 318 (1985): 162-163.

3 Elser, V., and R.C. Haddon, "Icosahedral C_{60}: an aromatic molecule with a vanishingly small ring current magnetic susceptibility," *Nature* 325 (1987): 792-794.

4 Haddon, R.C., "Measure of nonplanarity in conjugated organic molecules: which structurally characterized molecule displays the highest degree of pyramidalization?," *Journal of the American Chemical Society* 112 (1990): 3385-3389.

5 Fowler, P.W., P. Lazzeretti, and R. Zanasi, "Electric and magnetic properties of the aromatic sixty-carbon cage," *Chemical Physics Letters* 165 (1990): 79-86.

6 Haddon, R.C., and V. Elser, "Icosahedral C_{60} revisited: an aromatic molecule with a vanishingly small ring current magnetic susceptibility," *Chemical Physics Letters* 169 (1990): 362-364.

7 Schmalz, T.G., "The magnetic susceptibility of Buckminsterfullerene," *Chemical Physics Letters* 175 (1990): 3-5.

8 Taylor, R., J.P. Hare, A.K. Abdul-Sada, and H.W. Kroto, "Isolation, separation and characterisation of the fullerenes C_{60} and C_{70}: the third form of carbon," *Chemical Communications* (1990): 1423-1425 ; Hare, J.P., "A tale of two fullerenes," (unpublished, no date); Kroto, H.W., K. Prassides, R. Taylor, and D.R.M. Walton, "Separation and spectroscopy of fullerenes," *Physica Scripta* T45 (1992): 314-318.

9 Bethune, D.S., "The origins and early course of fullerene research at IBM Almaden Research Center," (unpublished, 1992).

10 Ajie, H., et al., "Characterization of the soluble all-carbon molecules C_{60} and C_{70}," *Journal of Physical Chemistry* 94 (1990): 8630-8633.

11 Haufler, R.E., et al., "Efficient production of C60 (Buckminsterfullerene), $C_{60}H_{36}$, and the solvated buckide ion," *Journal of Physical Chemistry* 94 (1990): 8634-8636.

12 Cox, D.M., et al., "Characterization of C_{60} and C_{70} clusters," *Journal of the American Chemical Society* 113 (1991): 2940-2944.

13 Johnson, R.D., G. Meijer, and D.S. Bethune, "C_{60} has icosahedral symmetry," *Journal of the American Chemical Society* 112 (1990): 8983-8984.

14 Fowler, P.W., P. Lazzaretti, M. Malagoli, and R. Zanasi, "Magnetic properties of C_{60} and C_{70}," *Chemical Physics Letters* 179 (1991): 174-180.

15 Haddon, R.C., et al., "Experimental and theoretical determination of the magnetic susceptibility of C_{60} and C_{70}," *Nature* 350 (1991): 46-47.

16 Taylor, R., J.P. Hare, A.K. Abdul-Sada, and H.W. Kroto, "Isolation, separation and characterisation of the fullerenes C_{60} and C_{70}: the third form of carbon," *Chemical Communications* (1990): 1423-1425.

17 Tycko, R., R.C. Haddon, G. Dabbagh, S.H. Glarum, D.C. Douglass, and A.M. Mujsce, "Solid-state magnetic resonance spectroscopy of fullerenes," *95* (1991): 518-520.

18 Yannoni, C.S., R.D. Johnson, G. Meijer, D.S. Bethune, and J.R. Salem, "^{13}C NMR study of the C_{60} cluster in the solid state: molecular motion and carbon chemical shift anisotropy," *Journal of Physical Chemistry* 95 (1991): 9-10.

19 Tycko, R., G. Dabbagh, R.M. Fleming, R.C. Haddon, A.V. Makhija, and S.M. Zahurak, "Molecular dynamics and the phase transition in solid C_{60}," *Physical Review Letters* 67 (1991): 1886-1889.

20 Fowler, P.W., and J. Woolrich, "p-systems in three dimensions," *Chemical Physics Letters* 127 (1986): 78-83; Shibuya, T., and M. Yoshitani, "Two icosahedral structures for the C_{60} cluster," *Chemical Physics Letters* 137 (1987): 13-16.

21 Yannoni, C.S., P.P. Bernier, D.S. Bethune, G. Meijer, and J.R. Salem, "NMR determination of the bond lengths in C_{60}," 113 *Journal of the American Chemical Society* (1991): 3190-3192.

CHAPTER 10

1 "Official Report Fifth Series Parliamentary Debates, Hansard," (House of Lords, 1991).

2 Snow, C.P., *The two cultures and the scientific revolution* (Cambridge: Cambridge University Press, 1959), p 36.

3 Some of the apparent ignorance on display can safely be put down to humorous self-deprecation. It is worth knowing, for example, that when Earl Russell talks of buckminsterfullerene as a molecule that "does nothing in particular and does it very well", he is taking out of context a fictional remark about the

very chamber in which he sits. In Gilbert and Sullivan's *Iolanthe*, Lord Mountararat sings: "When Wellington thrashed Bonaparte,/As every child can tell,/The House of Peers, throughout the war,/Did nothing in particular,/And did it very well ..." Gilbert, W.S., *Original plays* (Chatto and Windus, 1920) First series , p 273. The allusion would not have been lost on Earl Russell's auditors.

4 Passmore, J., *Science and its critics* (London: Duckworth, 1978), p 11.

5 Ibid., p 28.

6 One of Faraday's biographers makes this commonplace mistake. While exonerating him from the obligation to put his discoveries to immediate practical use, he all but credits Faraday with bringing in a modern technology: "for it was Michael who made floodlighting feasible". Kendall, J., *Michael Faraday man of simplicity* (London: Faber and Faber, 1954), p. 114.

7 *Science and its critics*, p 33.

8 Wolpert, L., *The unnatural nature of science* (London: Faber and Faber, 1992), p 41.

9 *Michael Faraday man of simplicity*, p 14.

10 Cantor, G.N., *Michael Faraday: Sandemanian and scientist* (Basingstoke: Macmillan, 1991), pp 155-156; *Michael Faraday man of simplicity*, p 105.

11 *Michael Faraday: Sandemanian and scientist*, p 109.

12 Bronowski, J., *Science and human values* (Harmondsworth, Middlesex: Penguin Books, 1964), pp 16-17.

13 *Michael Faraday man of simplicity*, p 125.

14 Bernal, J.D., *The social function of science* (London: Routledge, 1939), p 126.

15 Ibid., p 130.

16 Roberts, R.M., *Serendipity: accidental discoveries in science* (New York: John Wiley and Sons, 1989), pp 139-142; Schwartz, J., *The creative moment* (London: Jonathan Cape, 1992), p 94.

17 Wasson, T., ed., *Nobel prize winners* (New York: H.W. Wilson, 1987), p 824. The microwave spectroscopist Charles Townes (whose discovery of water and ammonia molecules in space was described in Chapter One) demonstrated the stimulated emission of microwave radiation by excited molecules of ammonia in 1954. Papers by John Polanyi and Theodore Maiman extending the principle to infrared and visible wavelengths – the optical maser or laser – were rejected by *Physical Review Letters* as lacking in scientific interest before

they were published elsewhere. The laser's main use at first was as a tool for spectroscopy. It was not really until the 1980s that lasers became widely used in printing, fibre-optic communications, medicine, industry, and a host of other fields.

18 Kroto, H.W., J.R. Heath, S.C. O'Brien, R.F. Curl, and R.E. Smalley, "C_{60}: Buckminsterfullerene," *Nature* 318 (1985): 162-163.

19 Smalley, R.E., letter to *Nature*, 1985.

20 Cox, D.M., et al., "Characterization of C_{60} and C_{70} clusters," *Journal of the American Chemical Society* 113 (1991): 2940-2944.

21 "The renaissance molecule," *Economist* 23 May 1992: 91-93.

22 Amato, I., "Doing chemistry in the round," *Science* 254 (1991): 30-31.

23 Beck, R.D., P.S. John, M.M. Alvarez, F.N. Diederich, and R.L. Whetten, "Resilience of all-carbon molecules C_{60}, C_{70}, and C_{84} – a surface-scattering time-of-flight investigation," *Journal of Physical Chemistry* 95 (1991): 8402-8409.

24 Cited in "Beam me up," *Nature* 356 (1992): 563.

25 Regueiro, M.N., P. Monceau, and J.L. Hodeau, "Crushing C_{60} to diamond at room temperature," *Nature* 355 (1992): 237-239.

26 Meilunas, R.J., R.P.H. Chang, S. Liu, and M.M. Kappas, "Nucleation of diamond films on surface using carbon clusters," *Applied Physics Letters* 59 (1991): 3461-3463.

27 Assink, R.A., J.E. Schirber, D.A. Loy, B. Morosin, and G.A. Carlson, "Intercalation of molecular species into the interstitial sites of fullerene," *Journal of Materials Research* 7 (1992): 2136-2143.

28 Arthur Hebard, interview with the author.

29 Tutt, L.W., and A. Kost, "Optical limiting performance of C_{60} and C_{70} solutions," *Nature* 356 (1992): 225-226.

30 This sections draws upon interviews with AT&T researchers, Walter Brown, Si Glarum, Art Ramirez, Thomas Palstra, Robert Fleming, Robert Haddon, and Arthur Hebard, and their review articles: Haddon, R.C., "Electronic structure, conductivity, and superconductivity of alkali metal doped C_{60}," *Accounts of Chemical Research* 25 (1992): 127-133; Hebard, A.F., "Superconductivity in doped fullerenes," *Physics Today* (November 1992): 26-32; Hebard, A.F., "Buckminsterfullerene," *Annual Review of Materials Science* 23 (1993).

31 Haddon, R.C., et al., "Conducting films of C_{60} and C_{70} by alkali-metal

doping," *Nature* 350 (1991): 320-322.

32 Hebard, A.F., M.J. Rosseinsky, R.C. Haddon, D.W. Murphy, S.H. Glarum, T.T.M. Palstra, A.P. Ramirez, and A.R. Kortan, "Superconductivity at 18 K in potassium-doped C_{60}," *Nature* 350 (1991): 600-601.

33 Rosseinsky, M.J., et al., "Superconductivity at 28 K in Rb_xC_{60}," *Physical Review Letters* 66 (1991): 2830-2832.

34 Holczer, K., O. Klein, S.M. Huang, R.B. Kaner, K.J. Fu, R.L. Whetten, and F. Diederich, "Alkali-fulleride superconductors: synthesis, composition, and diamagnetic shielding," *Science* 252 (1991): 1154-1157 ; Wang, H.H., et al., "Superconductivity at 28.6 K in a rubidium-C_{60} fullerene compound, Rb_xC_{60}, synthesized by a solution-phase technique," *Inorganic Chemistry* 30 (1991): 2839.

35 The anonymous interviewee in Marcson's survey of the conduct of science within applications-oriented companies might almost have had AT&T's experience in mind: "The trouble with these research people is that they go ahead and do research without any appreciation of the cost. There is a great deal of research done which should never have been taken on without a careful analysis of the product possibilities." Marcson, S., *The scientist in American industry* (Princeton, New Jersey: Princeton University Press, 1960), p 19.

36 Smalley has detected "borofullerenes" with atoms of boron, carbon's neighbour in the periodic table, replacing a few carbon atoms in the molecular framework. Boron and carbon's other neighbour, nitrogen, have bond strengths and directions that theoretically fit well with the requirements of the fullerene structure. The resulting compounds, $C_{59}B$, $C_{58}B_2$, etc, are presumed to be hollow spheres like their all-carbon parent and they too may go on to form endohedral complexes. Smalley has made compounds such as $K@C_{59}B$. These substitutions should provide reactive sites on the otherwise rather inert molecular surfaces. A number of highly substituted compounds have been proposed, from $C_{12}B_{24}N_{24}$ to $B_{30}N_{30}$, which would be exact electronic analogues of buckminsterfullerene with the same geometry but with a number of distinctive surface sites where reactions might occur. Smalley, R.E., "Doping the fullerenes," (ACS Symposium Series, 1991), 141-160; Smalley, R.E., "The third form of carbon," *Naval Research Reviews* .December (1991): 3-14; Amato, I., "CBN ball, anyone?," *Science* 255 (1992): 537.

37 Weiss, F.D., J.L. Elkind, S.C. O'Brien, R.F. Curl, and R.E. Smalley, "Photophysics of metal complexes of spheroidal carbon shells," *Journal of the American Chemical Society* 110 (1988): 4464-4465.

38 Bethune, D.S., R.D. Johnson, J.R. Salem, M.S. de Vries, and C.S. Yannoni, "Atoms in carbon cages: the structure and properties of endohedral fullerenes," *Nature* 366 (1993): 123-128.

39 Holloway, J.H., E.G. Hope, R. Taylor, G.J. Langley, A.G. Avent, T.J. Dennis, J.P. Hare, H.W. Kroto, and D.R.M. Walton, "Fluorination of buckminsterfullerene," *Chemical Communications* 14 (1991): 966-969.

40 Taylor, R., A.G. Avent, T.J. Dennis, J.P. Hare, H.W. Kroto, and D.R.M. Walton, "No lubricants from fluorinated C_{60}," *Nature* 355 (1992): 27-28.

41 These and other early explorations of fullerene chemistry are recounted in *Accounts of Chemical Research* 25 (1992).

42 Wudl, F., "The chemical properties if buckminsterfullerene (C_{60}) and the birth and infancy of fulleroids," *Accounts of Chemical Research* 25 (1992): 157-161; Fred Wudl, interview with the author.

43 "Doing chemistry in the round"; Baum, R.M., "Fullerenes broaden scientists' view of molecular structure," *Chemical and Engineering News* January 4, 1993: 29-34.

44 Friedman, S.H., D.L. DeCamp, R.P. Sijbesma, G. Srdanov, F. Wudl, and G.L. Kenyon, "Inhibition of the HIV-1 protease by fullerene derivatives: model building studies and experimental verification," *Journal of the American Chemical Society* 115 (1993): 6506-6509.

45 Schinazi, R.F., R. Sijbesma, G. Srdanov, C.L. Hill, and F. Wudl, "Synthesis and virucidal activity of a water-soluble, configurationally stable, derivatized C_{60} fullerene," *Antimicrobial Agents and Chemotherapy* 37 (1993): 1707-1710.

46 Aminoadamantane, for example, based on a fragment of the diamond lattice, is a "potent antiviral agent". Hammond, G.S., and V.J. Kuck, ed., *Fullerenes: synthesis, properties, and chemistry of large carbon clusters*, (Washington DC: American Chemical Society, 1992) 481, p xii.

47 Smalley, R.E., "Great balls of carbon, the story of buckminsterfullerene," *The Sciences* March-April 1991: 22-28; "The third form of carbon".

CHAPTER 11

1 Hoffmann, R., "How should chemists think?," *Scientific American* February 1993: 66-73.

2 Diederich, F., and R.L. Whetten, "Beyond C_{60}: the higher fullerenes," *Accounts of Chemical Research* 25 (1992): 119-126.

3 Bethune, D.S., R.D. Johnson, J.R. Salem, M.S. de Vries, and C.S. Yannoni, "Atoms in carbon cages: the structure and properties of endohedral fullerenes," *Nature* 366 (1993): 123-128.

4 Chibante, L.P.F., A. Thess, J.M. Alford, M.D. Diener, and R.E. Smalley,

"Solar generation of the fullerenes," *Journal of Physical Chemistry* 97 (1993): 8696-8700.

5 Taylor, R., G.J. Langley, H.W. Kroto, and D.R.M. Walton, "Formation of C_{60} by pyrolysis of naphthalene," *Nature* 366 (1993): 728-731.

6 This chapter draws on interviews with Rick Smalley and Gustavo Scuseria at Rice University, Daniel Ugarte, Harry Kroto, Robert Haddon, and Robert Whetten as well as on lectures given by Alan Mackay of Birkbeck College, and Thomas Ebbesen and Sumio Iijima of the NEC Fundamental Research Laboratories at the meetings, Fullerenes: The New Carbon Materials (Society of Chemical Industries, London, 15 July 1992) and A Postbuckminsterfullerene View of the Chemistry, Physics and Astrophysics of Carbon (Royal Society, London, 1-2 October 1992).

7 Iijima, S., "The 60-carbon cluster has been revealed!," *Journal of Physical Chemistry* 91 (1987): 3466-3467.

8 Iijima, S., "Helical microtubules of graphitic carbon," *Nature* 354 (1991): 56-58.

9 Ajayan, P.M., and S. Iijima, "Smallest carbon nanotube," *Nature* 358 (1992): 23.

10 Smalley, R.E., "From dopyballs to nanowires," (Fourth NEC Symposium on Physics and Chemistry of Nanometer-Scale Materials, Karuizawa, Japan, 1992).

11 "Helical microtubules of graphitic carbon."

12 Ajayan, P.M., T. Ichihashi, and S. Iijima, "Distribution of pentagons and shapes in carbon nano-tubes and nano-particles," *Chemical Physics Letters* 202 (1993): 384-388.

13 Ebbesen, T.W., and P.M. Ajayan, "Large-scale synthesis of carbon nanotubes," *Nature* 358 (1992): 220-222. The NEC team rapidly went on to produce samples of nanotubes with a thickness of a single carbon layer. These were important not only as the simplest examples of the type, and the possible nuclei for the concentric nanotubes, but also because such single tubes had been the basis of most theoretical calculations of huge tensile strength and other mechanical properties way in excess of those exhibited by present-day carbon fibres. Iijima, S., and T. Ichihashi, "Single-shell carbon nanotubes of known diameter," *Nature* 363 (1993): 603-605.

14 Mintmire, J.W., B.I. Dunlap, and C.T. White, "Are fullerene tubules metallic?," *Physical Review Letters* 68 (1992): 631-634.

15 Hamada, N., S. Sawada, and A. Oshiyama, "New one-dimensional conductors: graphitic microtubules," *Physical Review Letters* 68 (1992): 1579-1582;

Saito, R., M. Fujita, G. Dresselhaus, and M.S. Dresselhaus, "Electronic structure of chiral graphene tubules," *Applied Physics Letters* 60 (1992): 2204-2206.

16 "From dopyballs to nanowires."

17 Ajayan, P.M., and S. Iijima, "Capillarity-induced filling of carbon nanotubes," *Nature* 361 (1993): 333-334. Experiments with lead and carbon oxides by Iijima's group at by chemists at the University of Oxford soon showed that oxygen was indeed an important agent in uncapping the growing nanotubes. Ajayan, P.M., T.W. Ebbesen, T. Ichihashi, S. Iijima, K. Tanigaki, and H. Hiura, "Opening carbon nanotubes with oxygen and implications for filling," *Nature* 362 (1993): 522-524; Tsang, S.C., P.J.F. Harris, and M.L.H. Green, "Thinning and opening of carbon nanotubes by oxidation using carbon dioxide," *Nature* 362 (1993): 520-522.

18 Ugarte, D., "Curling and closure of graphitic networks under electron-beam irradiation," *Nature* 359 (1992): 707-709. Daniel Ugarte, interview with the author.

19 The phenomenon of approximately spherical particles in carbon science is nothing new. See, for example, Walker, P.L., ed., *Chemistry and physics of carbon*, (New York: Marcel Dekker, 1968) 4: pp 243-286. In these cases, particles such as "pitch spheres" are formed in an environment rich with solvents and other contaminants. They are of micron size, far bigger than even a seventy-shell fullerene which is about 50 nanometres in diameter. Their roundness arises not solely from the energetics of chemical bonding as it does in Ugarte's pure carbon particles, but in a spherulitic crystallization process dominated by surface tension considerations as the particles nucleate within a liquid medium.

20 Kroto, H.W., "Carbon onions introduce new flavour to fullerene studies," *Nature* 359 (1992): 670-671.

21 Mackay, A.L., and H. Terrones, "Diamond from graphite," *Nature* 352 (1991): 762.

22 Lenosky, T., X. Gonze, M. Teter, and V. Elser, "Energetics of negatively curved graphitic carbon," *Nature* 355 (1992): 333-335.

23 Fowler, P.W., and J. Woolrich, "p-systems in three dimensions," *Chemical Physics Letters* 127 (1986): 78-83.

24 See, for example, Walker, P.L., ed., *Chemistry and physics of carbon*, (New York: Marcel Dekker, 1965) 1: p 269 and elsewhere, which describe the carbon formed in various flames (ethylene and oxygen, propane and air, or whatever) as mixtures of hollow filaments, spheres, and twisted graphitic sheets. From pitch come roughly spherical particles between 10 and 200 nanometres in diameter, strung together like pearls on a necklace. X-ray and

electron diffraction of these particles shows that each contains thousands of crystallites made up of five to ten randomly stacked sheets of a hundred or so carbon atoms apiece like a disordered graphite. Little consideration appears to have been given to how or whether these club sandwich-like chunks of graphite are connected. The conventional wisdom appears to prefer broken and misaligned planes and long perimeters of dangling bonds (and to be fair, these conditions are generally rich in contaminants and in hydrogen from the hydrocarbons present which are available to satisfy these bonds) rather than admit for an instant the possibility that flat plane fragments might be joined along their edges by even a slightly curved carbon "hinge".

25 Hoffmann, R., "Under the surface of the chemical article," *Angewandte Chemie* 27 (1988): 1593-1602.

26 Smalley, R.E., "Self-assembly of the fullerenes," *Accounts of Chemical Research* 25 (1992): 98-105.

27 See, for example, Savage, G., *Carbon-carbon composites*, (London: Chapman & Hall, 1993).

28 Dresselhaus, M.S., "Down the straight and narrow," *Nature* 358 (1992): 195-196.

29 Drexler, K.E., *Engines of creation: the coming era of nanotechnology*, (Oxford: Oxford University Press, 1992). At intervals, Drexler mentions graphite and diamond fibres, carbyne cables, and other "strong, carbon-based materials" but does not elaborate on their exact molecular structures.

30 Though useful as a simile, bricklaying is hardly an accurate reflection of the contemporary building process. Commercial architecture at least has been transformed from a wet process of on-site assembly of very small components (bricks) into a sophisticated dry process where precision-made, large-scale modules are fabricated off-site and then brought together for final assembly. Buckminster Fuller agitated for just such a revolution. The chemical assembly of nanostructures should perhaps pursue a similar path, and move away from the manipulation of single atoms to the construction of molecular parts in one location for transport to the nanoscale building site for final assembly to form a complete molecular superstructure. I am indebted to Robert Whetten for drawing this intriguing parallel.

31 Zhang, X.B., et al., "The texture of catalytically grown helix-shaped carbon nanotubules," (unpublished, 1994).

EPILOGUE

1 The prospects of single-shell fullerenes and their analogues' being present in the interstellar medium, and providing an explanation for the diffuse interstellar absorption bands, the unidentified infrared emissions, and the 220

nanometre absorption feature are reviewed in Hare, J.P., and H.W. Kroto, "A postbuckminsterfullerene view of carbon in the galaxy," *Accounts of Chemical Research* 25 (1992): 106-112; Kroto, H.W., and M. Jura, "Circumstellar and interstellar fullerenes and their analogues," *Astronomy and Astrophysics* 263 (1992): 275-280.

2 The maverick astronomer, Fred Hoyle, was one of those to speculate in this area. He argued that some carbon species might survive the "cosmic turmoil" to emerge triumphant from a "Darwinian-style molecular evolution." Although silicates and ices of water and other small molecules have also been candidates, Hoyle and his colleague, Chandra Wickramasinghe, in 1962 proposed graphite as a main component of interstellar dust. They noted the "uncanny resemblance" of the 220 nanometre absorption band to that calculated for spherical graphite particles. Later, they began to doubt. "We now think, however, that the story is not so simple. Calculations show that the graphite particles' extinction feature is centred precisely on the desired wavelength (2200 Ångströms) only if the particles are sufficiently small (less than 300 Ångströms in radius) and have almost perfectly spherical shapes." Hoyle, F., and C. Wickramasinghe, *Lifecloud* (London: J.M. Dent and Sons, 1978), pp 81-85. Hoyle and Wickramasinghe dismiss carbon *per se* and propose that the feature may be due to organic (and hence, possible, prebiotic) molecules superimposed on graphite, a proposition they proceed to embellish in order to support their theory that life originated not on earth but elsewhere in space and was transported to earth. "It took very little time for us to confirm that bacteria are remarkably similar in size to the interstellar grains, and that furthermore, the particles in space had an abnormally low refractive index (a measure of the extent to which they scatter light), just as bacteria have when they are thoroughly dried. ... Hollow particles behave as if they have a very low refractive index..." Hoyle, F., *The intelligent universe* (London: Michael Joseph, 1983), pp 86-87. As we have seen elsewhere, however, any number of schemes can be devised to "explain" an interstellar spectrum (Chapter 8, note 23). The carbon onions are of course less than 300 ångströms in radius, almost perfectly spherical, and possibly hollow.

3 Buseck, P.R., S.J. Tsipursky, and R. Hettich, "Fullerenes from the geological environment," *Science* 257 (1992): 215-217.

4 Daly, T., P.R. Buseck, P. Williams, and C.F. Lewis, "Fullerenes from a fulgurite," *Science* 259 (1993): 1599-1601.

5 Becker, L., J.L. Bada, R.E. Winans, J.E. Hunt, T.E. Bunch, and B.M. French, "Fullerenes in the 1.85-billion-year-old Sudbury impact structure," *Science* 265 (1994): 642-645. In addition, Buseck has re-examined his studies done in 1981 of graphitic carbon in samples taken from the famous Allende meteorite and finds these to be much like Ugarte's carbon onions in appearance.

6 Heymann, D., L.P.F. Chibante, R.R. Brooks, W.S. Wolbach, and R.E. Smalley, "Fullerenes in the Cretaceous-Tertiary boundary layer," *Science* 265 (1994): 645-647.

INDEX

Page numbers in italics refer to diagrams